OCR
A LEVEL

2

BIOLOGY

Richard Fosbery
Adrian Schmit
Jenny Wakefield-Warren

HODDER
EDUCATION
AN HACHETTE UK COMPANY

Orders: please contact Hachette UK Distribution, Hely Hutchinson Centre, Milton Road, Didcot, Oxfordshire, OX11 7HH. Telephone: +44 (0)1235 827827. Email education@hachette.co.uk. Lines are open from 9 a.m. to 5 p.m., Monday to Friday. You can also order through our website: www.hoddereducation.co.uk

First published in 2016 by

Hodder Education,

An Hachette UK Company

Carmelite House

50 Victoria Embankment

London EC4Y 0DZ

www.hoddereducation.co.uk

Impression number   10 9 8 7 6 5

Year   2023 2022

Cover photo © kzww – Fotolia

Typeset in 10.5/12 pt Bliss Light by Integra Software Services Pvt., Pondicherry, India

Printed in CPI Group (UK) Ltd, Croydon, CR0 4YY

A catalogue record for this title is available from the British Library

ISBN 9781471827082

Contents

**Go to www.hoddereducation.co.uk/OCRABiology2
to find the following two chapters:**

Get the most from this book

Welcome to the **OCR A Level Biology 2 Student's Book**. This book covers Year 2 of the OCR A Level Biology specification.

The following features have been included to help you get the most from this book.

Prior knowledge

This is a short list of topics that you should be familiar with before starting a chapter. The questions will help to test your understanding.

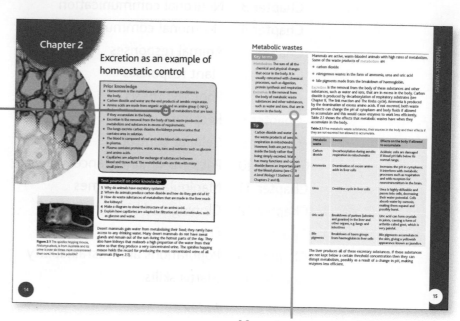

Key terms and formulae

These are highlighted in the text and definitions are given in the margin to help you pick out and learn these important concepts.

Examples

Examples of questions or calculations are included to illustrate chapters and feature full workings and answers.

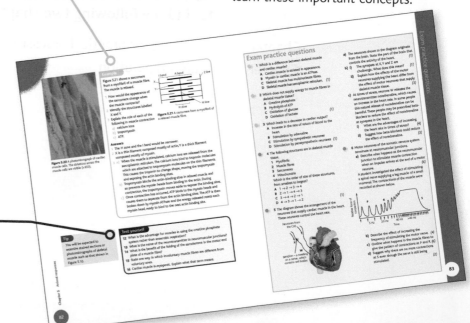

Test yourself questions

These short questions, found throughout each chapter, are useful for checking your understanding as you progress.

Tips

These highlight important facts, common misconceptions and signpost you towards other relevant chapters.

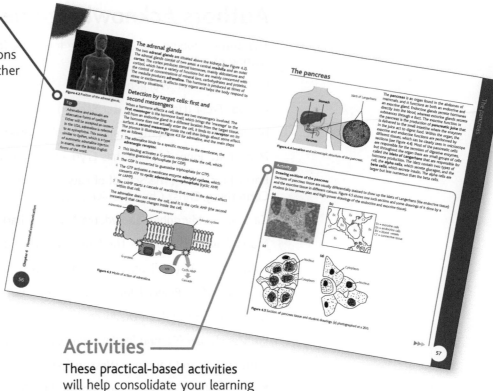

Activities

These practical-based activities will help consolidate your learning and test your practical skills.

Exam practice questions

You will find Exam practice questions at the end of every chapter. These follow the style of the different types of questions you might see in your examination, including multiple-choice questions, and are colour coded to highlight the level of difficulty. Test your understanding even further, with Maths questions and Stretch and challenge questions.

● Green – Basic questions that everyone should be able to answer without difficulty.

● Orange – Questions that are a regular feature of exams and that all competent candidates should be able to handle.

● Purple – More demanding questions which the best candidates should be able to answer.

● Stretch and challenge – Questions for the most able candidates to test their full understanding and sometimes their ability to use ideas in a novel situation.

Dedicated chapters for developing your **Maths** and **Practical skills** and **Preparing for your exam** can be found at the back of this book or online.

Authors Acknowledgements

The authors would like to thank the following for their help during the writing of this book:

Dr Clare van der Willigen

Gaynor Frost (Lincoln Minster School)

Maria Jose Rosello (The Grange School, Santiago, Chile)

Mary-Kate Jones (Cymdeithas Eryri - Snowdonia Society)

Barry Blackburn (The School of Law, Criminal Justice and Computing at Canterbury Christ Church University)

Carwyn ap Myrddhin (Snowdonia National Park Authority)

Maria Dayton (Edvotek - The Biotechnology Education Company)

Francesca Gale (Wellcome Trust Sanger Institute)

Professor Mark Reiger (University of Delaware)

Dr Paquita Hoeck (San Diego Zoo)

Ian Couchman (Cambridge International Examinations)

John Luttick and Laurence Wesson (James Allen's Girls' School)

Medina Valley Centre for Outdoor and Environmental Education (www.medinavalleycentre.org.uk)

Chapter 1

Communication and homeostasis

Prior knowledge

- Animals have two communication systems: the nervous system and the endocrine system.
- The nervous system transmits electrical impulses.
- The endocrine system produces hormones.
- Responses controlled by hormones are usually slower and longer lasting than responses controlled by the nervous system.
- The human body works to maintain steady levels of temperature, water and carbon dioxide and this is essential to life.
- The core temperature of the human body is maintained at 37 °C.
- Heat can be gained by respiration, shivering, exercise, and conserved by decreasing sweating, raising of hair/feathers or reducing blood flow near to the skin surface.
- Heat can also be gained or conserved by certain behaviours, e.g. exercise or (in humans) wearing thicker clothing.
- Heat can be lost by sweating, or by increasing blood flow near to the skin surface.
- Heat can also be lost by behaviour, e.g. panting or (in humans) wearing less clothing.

Test yourself on prior knowledge

1 Which two organ systems are involved in communication in an animal?
2 Why is the maintenance of a relatively constant core temperature an advantage to organisms?
3 Explain how sweating can lower the temperature of the blood.
4 How does the body control blood flow to the body surface?

Introduction

Although you may often wish you did not sweat as it can feel unpleasant (see Figure 1.1), it is vital for keeping you alive. A lot of people think that sweating cleanses your body of toxins, but in the scientific community that is widely regarded as a myth. However, a scientific study in the *Archives of Environmental and Contamination Toxicology* in 2010 seemed to indicate that heavy metal toxins were preferentially excreted in sweat. These results have not yet been widely replicated, but in any case we know that sweating is essential as a means of cooling you down and keeping your body temperature constant.

Figure 1.1 Next time you have sweat dripping into your eyes, remember it is helping to keep you alive.

As multicellular organisms have increased in size and complexity over the course of evolution, they have developed specialised cells and tissues. These structures are found in organs in specific parts of the body, yet each organ performs functions that relate to the whole body. These different specialised structures may also have roles that interact with one another. For these reasons it is essential that the different cells, tissues and organs in the body are able to communicate with one another. In addition to internal coordination, animals and plants need to respond to changes in the environment, yet the cells that bring about the response are nearly always different from those that detect the change, and communication is needed to transfer information about the stimulus to the responding cells.

Animals show a wider range of responses than plants and generally have a greater variety of specialised cells. They have two communication systems: a nervous system, which is capable of rapid responses, and an endocrine system, which tends to carry out longer-lasting communication.

Both hormones and nerves are examples of **cell signalling**, by which individual cells communicate with others in a different location in the body, but the method by which hormones transmit those signals is very different from that of nerves. The nature of the signal is different as it is chemical rather than electrical. The method of transfer is different too; via the bloodstream rather than via nerves. The transmission via the blood is significant, because as the blood system penetrates all parts of the body, a hormone in the blood can reach and affect many different organs and tissues. Although this is useful, it demands a system by which the action of the hormone can be restricted only to relevant cell types. This is done by hormone-specific receptors on the plasma membranes or in the cytoplasm of **target cells**, which are absent on others. Hormones are only removed from the blood when they are used by target cells, and so action over a long period is possible, especially as hormone levels can, if necessary, be continually 'topped up' by the relevant endocrine gland.

The ability of hormones to act over a prolonged period of time is significant for homeostasis. The human body, like that of all mammals, maintains a nearly constant body temperature. The concentration of the blood, in terms of both glucose and water content, is kept within a slightly greater but restricted range, as are the levels of oxygen and carbon dioxide. In order to do this, it must be possible to detect changes in the factor to be controlled. This is done by receptors. The receptors then send a signal, either directly or indirectly, to organs and tissues that actually do the controlling, and these are the effectors. The transfer of information may be either via hormones or nerves.

Key terms

Endocrine gland A gland that secretes its product directly into the bloodstream, rather than via a duct.

Homeostasis The maintenance of a condition of equilibrium or of near-constant internal conditions.

Receptor A structure in the body that can detect changes in its environment and react to stimuli.

Effector A structure in the body that responds to a stimulus and brings about a response. In animals, muscles and glands are common effectors.

Cell signalling

If a multicellular organism is to function in a coordinated way, the cells of which it is made must communicate with each other. This is called cell signalling, and it may occur between cells that are adjacent (or very close to each other) or between cells in different parts of the body.

In order for cells to communicate, one cell must produce some sort of signal that can be detected by another cell and induce a response. This signal is nearly always a chemical (although some nerve cells communicate by electrical impulses at electrical synapses) and the receiving cell has special receptors in its cell membrane, which can detect that chemical.

Cell signalling pathways in animals are categorised into two types, according to the distance the signal travels

- **paracrine signalling** occurs between cells that are close together
- **endocrine signalling** involves signalling over longer distances, with the signalling molecule transported in the circulatory system.

Signalling molecules produced by cells can belong to a variety of chemical groups, including proteins, amino acids, lipids, glycoproteins and phospholipids. In endocrine signalling the signalling molecules are hormones, which will be discussed in more detail in Chapter 4.

Receptor molecules are proteins or glycoproteins, usually on or in the cell surface membrane, although some (e.g. oestrogen receptors) are actually in the cytoplasm, as steroid hormones can diffuse through the cell membrane. The signalling molecule binds to the receptor and causes specific changes in the receiving cell.

An example of cell signalling is the action of neurotransmitters: chemicals that transfer an impulse from one nerve cell across a small gap (the synapse) to another nerve cell. You will learn more about neurotransmitters in Chapter 3. The action of all hormones is another example of cell signalling. A further example is the production of histamine, which is best known as a cause of allergic reactions. Its normal function is to help to produce the inflammatory response to infections or parasites. White blood cells called mast cells produce histamine in response to a chemical signal, an antibody called immunoglobulin E.

Cell signalling is not exclusive to animals. Plant hormones act as signalling molecules; for example, ethylene, which promotes fruit ripening, is detected by protein receptors, resulting in the activation of genes that cause ripening.

Key terms

Oestrogen A female sex hormone (although also present in small quantities in males) that is a steroid and plays a role in the female reproductive cycle and the development of female secondary sex characteristics.

Synapse A junction between two nerve cells, consisting of a minute gap across which impulses pass either by an electrical current (electrical synapses) or, more often, by diffusion of a chemical neurotransmitter (chemical synapses).

The principles of homeostasis

We have seen that homeostasis is the maintenance of a near-constant internal environment within narrow limits. In order to survive, all organisms must do this to some extent. Depending on the organism, there may be internal mechanisms to maintain a steady state, or the behaviour of the organism might assist in keeping factors constant. In mammals, the main components controlled are internal temperature, blood glucose concentration and water content. All the processes of life depend upon chemical reactions controlled by enzymes, which are affected by temperature. If the temperature is so low as to make enzymes inactive, or high enough to denature them, the organism will certainly die. Even if body temperature is slightly too high or too low the enzymes will not work at their optimum rate, which will have serious consequences. The concentrations of body fluids play a part in water potential gradients that are essential for some living processes. Although the body can tolerate slight variations in blood concentrations, larger changes will be harmful. Glucose is a blood solute, and so has an effect on water potentials, but it is also a vital supplier of energy. However, it can cause damage if present in high concentrations.

To keep internal conditions constant the body must be able to detect any deviations and respond in an appropriate way to reverse them. One of the key principles of such systems is the process of negative feedback.

Negative and positive feedback

Negative and positive feedback are the mechanisms by which the internal conditions in the body are maintained or adjusted. Of the two, negative feedback is the most common and indeed it can be found in most homeostatic mechanisms.

Negative feedback

In negative feedback mechanisms a change initiates a series of events which then reverse that change. This is easiest to see in an example (see Figure 1.2). It is important that the body keeps its metabolic rate relatively constant, and this is done by hormones produced by the thyroid gland which boost the metabolic rate if it drops. The starting point of the process that results in the release of thyroid hormones is the release of a hormone, thyrotropin-releasing hormone (TRH), by the hypothalamus in the brain. The thyroid hormones, as well as boosting metabolism, also 'switch off' the production of TRH, and so the levels of thyroid hormones (and metabolic rate) are kept within a certain range, preventing metabolic rate from going too high. So, the production of TRH in the hypothalamus brings about events which ultimately result in that production being switched off. This is negative feedback.

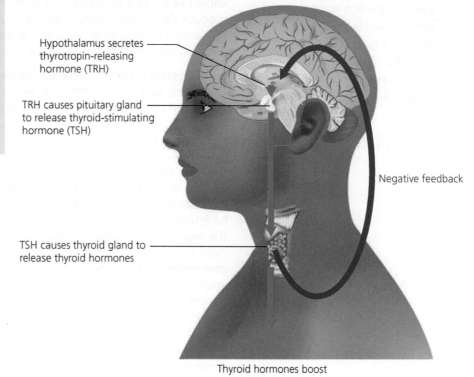

Hypothalamus secretes thyrotropin-releasing hormone (TRH)

TRH causes pituitary gland to release thyroid-stimulating hormone (TSH)

Negative feedback

TSH causes thyroid gland to release thyroid hormones

Thyroid hormones boost metabolic rate

Figure 1.2 Negative feedback control of the production of thyroid hormones.

Positive feedback

Negative feedback mechanisms automatically lead to control and stabilisation, and so are used in homeostasis. There is such a thing as **positive feedback**, which causes proliferation rather than control, and so is not involved in maintaining a steady state in the body. However, hormones can be involved in positive feedback mechanisms. An example is the hormone oxytocin, which intensifies uterine contractions in childbirth. Contractions actually stimulate the release of oxytocin, so the increase it causes results in even more oxytocin being released. Birth itself stops the release of oxytocin and so ends this positive feedback loop. Unrestricted positive feedback mechanisms would result in the level of products, for example oxytocin, getting out of control, so there is always an outside factor (in this case, the birth itself) which can bring the positive feedback loop to an end.

Small molecules of RNA are synthesised in animals and plants as cell signalling compounds. MicroRNAs (miRNAs) are a group of these small RNA molecules that have roles in coordinating the responses of plants to ion deficiencies.

Leaves that are deficient in phosphate ions release specific miRNA molecules that are transported in the phloem to root cells. Here they have the effect of increasing the absorption of phosphate ions. It is believed that they do this by interfering with the synthesis of proteins that recycle and break down carrier proteins from the cell surface membrane.

1 Why are miRNAs transported from shoots to roots in the phloem and not in the xylem?

2 miRNAs are non-coding forms of RNA that are 21–24 nucleotides in length. Explain the meaning of **non-coding**. State two other examples of non-coding RNA.

3 Suggest how miRNAs can interfere with protein synthesis in root cells.

4 Explain why carrier proteins are required for the uptake of phosphate ions.

5 miRNAs are also involved in responses to deficiencies of other ions. Explain why the responses of plants to ion deficiencies are examples of homeostasis and negative feedback.

Answers

1 miRNAs are transported from shoots to roots in the phloem and not in the xylem because phloem conducts from leaves (source) to roots (sink) and xylem conducts only upwards, from roots to leaves.

2 Non-coding means that miRNAs are not used in the synthesis of polypeptides or proteins, so their sequence of nucleotides is not translated. Two other examples of non-coding RNA are transfer RNA (tRNA) and ribosomal RNA (rRNA).

3 miRNAs can interfere with protein synthesis in root cells because the sequence of bases on the miRNA may be complementary to the sequence of bases on one or more molecules of mRNA. If miRNA pairs with mRNA then translation cannot occur as there are no exposed codons to form base pairs with anticodons on tRNA. Therefore no polypeptide or protein is produced. Although miRNA molecules are short chains of nucleotides they will stop translation of much longer mRNA molecules.

4 Carrier proteins are required for the uptake of phosphate ions because phosphate ions are absorbed by active transport against a concentration gradient. There is a lower concentration of phosphate ions in the soil than in the root. In addition, phosphate ions are charged so cannot pass through the hydrophobic region of the phospholipid bilayer of membranes. Ions have to pass through a channel protein or carrier.

5 The responses of plants to ion deficiencies are examples of homeostasis because homeostasis is the process of keeping internal conditions (near) constant. Phosphate ions are very important for the synthesis of ATP, DNA and RNA and cells require a constant supply of ions. If the supply decreases the plant can respond by absorbing more ions from the soil. This is achieved by sending signalling compounds (miRNA) from the leaf where the deficiency is detected to the root. Root cells are the effector because these cells respond by increasing the uptake of phosphate ions. This control is negative feedback because a change in the plant sets in motion a corrective action to restore the supply of phosphate ions.

Control of body temperature: overview

The human body keeps its core temperature within a very restricted range. Although 37 °C is normally thought of as human body temperature, it actually falls within a range of 36.8 ± 0.5 °C and tends to be slightly higher in the daytime than at night. Temperatures of 38 °C or higher and 35 °C or lower indicate fever or hypothermia respectively. Human enzymes have evolved alongside the development of a body temperature of around 37 °C, so that is their optimum temperature. Any significant variation from that temperature affects the rate of enzyme-controlled reactions, which are vital to life. The control of temperature is called **thermoregulation**.

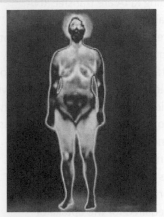

Figure 1.3 Thermal imaging shows the main areas of heat loss from the body. The yellow areas show the greatest heat loss, the blue areas the least. Clothing will reduce this heat loss considerably.

Mammals and birds are endotherms, meaning that they have physiological mechanisms to control their body temperature. All other members of the animal kingdom are ectotherms and they have to rely on behavioural adaptations to ensure that their core temperature does not sink too low or rise too high.

Heat loss from the body

Heat can be lost from the body in four ways (see Figure 1.3).

- **Radiation** is the loss of heat (in the form of electromagnetic radiation) from hot objects into cooler surroundings. It is the main way in which the human body loses heat, but is also the way we gain heat from external sources, like the Sun or a fire.

- **Convection** is the movement of currents or warm air upwards, because the density of air decreases when it is warm.

- **Conduction** is the transfer of heat energy from a warmer material to a cooler one. The body can lose heat to the air by conduction, but, because air is a good insulator, if a layer of air can be trapped around the body it will reduce further heat loss by radiation. This is the reason why wearing thick clothes made of materials like wool (which traps air) keeps you warm in cold conditions.

- **Evaporation** of water from the skin cools the surface, because the heat necessary to change the water into water vapour is extracted from the skin.

Temperature control in ectotherms

Aquatic ectotherms have relatively few problems in maintaining a stable body temperature. Water temperatures do not fluctuate nearly so much as those on land, due to the high specific heat capacity of water.

On land, however, environmental temperatures can vary a lot at different times of the day and in different seasons. Ectothermic animals must make sure they do not expose themselves to extremes of temperature, and must warm their bodies up rapidly if they get cold (which is often unavoidable during the night; Figure 1.4). Low body temperatures mean that their movements will be sluggish, making it difficult for ectotherms to catch prey or to escape predators.

If they are in danger of overheating, ectotherms will seek shade or water so that they can cool down. There are many other adaptations that help ectotherms avoid large changes in their body temperature, but these vary from group to group. For example, many fish have blood vessels bringing cold blood from the body surface that pass very close to those bringing warm blood from the internal organs and muscles, and this evens out the temperatures. Ectotherms living in cold conditions are often dark in colour, which allows their body to retain heat better. Ectotherms also tend to be able to tolerate a much greater range of body temperatures than endotherms; for example, some earthworms have been shown to survive at temperatures down to −20 °C. Despite these apparently rather haphazard and laborious methods, ectotherms can be surprisingly successful at keeping their core body temperature within a limited range. The body temperature of some large ectotherms only seems to vary about 2 °C during the course of a day.

Figure 1.4 After a cool night, this lizard is sitting on a warm and sunny rock to raise its body temperature before it can be fully active.

Ectotherms suffer in comparison to endotherms because their behaviour is more restricted by environmental temperature, and they cannot easily colonise very hot or cold habitats. However, endotherms need a lot of energy to maintain their body temperature, which means they have a higher metabolic rate and need more food. Ectotherms can survive better in situations where food is limited because they need less and are able to go for longer periods without eating.

> **Test yourself**
>
> 1 Exocrine glands secrete a product via a duct, whereas endocrine glands secrete products directly into the blood. Suggest why hormones are produced in endocrine glands and never by exocrine.
> 2 What are the advantages of having two coordination systems (nerves and hormones) rather than just one?
> 3 Explain why a positive feedback mechanism needs some sort of additional control.
> 4 Suggest a reason why large ectotherms show less variation in body temperature than small ones.
> 5 Suggest a reason why ectotherms may grow more rapidly than endotherms.

Temperature control in endotherms

Figure 1.5 The Arctic fox, *Vulpes lagopus*, can maintain its core temperature even when the temperature is as low as −40°C.

> **Tip**
>
> A common mistake in answers to questions about temperature control is to say that, in hot conditions, the blood capillaries move nearer the surface of the skin. The capillaries stay where they are: they have no way of moving. More blood flows nearer the surface of the skin, because of vasodilation of the arterioles.

To control their body temperature, endothermic animals have mechanisms to increase their core temperature in cold conditions and lower it when it is hot. In order to do this, they must also have a method of detecting the temperature of the blood.

Cooling mechanisms

Vasodilation of skin capillaries

Heat exchange, whether it results in heating or cooling, takes place at the body surface, where the blood is in close proximity to the environment. It would have to be extremely hot for the blood to actually **gain** heat from the environment, although this does happen. However, even if the temperature of the air is lower than that of the blood, the warmer it is the less heat will be lost from the surface (see Figure 1.6).

Given that the blood generally loses heat to the air when it travels near the surface of the skin, one way to lose more heat is to get a greater volume of blood to the capillaries in the skin. Arterioles have muscles in their walls which can contract or relax to allow less or more blood through them. In warm conditions, the heat loss is maximised by dilating the arterioles near to the skin, so that more blood flows though capillaries. This is referred to as **vasodilation** (see Figure 1.6). This works provided the air temperature is lower than the blood temperature, which it almost always will be. This explains why people with pale skin go red when they are hot: there is a lot of blood in the vessels under the skin and the red colour of the blood can be seen. Note that vasodilation involves muscles in the arterioles, not the capillaries, which have no muscle in their walls.

7

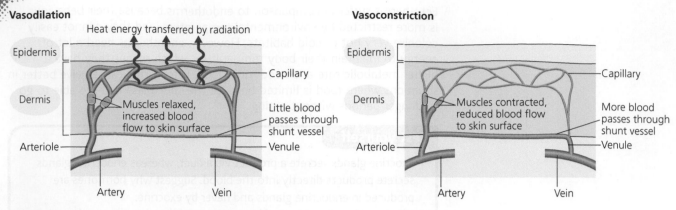

Figure 1.6 Vasodilation and vasoconstriction in the skin.

Sweating

The skin has **sweat glands** which produce sweat in hot conditions (see Figure 1.7). In fact, they produce a little sweat at all times, but the secretion is greatly increased when the temperature rises.

Structure of the skin

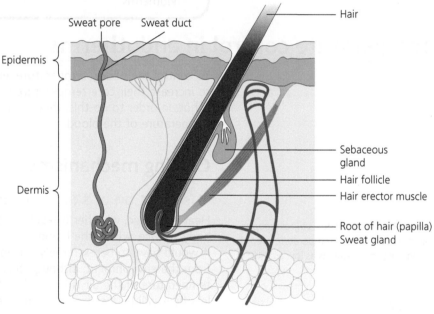

Figure 1.7 Structure of the skin.

The main function of sweating is to cool the skin. It is not the sweat itself which has the cooling effect – it will be at more or less the same temperature as the body – but the evaporation of that sweat. We have seen that evaporation extracts heat from the skin to convert water into water vapour, and so the skin is cooled. This means that sweating is much less effective in humid conditions, as damp air is less effective at evaporating water (due to a reduced or unfavourable concentration gradient). See Figure 1.8.

Flattening of the hair

Air trapped between the hairs on the skin forms an insulating layer. The **hair erector muscles** (effectors) in the skin (see Figure 1.7) can raise the hairs by contracting, and lower them again when they relax. In warm conditions these muscles relax and so the insulating layer of air on the skin is thinner, allowing more heat to be lost.

Figure 1.8 Sweating cools you down, but only if the air is dry enough to evaporate it.

Warming mechanisms

Boosting metabolic rate

The great majority of the chemical reactions in the body are exothermic, i.e. they produce heat. This is why the body is warm. In cold conditions, one of the hormones, thyroxine, is released from the thyroid gland just in front of the larynx and this hormone boosts the basal metabolic rate (BMR) and so increases heat production. Another hormone, adrenaline, also boosts metabolism but its action is short-term. The liver, where a huge number of chemical reactions take place, plays a big role in generating heat for the body.

Shivering

Shivering is a reflex action in response to a slight drop in core body temperature, and so is a nervous rather than a hormonal mechanism. The effectors are the muscles, and the rapid and regular muscle contractions which comprise shivering generate heat, which warms the blood. This heat is generated by the metabolic reactions going on in the muscle not, as is commonly thought, by friction.

Vasoconstriction

In the same way that hot conditions affect the blood supply to the skin, so does a drop in temperature. The response is the reverse: the arterioles constrict so that not so much blood reaches the capillaries near the surface. This is called **vasoconstriction**. The blood is diverted through **shunt vessels**, which are deeper in the skin and so do not lose heat to the surroundings (see Figure 1.6). This is not strictly a 'warming' mechanism, as it does not raise the temperature of the blood, but it does reduce the normal cooling effect that happens when blood flows through the skin.

Erection of hairs

As mentioned above, each hair has an erector muscle attached to it, and when this contracts the hair stands on end. This allows the hair to trap a thicker layer of air, which insulates the skin and reduces heat loss (once again, not actually warming the blood). This can be effective in animals with a thick covering of fur, but has minimal effect in humans, as the hair on most of our skin is very sparse.

Behavioural homeostatic responses

Although endotherms have internal homeostatic mechanisms, they also exhibit similar behavioural responses to ectotherms. They will seek shade or cool places when the environment is particularly hot, and will take measures to warm themselves in extreme cold. For example, penguins huddle together to withstand the extreme Antarctic winters, and humans build fires or wear thick clothing. An example of a behavioural response is shown in Figure 1.9.

In order for these behaviours to take place, the animal must have a means of detecting external temperatures. This is done by **peripheral receptors**, thermoreceptors found in the skin and the mucous membranes. There are receptors for both heat and cold, but exactly how they work is not yet fully understood. As well as behavioural responses, they communicate with the hypothalamus and so may play a part in physiological mechanisms.

Figure 1.9 The hippopotamus is a mammal, and so is endothermic. Nevertheless, living in a hot climate, it stays in water throughout the day to keep cool, only emerging to feed at night.

Example

Diurnal temperature variation in humans

Human body temperature is kept within a very restricted range, but does vary slightly during 24 hours. This variation is shown in the graph in Figure 1.10.

1 By how many degrees Celsius does the body temperature vary during the course of 24 hours?

2 Suggest a reason why the body temperature rises slightly in the day time.

3 When the body temperature is at its highest, sweat may be produced. Explain how this can lower the body temperature.

4 The body temperature is controlled by a negative feedback mechanism. Explain the term *negative feedback*.

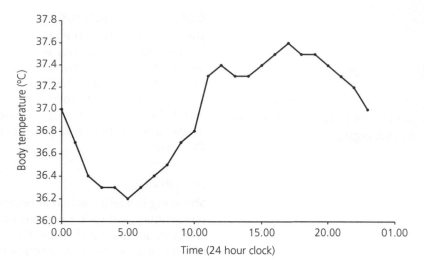

Figure 1.10 Diurnal variation in body temperature.

Answers

1 It varies by 1.4 °C (from 36.2 to 37.6 °C).

2 This is due to increased activity in the day, which will mean that the muscles will generate more heat than at night, when the body is sleeping.

3 The water in the sweat will evaporate. Heat energy from the skin is used to vaporise the water, and so the skin is cooled.

4 Negative feedback occurs when a change causes a series of events which result in a reversal of that change.

Monitoring temperature in endotherms

To control their body temperature endotherms need to have a means of monitoring it and then coordinating corrective measures. The centre for temperature control is located in the **hypothalamus** in the brain, which monitors the core blood temperature as it passes through. As well as **temperature receptors**, it contains two control centres: the **heat loss centre** and the **heat gain centre**. These centres send out both nervous and hormonal signals which bring about the various actions related to temperature control. Sweating, shivering, vasoconstriction and vasodilation are controlled by nerve impulses via the autonomic nervous system. Thyroxine is released from the thyroid gland and is controlled by thyroid-stimulating hormone (TSH), which is released from the pituitary gland in response to hormones released by the hypothalamus, which is just above it. Adrenaline is released from the adrenal glands as a result of nervous stimulation. Negative feedback mechanisms reverse these actions when the core body temperature returns to normal. The control process is summarised in Figure 1.11.

Key term

Autonomic nervous system The part of the nervous system that controls automatic responses. It consists of the sympathetic and parasympathetic nervous systems.

Test yourself

6 Explain why hot and humid conditions feel much more uncomfortable than equally hot, but dry conditions.

7 Explain why boosting the metabolic rate raises the body temperature.

8 What is the role of the shunt vessels in the skin?

9 Explain why people who are cold go pale or 'blue'.

10 Explain the role of the hypothalamus in thermoregulation.

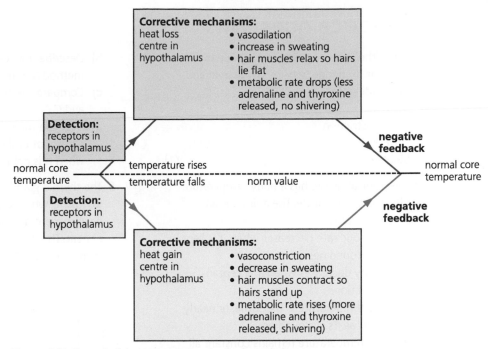

Figure 1.11 Control of thermoregulation by the hypothalamus.

Activity

Oxygen consumption and temperature

A student determined the respiratory rates of a mouse (an endotherm) and a lizard (an ectotherm) at different environmental temperatures using an open-flow respirometer (shown in Figure 1.12).

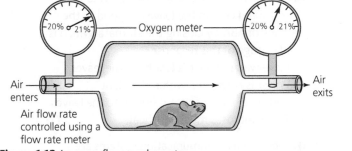

Figure 1.12 An open-flow respirometer.

The student directed the air at a standard flow rate into a temperature-controlled chamber containing the experimental animal. She measured the concentration of oxygen in the air before entering the chamber, and when it left. She calculated the oxygen consumption of the animals using the difference in the concentrations and the rate of air flow. Her results are shown in the graph in Figure 1.13.

1 Why is it important to control the flow rate through the chamber?
2 Describe the differences between the data for the mouse and the lizard.
3 Explain the pattern of oxygen consumption seen in the mouse.
4 Explain the difference between the patterns seen in the mouse and the lizard
   a) between 5 °C and 20 °C
   b) between 30 °C and 40 °C.
5 Suggest a possible source of inaccuracy in this experiment.

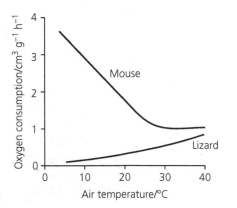

Figure 1.13 The rate of oxygen consumption of a mouse and a lizard over a range of temperatures.

Exam practice questions

1 Which of the following does not reduce the rate of heat exchange between an animal and its surroundings?
   A feathers
   B vasoconstriction
   C shivering
   D blubber (1)

2 A large tropical reptile has a body temperature that fluctuates very little. The animal is an ectotherm because:
   A its metabolic rate decreases when it is kept in warm surroundings
   B its activity decreases when it is kept in cold conditions
   C its body temperature is the same, or nearly the same, as its surroundings
   D its body temperature remains constant all the time. (1)

3 In dry conditions, root cells release abscisic acid (ABA), which travels through the xylem to the leaves. Guard cells in the leaves respond to the presence of ABA by closing the stomata. Which is an advantage of this response?
   A More water vapour is lost from the leaves.
   B Water is conserved within the plant.
   C Carbon dioxide cannot diffuse into the leaves.
   D The plant loses much less heat. (1)

4 a) Explain why communication systems are needed by multicellular animals. (5)

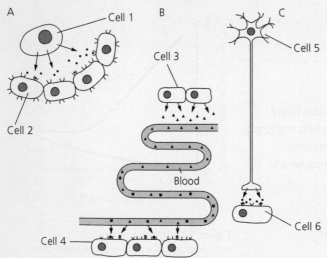

The diagram shows different ways in which cells in multicellular animals signal to each other. In the questions which follow you can refer to the cells by their numbers.

b) Describe how cell signalling is achieved in method A in the diagram. (5)
c) Compare the two methods of signalling in B and C. (5)
d) Describe the roles of cell signalling in the control of body temperature in mammals when it is cold. (6)

5 The thyroid gland secretes the hormone thyroxine, which controls growth and development and also stimulates the production of heat. The diagram shows how the release of thyroxine is controlled.

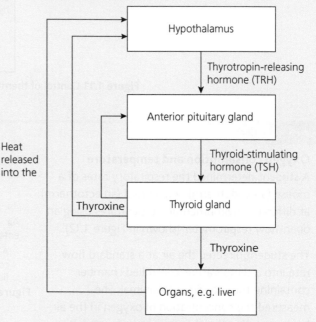

a) State what will happen to the release of thyroxine if the secretion of
   i) TSH increases
   ii) TRH decreases. (2)
b) Use the diagram to suggest and explain what happens to the control of thyroxine release if the body temperature increases. (4)
c) Thyroid hormones, such as thyroxine, cross cell membranes to enter the nucleus and bind to receptors. Suggest how thyroxine stimulates heat production in liver cells. (4)
d) Explain the term *negative feedback* as applied to thermoregulation in mammals. (4)

Stretch and challenge

6 T.H. Benzinger led a research team at the US Naval Medical Research Institute in Maryland, USA. The team carried out a series of experiments to investigate thermoregulation in humans by using a man who lay naked, suspended and totally enclosed in a specially constructed chamber. This chamber was designed to take simultaneous measurements of the man's core body temperature, his rate of sweating and the heat given off from his body.

The graph below shows the results from one of Benzinger's experiments. The man rested in the temperature-controlled chamber which was kept at 45 °C. At the times indicated on the graph the man ate a quantity of crushed ice. His core body temperature was measured by placing a thermocouple near his eardrum. The artery that supplies blood to the eardrum also supplies the brain.

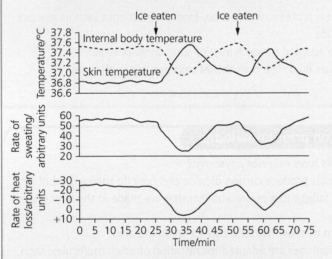

a) Comment on the results shown in the graph, explaining what they show about the location and functioning of the body's thermostat.

Two further investigations were carried out using the same chamber.

The man's rate of sweating and rate of heat production were recorded over 2 days. During this time, the man's core body temperature was caused to fluctuate between 36.3 °C and 37.6 °C. The air in the chamber was kept constant at different temperatures. This meant that the man's skin temperature remained constant.

In the first investigation (**A**) the man's rate of sweating was measured as the core body temperature changed. The investigation was carried out at a skin temperature of 33 °C and then repeated at 29 °C.

In the second investigation (**B**) the rate of heat production in the body was measured at different core body temperatures. The investigation was carried out at a skin temperature of 20 °C and then again at 31 °C.

The results are shown in the two graphs that follow.

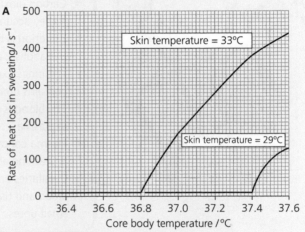

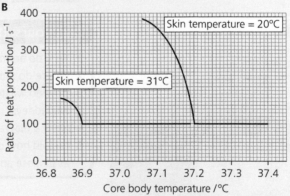

b) Describe the effect of changing core body temperature on the rates of sweating and heat production in the different air temperatures.

c) Explain what the results in **A** and **B** show about the set point for core body temperature.

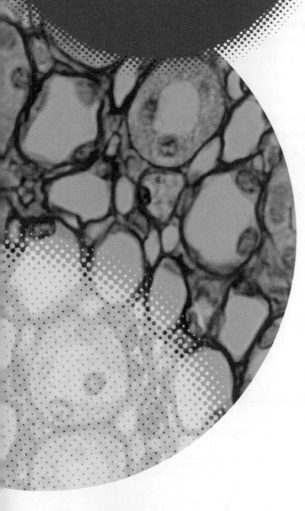

Chapter 2

Excretion as an example of homeostatic control

Prior knowledge
- Homeostasis is the maintenance of near-constant conditions in the body.
- Carbon dioxide and water are the end products of aerobic respiration.
- Amino acids are made from organic acids and an amine group ($-NH_2$).
- Carbon dioxide and urea are waste products of metabolism that are toxic if they accumulate in the body.
- Excretion is the removal from the body of toxic waste products of metabolism and substances in excess of requirements.
- The lungs excrete carbon dioxide; the kidneys produce urine that contains urea in solution.
- The blood is composed of red and white blood cells suspended in plasma.
- Plasma contains proteins, water, urea, ions and nutrients such as glucose and amino acids.
- Capillaries are adapted for exchange of substances between blood and tissue fluid. The endothelial cells are thin with many small pores.

Test yourself on prior knowledge

1 Why do animals have excretory systems?
2 Where do animals produce carbon dioxide and how do they get rid of it?
3 How do waste substances of metabolism that are made in the liver reach the kidneys?
4 Make a diagram to show the structure of an amino acid.
5 Explain how capillaries are adapted for filtration of small molecules, such as glucose and water.

Figure 2.1 The spinifex hopping mouse, *Notomys alexis*, is from Australia and its urine is over six times more concentrated than ours. How is this possible?

Desert mammals gain water from metabolising their food; they rarely have access to any drinking water. Many desert mammals do not have sweat glands and remain out of the sun during the hottest parts of the day. They also have kidneys that reabsorb a high proportion of the water from their urine so that they produce a very concentrated urine. The spinifex hopping mouse holds the record for producing the most concentrated urine of all mammals (Figure 2.1).

Metabolic wastes

Key terms

Metabolism The sum of all the chemical and physical changes that occur in the body. It is usually concerned with chemical processes, such as digestion, protein synthesis and respiration.

Excretion is the removal from the body of metabolic waste substances and other substances, such as water and ions, that are in excess in the body.

Tip

Carbon dioxide and water are the waste products of aerobic respiration in mitochondria. However, both are put to use inside the body rather than being simply excreted. Water has many functions and carbon dioxide forms an important part of the blood plasma (see *OCR A level Biology 1 Student's Book* Chapters 2 and 8).

Mammals are active, warm-blooded animals with high rates of metabolism. Some of the waste products of metabolism are

- carbon dioxide

- nitrogenous wastes in the form of ammonia, urea and uric acid

- bile pigments made from the breakdown of haemoglobin.

Excretion is the removal from the body of these substances and other substances, such as water and ions, that are in excess in the body. Carbon dioxide is produced by decarboxylation of respiratory substrates (see Chapter 8, The link reaction and The Krebs cycle). Ammonia is produced by the deamination of excess amino acids. If not excreted, both waste products can change the pH of cytoplasm and body fluids if allowed to accumulate and this would cause enzymes to work less efficiently. Table 2.1 shows the effects that metabolic wastes have when they accumulate in the body.

Table 2.1 Five metabolic waste substances, their sources in the body and their effects if they are not excreted but allowed to accumulate.

Metabolic waste	Source	Effects on the body if allowed to accumulate
Carbon dioxide	Decarboxylation during aerobic respiration in mitochondria	Acidosis: cells are damaged if blood pH falls below its normal range.
Ammonia	Deamination of excess amino acids in liver cells	Increases the pH in cytoplasm; it interferes with metabolic processes such as respiration and with receptors for neurotransmitters in the brain.
Urea	Ornithine cycle in liver cells	Urea is highly diffusible and passes into cells, decreasing their water potential. Cells absorb water by osmosis, making them expand and possibly burst.
Uric acid	Breakdown of purines (adenine and guanine) in the liver and other organs, e.g. lungs and intestines	Uric acid can form crystals in joints, causing a form of arthritis called gout, which is very painful.
Bile pigments	Breakdown of haem groups from haemoglobin in liver cells	Bile pigments accumulate in the skin, giving a yellowish appearance known as jaundice.

The liver produces all of these excretory substances. If these substances are not kept below a certain threshold concentration then they can disrupt metabolism, possibly as a result of a change in pH, making enzymes less efficient.

The liver

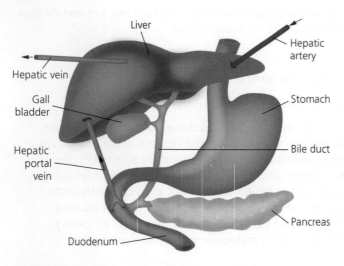

Liver
Hepatic vein
Gall bladder
Hepatic portal vein
Duodenum
Hepatic artery
Stomach
Bile duct
Pancreas

Figure 2.2 The gross structure of the liver, its blood supply and its connection to the duodenum. The hepatic artery is a branch of the aorta and blood in the hepatic vein drains into the vena cava.

You have seen that the liver has important roles as an effector in the maintenance of body temperature (Chapter 1) and it also has a similar role in the control of glucose concentration in the blood (Chapter 4). To carry out its roles, including the breakdown of unwanted substances and the production of excretory waste, it needs a good supply of blood. The liver is unusual as it is supplied with blood from two sources: as you can see in Figure 2.2, oxygenated blood flows from the heart in the hepatic artery and deoxygenated blood flows from the digestive system in the hepatic portal vein. The liver absorbs and metabolises much of the nutrients that are absorbed in the small intestine. Deoxygenated blood flows back towards the heart in the hepatic vein. The liver makes bile salts, which help digest fats, and bile pigments, which are a waste product. These are stored in the gall bladder as part of the bile and released into the duodenum via the bile duct.

The liver has a rather simple internal structure. It is divided into many lobules that are separated from each other by connective tissue (Figure 2.3). The liver does not have cells specialised for different functions. Almost all of the functions occur in hepatocytes (literally, liver cells). Each lobule is supplied with blood from the branches of the hepatic artery and the hepatic portal vein. This blood flows through wide capillaries known as sinusoids that are lined by an incomplete layer of endothelial cells which allows blood to reach hepatocytes, making it easy for the exchange of substances between blood and cells. The liver carries out a huge number of chemical changes. Each **hepatocyte** has a large surface area in contact with blood for exchange of substances. A branch of the hepatic vein drains blood away from each lobule. All the functions of the organ occur within each lobule, so the lobule is referred to as the functional unit of the liver.

Key terms

Connective tissue A tissue that consists of cells that secrete an extracellular matrix, such as collagen fibres, bone or cartilage.

Endothelial The adjective from endothelium. The squamous cells lining all the blood vessels (including sinusoids in the liver), heart and the lymphatic vessels are known as endothelial cells.

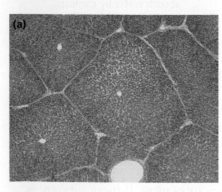

(a)

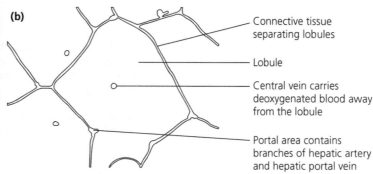

(b)

Connective tissue separating lobules

Lobule

Central vein carries deoxygenated blood away from the lobule

Portal area contains branches of hepatic artery and hepatic portal vein

Figure 2.3 (a) A low-power photomicrograph of some lobules from pig's liver (x30) and (b) a plan drawing made from the photomicrograph.

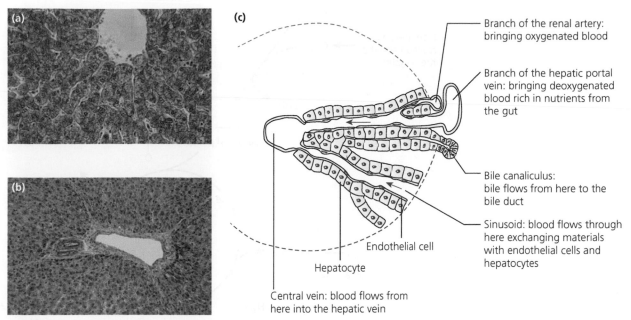

Figure 2.4 (a) hepatocytes around the central vein of a lobule (x200); (b) portal area (x40) and (c) the arrangement of blood vessels, sinusoids, hepatocytes and bile canaliculi inside a liver lobule.

Hepatocytes

Hepatocytes store glucose as the polysaccharide glycogen. They also make bile, which is a digestive secretion that is stored in the gall bladder and enters the duodenum. It contains bile pigments, which are excretory products made from haemoglobin, and bile salts, for the emulsification of fats. These substances enter the little channels (canaliculi) that join together to form the bile duct, which drains into the gall bladder and the duodenum. The blood flows along the sinusoids from the branches of the hepatic artery and the hepatic portal vein to drain into a branch of the hepatic vein (Figure 2.4). This deoxygenated blood flows back to the heart through the vena cava.

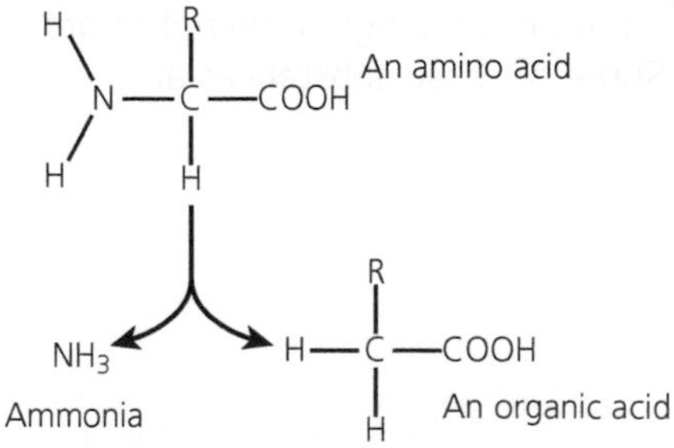

Protein in the diet is digested into amino acids, which are absorbed into the blood and taken directly to the liver. Excess amino acids are not excreted. As they are good sources of energy, the amine group ($-NH_2$) is removed so that the rest of the molecule can be used. Each amino acid is deaminated by removal of the $-NH_2$ to form ammonia (NH_3) which in cytoplasm forms the ammonium ion (NH_4^+). What remains after removal of the amino group in **deamination** is an organic acid that may be respired in the Krebs cycle in mitochondria (see Chapter 8) or used in the synthesis of other compounds (Figure 2.5).

Ammonia is made less harmful by a series of reactions that occur in liver cells. These reactions form a cycle: the **ornithine cycle** or urea cycle. Figure 2.6 shows an outline of the ornithine cycle. The production of carbamyl phosphate and citrulline occur in mitochondria using the ammonia released from deamination. The other reactions occur in the cytosol. Aspartate is formed from excess amino acids. The ornithine cycle produces one molecule of urea from one molecule of carbon dioxide and two amino groups from two amino acids. The advantage of a cycle is that only small quantities of the intermediate compounds, such as ornithine and citrulline, are needed to process large quantities of the waste amino groups and carbon dioxide.

Figure 2.5 Deamination. Most amino acids are deaminated in mitochondria. This means that ammonia is held in a compartment separate from the rest of the cytoplasm.

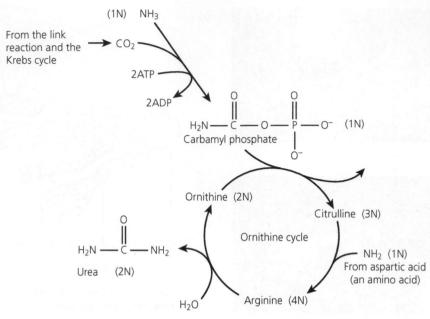

Figure 2.6 The ornithine cycle. The numbers of nitrogen atoms in each compound are shown in parentheses.

Urea is soluble in water and less toxic than ammonia. Urea diffuses readily through the phospholipid bilayer of the membranes and so leaves hepatocytes and is transported to the kidneys dissolved in the blood plasma. We also excrete very small quantities of ammonia. There are two other nitrogenous waste products

● uric acid, which is made from excess purine bases (see page 50 in *OCR A level Biology 1 Student's Book*)

● creatinine, which is made from creatine phosphate in muscle (see Chapter 5 in this book).

Detoxification

The liver breaks down many substances that are no longer required or are toxic. This is known as **detoxification**. Some of these substances are

● lactate

● alcohol

● hormones

● medicinal drugs.

Lactate metabolism

Lactate is the end product of anaerobic respiration. It is produced by skeletal muscles during strenuous activity when there is insufficient oxygen supply in the blood (see Chapter 8, Anaerobic respiration, for more details). Lactate molecules diffuse out of muscles into the blood. Lactate is an energy-rich compound which is respired by cardiac muscle and some other tissues. The

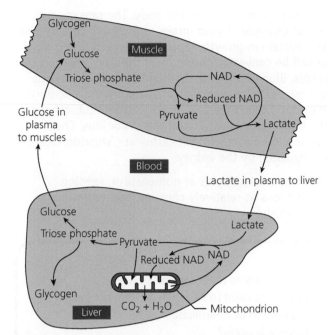

Figure 2.7 Lactate is produced in skeletal muscle tissue during strenuous exercise, and then transported in the blood to the liver where it is metabolised.

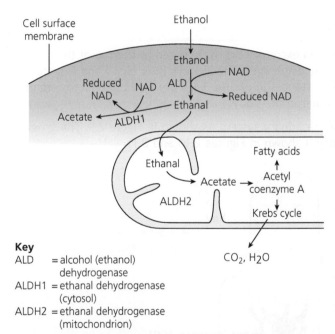

Key

ALD = alcohol (ethanol) dehydrogenase

ALDH1 = ethanal dehydrogenase (cytosol)

ALDH2 = ethanal dehydrogenase (mitochondrion)

Figure 2.8 The metabolism of alcohol (ethanol) in the liver.

rest is absorbed by liver cells and metabolised. Lactate is converted to pyruvate, some of which enters mitochondria to be respired aerobically to provide energy for converting the rest of the lactate to glucose. Some of the glucose produced is stored as glycogen and the rest enters the blood to help restore the normal concentration in the blood.

The reaction in which lactate is converted to pyruvate is an oxidation reaction that is coupled with the reduction of the coenzyme NAD (see Figure 2.7). The reduced coenzyme is a source of energy (see Chapter 8 for how this happens).

Alcohol metabolism

Alcohol (ethanol) is readily absorbed in the stomach. It is very quickly distributed throughout the body and is absorbed by hepatocytes. Like lactate, ethanol is a good source of energy and is respired by hepatocytes in preference to fat. Most is converted into ethanal (CH_3CHO) and then into acetate (CH_3COO^-), which can either be respired or used to make other compounds such as fatty acids. The rest is oxidised by enzymes in the smooth endoplasmic reticulum.

You can see in Figure 2.8 that the metabolism of alcohol is linked with the reduction of NAD, which is recycled by oxidising it in mitochondria. This in turn generates ATP for the cell, as you will discover in Chapter 8 (see Ethanol fermentation). Since the metabolism of ethanol generates plenty of ATP, liver cells do not use as much fat as usual so this gets stored within the cells giving rise to the condition known as fatty liver. The liver cells of binge drinkers show this condition for several days after a bout of heavy drinking. The fat stored in the liver reduces the efficiency of the hepatocytes in carrying out their many functions. Fatty liver can lead to life-threatening conditions such as cirrhosis, a condition that is on the increase in the UK among young people who misuse alcohol.

Other drugs, such as paracetamol, steroids and antibiotics, are also broken down in the liver.

Hormones

Hormones are also removed from the circulation and metabolised. Protein hormones, such as insulin and glucagon, and peptide hormones, such as anti-diuretic hormone, are hydrolysed into amino acids. Steroid hormones, such as oestrogen and testosterone, are also inactivated by conversion to other compounds, which are excreted in the urine (see Urine tests, in this chapter).

Metabolic wastes are produced continually in the body. The rate of production can change. For example, if your intake of protein increases and you do not use the extra protein in growth, repair or muscle building, then the excess amino acids will be converted to urea and the concentration of urea in the blood increases. Blood is constantly filtered by the kidneys to remove any metabolic wastes. This prevents their concentrations increasing to levels that will be harmful. The kidney also removes useful substances that are in excess in the body, such as water and the major ions. The concentrations of potassium, sodium, hydrogen carbonate, chloride, calcium and phosphate ions are regulated by the kidney.

These functions of the kidney are examples of homeostasis: keeping excretory wastes, water and ions at relatively constant concentrations so that they do not become harmful.

Test yourself

1 List the metabolic waste substances that mammals excrete.
2 Explain why excretory products need to be removed from the body.
3 Explain why amino acids have to be deaminated.
4 Where does the ornithine cycle occur?
5 One of the functions of the liver is detoxification. List the substances that the liver destroys or makes harmless.
6 Name the polysaccharide that is stored in hepatocytes.

The kidney

The kidneys are the main excretory organs situated at the top of the abdomen just below the diaphragm. The kidneys filter blood and excrete waste products and substances in excess. They are also the effector organ in the regulation of water in the body.

The kidney is part of the urinary system as shown in Figure 2.9. The gross structure of the kidney is shown in Figures 2.10 and 2.11.

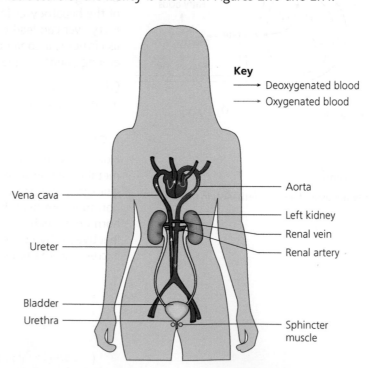

Key
→ Deoxygenated blood
→ Oxygenated blood

Vena cava
Aorta
Left kidney
Renal vein
Ureter
Renal artery
Bladder
Urethra
Sphincter muscle

Figure 2.9 The urinary system. The kidneys are situated very close to the heart.

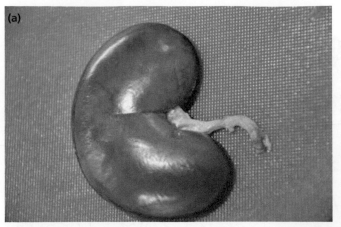

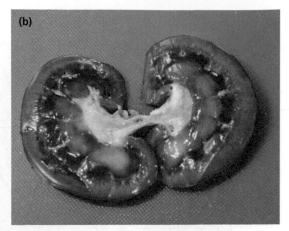

Figure 2.10 The gross structure of a kidney. (a) An external view of a kidney with a ureter attached. (b) A kidney in vertical section.

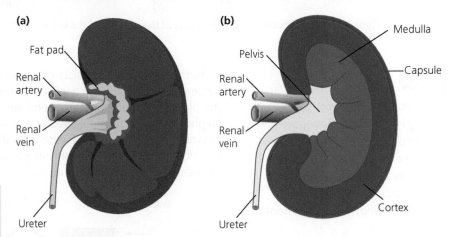

Figure 2.11 The gross structure of a kidney. (a) An external view and (b) a vertical section.

Key term

Nephron The functional unit of the kidney that filters blood, reabsorbs useful substances and forms urine. Each nephron consists of a glomerulus and a tubule and is associated with many blood capillaries.

The kidneys are full of many tubules known as nephrons. It is not possible to see complete nephrons in any section of a kidney because they are not regular structures that lie in any one plane within the kidney. Sections like those in the photomicrographs in Figures 2.12 and 2.13 show the parts of many adjacent nephrons.

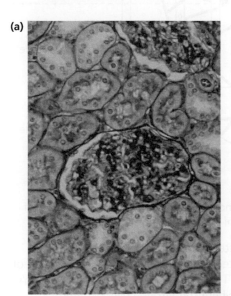

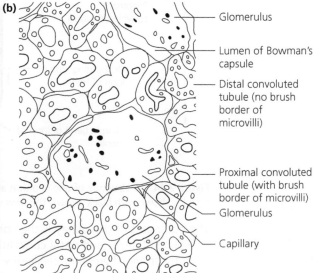

Figure 2.12 (a) A photomicrograph of the cortex of the kidney (x200) and (b) a drawing of part of the photomicrograph.

(a)

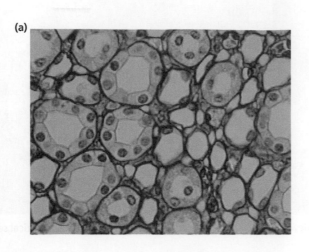

(b)

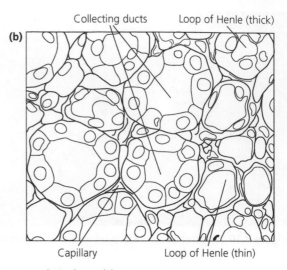

Figure 2.13 (a) A photomicrograph of part of the medulla of the kidney (x600) and (b) a drawing of part of the photomicrograph.

Tip

In your practical work you may have to make low-power plan drawings and high-power drawings like those in Figures 2.12(b) and 2.13(b) from microscope slides. See *OCR A level Biology 1 Student's Book* Chapter 17.

At birth, each kidney consists of around a million nephrons. The number decreases with age. Each nephron filters blood to produce a filtrate from the plasma. Filtrate contains useful substances as well as waste substances. Nephrons reabsorb useful substances into the blood and control the volume of water lost in the urine. Figure 2.14 shows the structure of a nephron. The urine formed in the nephrons passes into collecting ducts. It is usual to consider the collecting duct as part of a nephron.

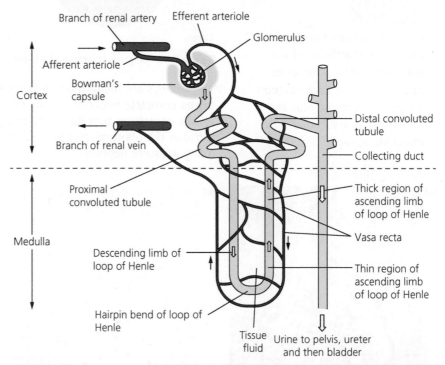

Figure 2.14 The structure of one nephron and its associated blood vessels. The flow of blood is shown with black arrows, the flow of filtrate and urine with white arrows.

Tip

Lobules are the functional unit structures of the liver; nephrons are the functional unit structures of the kidney.

The outer part of the kidney is the cortex and contains glomeruli, proximal convoluted tubules and distal convoluted tubules. Each glomerulus consists of a tightly arranged group of capillaries. These capillaries sit inside a Bowman's capsule, which is a cup-like structure where the filtrate collects before entering the rest of the nephron.

The inner part of the kidney is the pelvis, where urine collects. Between it and the cortex is the medulla, which contains loops and collecting ducts. These are visible in Figure 2.13 in transverse section. Table 2.2 summarises the structural features of the main regions of the kidney that you can see with the light microscope, and their functions.

Table 2.2 Regions of the kidney, their structural features and functions.

Region of the kidney	Structural features	Function
Capsule	Thin layer of connective tissue with collagen fibres	Protection
Cortex	Glomeruli, proximal and distal convoluted tubules, blood vessels	Ultrafiltration in glomeruli; selective reabsorption in proximal and distal convoluted tubules
Medulla	Loops of Henle and collecting ducts with associated capillaries (vasa recta)	Loops of Henle form tissue fluid with a low water potential. Collecting ducts may reabsorb water from urine by osmosis
Pelvis	Urine-filled space surrounded by white fibrous tissue that connects to the ureter	Collects urine from collecting ducts
Ureter	Thick-walled, muscular tube (smooth muscle)	Moves urine to the bladder by peristalsis

Tip

As you read about the structure and function of each part of the kidney, look carefully at the photomicrographs and drawings in Figures 2.12 and 2.13 as well as the diagram of the nephron in Figure 2.14.

Test yourself

7 List the structures through which urea travels from where it is produced to where it leaves the body.
8 A kidney is cut in vertical section. Name the regions of the kidney that you can see without the aid of a microscope or hand lens.
9 Nephrons are the functional units of the kidney. What does this mean?
10 Name the parts of a nephron.
11 Calculate the largest diameter of the glomerulus shown in Figure 2.12(a).

Features of the different parts of the nephron that can be seen with the light microscope are described below.

Each **glomerulus** is composed of a tight knot of capillaries. Surrounding the capillaries are cells with prominent nuclei. These cells are **podocytes** (literally, foot cells) that have branching extensions, as shown in Figure 2.15. These cells suspend the capillaries within the **Bowman's capsule**, which is the fluid-filled space around the glomerulus. The Bowman's capsule is lined by a single layer of squamous epithelial cells (indicated by yellow in Figure 2.14).

The filtrate collects in the Bowman's capsule before entering the **proximal convoluted tubule** (PCT). The tubule is formed from cuboidal epithelial cells. These cells are lined by a brush border made up of many microvilli and contain many mitochondria (see Figures 2.12 and 2.16).

The volume of the filtrate decreases as it moves along the PCT as substances and much water are reabsorbed into the blood. The filtrate continues into the **loop of Henle**, which is narrower in cross-section, with a squamous epithelium. The descending limb of the loop enters the medulla and turns a hairpin bend, with the ascending limb returning towards the cortex. The ascending limb has thicker, cuboidal epithelial cells with no brush border, but many mitochondria. The loops are surrounded by the **vasa recta**: capillaries that also dip down into the medulla and return to the cortex.

After leaving the loop, the filtrate enters the **distal convoluted tubule** (DCT), which is in the cortex. The DCT has cuboidal cells, but they are not as wide as the cells of the PCT. The cells have few microvilli and therefore no brush border.

Several nephrons drain into a **collecting duct** in the medulla. Transverse sections of these ducts are wider than those of the other parts of the nephron. The epithelial cells are like those of the DCT: cuboidal without a brush border. The liquid that flows out of the DCT is urine. It may flow unchanged all the way through the collecting duct into the pelvis, but when water is in short supply the collecting ducts become permeable to water and reabsorb water from the urine into the blood. This helps to maintain the water balance of the blood by producing urine that is more concentrated.

Ultrafiltration

Blood enters the glomerulus from a branch of the renal artery at high pressure. Notice in Figure 2.14 that the diameter of the efferent arteriole is narrower than the diameter of the afferent arteriole; this builds up a head of pressure to force small molecules into the Bowman's capsule. This is **pressure filtration** that occurs in all capillaries but there are structural adaptations in the glomerulus to make this even more effective at removing substances from the blood. Filtration here is known as **ultrafiltration**. The hydrostatic pressure of the blood which forces fluid out of the capillaries is opposed by the oncotic pressure of the proteins in the plasma (see page 151 in the *OCR A level Biology 1 Student's Book*). The filtrate also has a hydrostatic pressure and an oncotic pressure, although both of these are very low. The net effect of these four pressures is an overall pressure forcing substances from the blood into the filtrate.

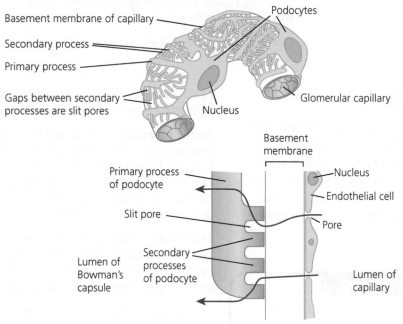

Figure 2.15 Ultrafiltration in the glomerulus. Endothelial cells forming the capillary walls have many pores in them and podocyte cells have slit pores to reduce the resistance to the flow of filtrate.

This region is well adapted for filtration because the endothelial cells lining the capillaries have pores in them that allow substances to leave the blood. On the outside of these cells is a basement membrane made of glycoproteins. This acts as a sieve retaining all the blood cells and platelets. The basement membrane allows substances with a relative molecular mass (RMM) of less than 69 000 through the glomerular capillaries into the filtrate. Most of the proteins in the plasma are larger than this so they are retained in the blood. The capillaries in the glomerulus are supported by podocyte cells that form an incomplete layer so they do not offer any resistance to the flow of filtrate.

Example

1 Outline the role of the kidneys in homeostasis.

2 The total blood volume in a human is about 5.5 dm³. At rest the kidneys receive 25% of the blood pumped out by the heart. The cardiac output at rest is 5.6 dm³ min⁻¹. In a healthy person, the glomerular filtration rate (GFR) remains constant all the time at 125 cm³ min⁻¹. The volume of urine produced each day varies between 1.2 and 2.0 dm³.

   a) Calculate the volume of blood that flows through the kidneys each minute.

   b) Use the figure you have calculated in (a) to determine the percentage of the blood flow through the kidneys that becomes filtrate.

   c) A person produces 1.5 dm³ of urine in a day. Calculate the percentage of the filtrate that is lost in the urine during the course of the day.

3 Explain how the hydrostatic pressure and oncotic pressure of the blood and filtrate influence the GFR.

Answers

1 The kidneys are the body's main excretory organ. They keep metabolic wastes, such as urea, below harmful concentrations. The kidneys are the effector organs in regulating the salt and water balance of the blood. They also help to maintain the blood pH. If these factors change they would interfere with the action of enzymes and so metabolism would not function efficiently. Cells might also swell and burst if blood became too dilute.

2 a) The volume of blood pumped out of the heart = 5.6 dm³ min⁻¹. The volume of blood that enters the kidneys is 25% of this

$$= \frac{25}{100} \times 5.6 = 1.4 \, dm^3 \, min^{-1} = 1.4 \times 1000$$
$$= 1400 \, cm^3 \, min^{-1}$$

b) The volume of filtrate formed = 125 cm³ min⁻¹. The percentage of the blood that flows through the kidney that becomes filtrate

$$= \frac{125}{1400} \times 100 = 8.93\%.$$

c) The volume of filtrate produced in a day
$$= 125 \, cm^3 \, min^{-1} \times 60 \times 24$$
$$= 180 \, dm^3.$$ The volume of urine as a percentage of the volume of filtrate $= \frac{1.5}{180} \times 100 = 0.83\%.$

3 The hydrostatic pressure of the blood is blood pressure created by the contraction of the heart. This pressure is high in the capillaries of the glomerulus as the efferent arteriole is narrower than the afferent arteriole. The filtrate has a hydrostatic pressure but that is much lower. Oncotic pressure is the osmotic effect of large molecules, especially the protein albumen, in the plasma. The oncotic pressure of blood plasma is greater than the oncotic pressure of filtrate because very few proteins pass through the basement membrane. Hydrostatic pressure of the blood forces blood out of the capillaries. The hydrostatic pressure of filtrate counteracts this pressure, but does not equal it. The oncotic pressure of blood plasma opposes the movement of filtrate into Bowman's capsule. The oncotic pressure of filtrate has a negligible effect.

Selective reabsorption

The filtrate collects in the Bowman's capsule and passes into the proximal convoluted tubule (PCT). The cells that line the PCT are specialised for reabsorption of useful substances from the filtrate, including

- glucose
- ions, e.g. sodium ions
- amino acids
- water.

Absorption of glucose and amino acids requires a supply of ATP from respiration so this is active uptake. ATP is used to provide energy for sodium/potassium (Na⁺/K⁺) protein pumps on the lateral and basal membranes of the PCT cells. These create a low concentration of sodium ions inside the cytoplasm. This means there is a concentration gradient for sodium ions from the filtrate into the cytoplasm, which is used to drive the uptake of other molecules, such as glucose and amino acids.

Sodium ions can only diffuse through specialised channel proteins or carriers. The carrier protein on the luminal membrane of the PCT cells facing the filtrate is a symport (or **co-transporter protein**) that has binding sites for sodium ions and glucose. When both of these binding sites are filled the carrier changes shape to deliver the sodium ions and glucose into the cytoplasm. This gives a high concentration of glucose inside the cell and glucose molecules diffuse out through carrier proteins in the basal and lateral membranes into the blood of the vasa recta. The absorption of glucose and amino acids in this way is an example of **indirect active transport** as the molecules themselves diffuse into and out of the cell, but this movement is driven by the active transport of sodium ions and the presence of the co-transporter (symport) proteins in the luminal membrane. There are similar methods for reabsorbing amino acids.

Figure 2.16 shows the structure of a PCT cell and how it reabsorbs sodium ions and glucose. Urea also diffuses across the cells back into the bloodstream by diffusion. The movement of solutes from the filtrate to the blood gives the blood a lower water potential than the filtrate. As a result, water diffuses by osmosis from the filtrate back into the blood. The reabsorption of urea and water is passive.

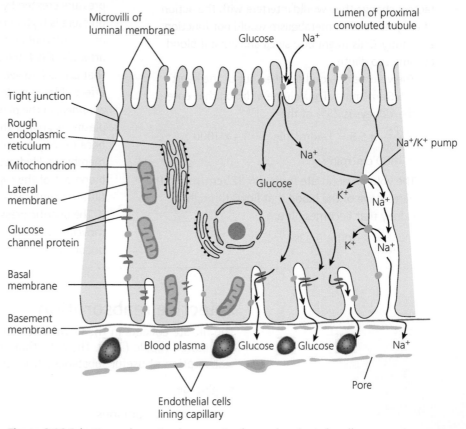

Figure 2.16 Selective reabsorption by a proximal convoluted tubule cell.

These PCT cells are a very good example of the relationship between structure and function:

- Tight junctions, which are like Velcro, between cells ensure that movement occurs *through* the cells and not *between* them.

- Microvilli provide a large surface area, allowing many symport carrier proteins to line the tubule (see Figure 2.17).

- Many mitochondria produce ATP for active transport by Na$^+$/K$^+$ protein pumps.

- Infoldings of the basal membrane give a large surface area for protein pumps and carrier proteins for glucose.

- Rough endoplasmic reticulum makes proteins for Na$^+$/K$^+$ pumps, symport carrier proteins and carrier proteins for glucose and amino acids.

By the time filtrate has reached the end of a PCT a large proportion of the solutes and much of the water have been reabsorbed. The volume has decreased considerably. The remaining filtrate passes into the descending loops.

Making a water potential gradient

The main function of the rest of the nephron is to regulate the concentration of the blood and determine the concentration of the urine. Mammals have two types of nephron. Nephrons can either have short loops that do not extend very far, if at all, into the medulla or long loops that extend to the tip of the medulla. Human kidneys have a mixture of long and short loops with about 15% of the loops extending to the tip of the medulla.

We are able to concentrate urine by a factor of four compared with the concentration of blood plasma. This means that when dehydrated we can conserve water rather than letting it go to waste in the urine. When required, water can be conserved by reabsorbing it from the urine that travels through the collecting ducts. Concentrating the urine is achieved by making the tissue fluid rich in ions and urea so it has a lower water potential than the urine in the collecting ducts. Remember that water diffuses by osmosis down its water potential gradient from a region with a high water potential to a region with a low water potential.

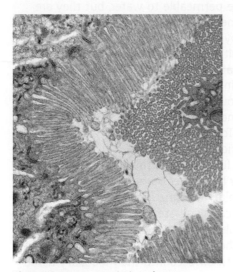

Figure 2.17 A transmission electron micrograph of brush borders of cells from the proximal convoluted tubule (x6000).

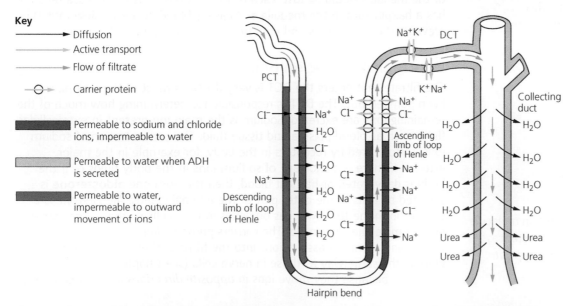

Key

→ Diffusion
→ Active transport
→ Flow of filtrate
—○→ Carrier protein

Permeable to sodium and chloride ions, impermeable to water

Permeable to water when ADH is secreted

Permeable to water, impermeable to outward movement of ions

Figure 2.18 Loops of Henle and collecting ducts determine the volume and concentration of urine produced.

As filtrate flows down the descending loops of Henle it becomes more concentrated. Cells lining the walls are permeable to water, but they are not permeable to the outward movement of ions and urea. Water diffuses out into the surrounding tissue fluid by osmosis. Sodium ions and chloride ions diffuse into the filtrate. As a result of the loss of water and gain of ions, the filtrate increases in concentration as it flows down towards the hairpin bend. The walls of the ascending limbs are impermeable to water, but permeable to ions, which diffuse out into the tissue fluid. As filtrate flows up the ascending loops it becomes less concentrated as a result of the loss of sodium and chloride ions. Towards the top of the ascending loops the walls are thicker with cells rich in mitochondria. These cells pump ions out of the filtrate into the tissue fluid and it is this active transport that is mainly responsible for the very low water potential of the tissue fluid in the medulla. Mammals with relatively long loops have space for many of these cells and therefore can generate a much lower water potential in the medulla than mammals with relatively short loops. When water is in short supply in the body, the cells at the base of the collecting ducts become permeable to urea, which diffuses into the medullary tissue fluid to contribute to decreasing the water potential so more water diffuses out of the urine before it enters the pelvis.

Countercurrent exchange

You will have noticed from Figure 2.14 that loops, vasa recta (capillaries) and collecting ducts are all parallel to one another. This arrangement produces a concentrated tissue fluid in the medulla. The descending and ascending limbs of the loops are all close together and they exchange substances through the tissue fluid. The nephrons concentrate the sodium and chloride ions in the lower part of the medulla because there is diffusion of ions from ascending to descending limbs. The same principle applies to the capillaries in this region, which lose water and gain sodium ions and chloride ions as the blood flows down into the medulla. The reverse exchanges occur as the blood flows up the medulla. This countercurrent exchange ensures that the ions in the medullary tissue do not 'leak' away in the blood leaving the kidneys. If there were no countercurrent exchange, solutes in the medullary tissue fluid would diffuse into the blood. The blood would then transport them out of the kidneys and the concentrating effect of the medulla would be lost. Each of the blood vessels in the vasa recta has a hairpin turn in the medulla and carries blood at a very slow rate; this prevents the loss of ions and urea from the tissue fluid in the medulla.

Controlling excretion of ions

The filtrate that enters the DCT is very dilute as most of the ions have been pumped out. The DCT is responsible for determining how much of the remaining ions are excreted. Sodium is the main ion present in extracellular fluids such as blood plasma and tissue fluid. The concentration of sodium ions is monitored by receptors in the body, for example in the major arteries. If there is a shortage of sodium ions in the body fluids or if the body is dehydrated or has lost blood, then the hormone **aldosterone** is secreted by cells in the outer part of the adrenal glands. Aldosterone is a steroid hormone that switches on genes in DCT cells so that more carrier protein molecules are made. The carriers pump sodium ions out of the filtrate and pump potassium ions into the filtrate. These carrier proteins work in the same way as those in nerve cells (see Chapter 3) and are antiports because they move ions in *opposite directions across membranes*.

Tip

The mechanism that drives the high concentration of solutes in the medullary tissue fluid is the active transport of sodium and chloride ions out of the filtrate that occurs in the thick ascending limbs of the loops of Henle. The hairpin bends of the loops and the vasa recta maintain this high concentration.

Key term

Antiport A carrier protein molecule that transports two substances across a membrane *in opposite directions*.

Regulation of the water content of the blood

We have already seen how body temperature is controlled (see Chapter 1). The water potential of the blood is another physiological factor that is kept within narrow limits. The hypothalamus is the control centre and contains osmoreceptors that detect changes in the water potential of the blood. Cells do not function efficiently if their water content varies. If the water potential of the blood is too high cells absorb water and swell, and maybe even burst. If the water potential of the blood is too low water will diffuse out of the cells by osmosis and they will shrink. Figure 2.19 shows what happens in conditions of water stress when the body is dehydrated.

The axons of neurosecretory neurones from the hypothalamus extend into the posterior pituitary gland. Instead of forming synapses with other neurones they terminate near blood capillaries. When they are stimulated, impulses travel down the axons to release molecules of a small peptide hormone known as **anti-diuretic hormone** (ADH) by exocytosis of vesicles. The target cells of ADH are the cells of the DCT and collecting ducts. Figure 2.20 shows what happens to these cells when ADH attaches to receptors on their surfaces. **Aquaporins** are water channels that are inserted into the cell surface membrane. When open, 3 billion molecules of water a second move through each aquaporin.

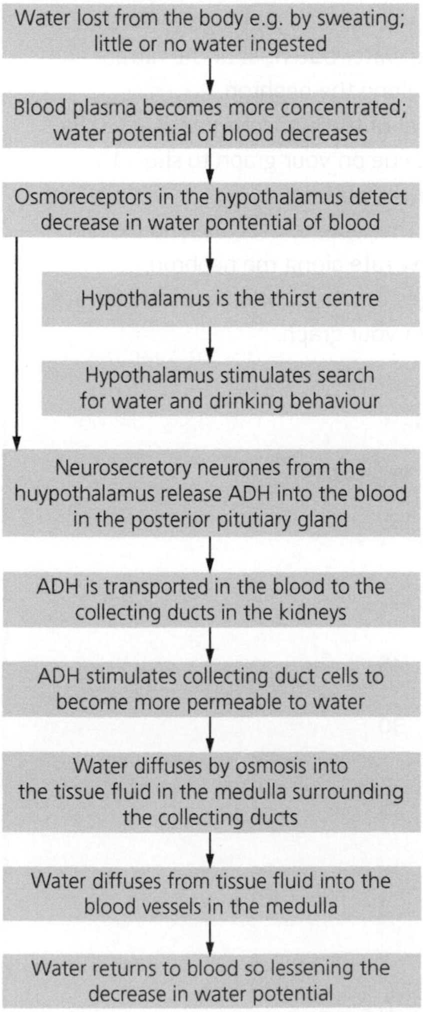

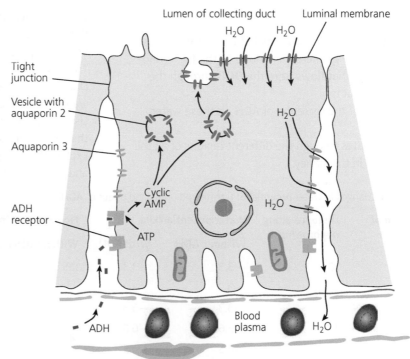

Figure 2.20 The changes that occur in a collecting duct cell in response to stimulation by ADH (aquaporin 1 is found on cells of the PCT and the descending limb of the loop of Henle).

When the water potential of the blood is higher than the set point, the osmoreceptors are not stimulated and ADH is not secreted. ADH also stimulates the movement of channel proteins into cells at the base of the collecting duct. These channels allow urea to diffuse down its concentration gradient into the surrounding tissue fluid to decrease the water potential. In the absence of ADH, aquaporins are taken back into the cytoplasm by endocytosis at the luminal surface. This means that the membrane becomes impermeable to water and no water is reabsorbed by osmosis. Urine flows straight through the collecting ducts without any loss in water. Under these conditions large

Water lost from the body e.g. by sweating; little or no water ingested

↓

Blood plasma becomes more concentrated; water potential of blood decreases

↓

Osmoreceptors in the hypothalamus detect decrease in water pontential of blood

↓

Hypothalamus is the thirst centre

↓

Hypothalamus stimulates search for water and drinking behaviour

↓

Neurosecretory neurones from the huypothalamus release ADH into the blood in the posterior pituitary gland

↓

ADH is transported in the blood to the collecting ducts in the kidneys

↓

ADH stimulates collecting duct cells to become more permeable to water

↓

Water diffuses by osmosis into the tissue fluid in the medulla surrounding the collecting ducts

↓

Water diffuses from tissue fluid into the blood vessels in the medulla

↓

Water returns to blood so lessening the decrease in water potential

Figure 2.19 This flow chart shows how the kidney is involved in osmoregulation.

volumes of dilute urine are produced. Control of water potential of the blood is another example of negative feedback (see Chapter 1).

12 What is ultrafiltration?

13 State three ways in which the glomerulus is adapted for efficient ultrafiltration.

14 Describe three ways in which the structure of a PCT cell is adapted for reabsorption.

15 Why do the microvilli on either side of the lumen in Figure 2.17 look different?

16 What is the role of loops of Henle in the medulla?

17 What is the role of aldosterone?

18 Suggest how the spinifex hopping mouse is able to produce urine 25 times more concentrated than its blood plasma.

Activity

Changes along the nephron

Table 2.3 shows

- changes in the water potential of the filtrate that is formed in the glomeruli and flows to the DCTs, and of the urine that flows from the DCTs through the collecting ducts
- the effect of the secretion of ADH on these water potentials
- the flow rates through the different regions of all the nephrons in the kidney.

1 Use the figures in the table to plot a graph to show the changes in the water potential of the filtrate and urine as it passes along the nephron.

2 The water potential of the blood plasma is −3.3 kPa. Draw a horizontal line on your graph to show the water potential of the plasma.

3 Use the figures in the table to describe and explain the changes in flow rate along the nephron.

4 Explain the changes in water potential of the filtrate and urine shown in your graph.

Table 2.3 Changes in water potential of filtrate and urine, and effect of ADH.

Region of nephron	Distance along nephron/mm	Water potential/kPa		Flow rate/cm³ min⁻¹	
		Without ADH	**With ADH**	**Without ADH**	**With ADH**
PCT	0	−3.3	−3.3	125	125
	7	−3.3	−3.3		
	14	−3.3	−3.3		
Descending loop	16	−6.6	−6.6	45	45
	19	−10.5	−10.5		
Hairpin bend	20	−14.0	−14.0	30	30
Ascending loop	21	−12.0	−12.0	25	25
	23	−9.0	−9.0		
	25	−1.5	−1.5		
DCT	26	−1.0	−3.0	15	8
	27	−0.8	−7.8		
Collecting duct	28	−0.5	−9.0	20	0.2
	30	−0.3	−11.0		
	32	−0.3	−14.0		

Urine tests

You may well have had a medical examination and been asked for a sample of urine. The most common urine test is for 'sugar' (glucose). As we have seen, all the glucose in the filtrate should be reabsorbed by the PCT. If the blood glucose concentration increases too much, then the PCT cells cannot reabsorb all of the glucose. The limit is the **renal threshold**, which is 180 mg 100 cm^{-3} blood. If glucose is present in the urine, then this may indicate that there is something wrong with the homeostatic control of glucose, especially the function of insulin. There is more about this in Chapter 4. A doctor may decide to give a person a glucose tolerance test, which is also described in Chapter 4. Ketones are produced by the metabolism of people who have diabetes. The presence of ketones (acetone (propanone) and acetoacetate) in the blood is an indicator that a person has diabetes mellitus.

Urine is also tested for protein. This should not be present in filtrate, let alone urine, as most plasma proteins have relative molecular masses (RMM) greater than 70 000. Serum albumin (also known as albumen) is the most common plasma protein. The RMM of albumin is 69 000, but as it is negatively charged, only a very small quantity is filtered. Any filtered albumin is reabsorbed by endocytosis in the PCT, broken down and the amino acids absorbed into the blood. The presence of protein in urine may indicate that the blood pressure is too high, or it may indicate a kidney infection or that there is something wrong with the filtration mechanism. Positive results for albumin are common in pregnancy, but usually become negative after delivery.

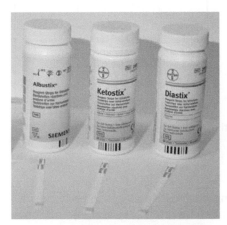

Figure 2.21 These test strips are used to test urine for glucose, protein and ketones.

Positive tests for nitrite ions indicate a bacterial infection in the urinary tract.

Rather than use test strips for single substances, urinalysis test strips test for 10 different factors including those listed above and the pH of the urine. These test strips contain chemicals that react with the test substance and give a colour change to indicate positive results (Figure 2.21).

Tip

As part of your practical work you may use test strips to test fake urine for glucose, protein, ketone bodies and nitrite ions.

Urine samples are also used for pregnancy testing (Figure 2.22). These testing sticks contain molecules of a monoclonal antibody that are specific to the hormone human chorionic gonadotrophin (hCG), which is secreted by the early embryo shortly after implantation in the uterus (Figure 2.23). The antibodies all originate from one clone of B lymphocytes that all produce the same antibody specific to hCG. This reduces the chance of getting false results.

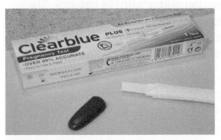

Figure 2.22 A pregnancy testing kit. There are similar kits that detect the release of the hormone that stimulates ovulation: women use these to help them become pregnant.

Athletes are tested regularly to check that they have not been taking anabolic steroids to build up their muscle mass. Anabolic steroids stimulate protein synthesis. They are detected in the urine by gas chromatography or mass spectrometry. These drugs are based on steroid hormones and may have a half-life of 16 hours so it is important that samples are taken close to the event. The half-life of a drug is the time taken for its concentration in the blood to decrease to half its initial value.

Test yourself

19 Name the substance in urine that is detected by a pregnancy testing kit.

20 Explain, briefly, why there are two types of monoclonal antibody in a pregnancy testing kit like that in Figure 2.22.

21 Explain why some athletes and body builders take anabolic steroids.

22 How are anabolic steroids detected in the urine of athletes?

23 a) What is meant by the term *renal threshold*?
   b) Explain the importance of the renal threshold for glucose.

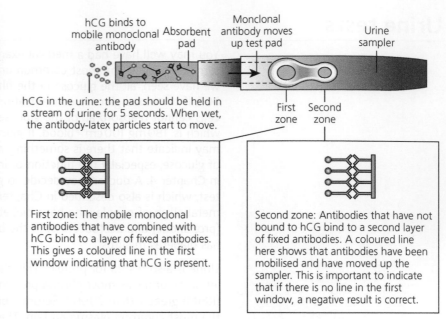

hCG binds to mobile monoclonal antibody Absorbent pad Monclonal antibody moves up test pad Urine sampler

hCG in the urine: the pad should be held in a stream of urine for 5 seconds. When wet, the antibody-latex particles start to move.

First zone Second zone

First zone: The mobile monoclonal antibodies that have combined with hCG bind to a layer of fixed antibodies. This gives a coloured line in the first window indicating that hCG is present.

Second zone: Antibodies that have not bound to hCG bind to a second layer of fixed antibodies. A coloured line here shows that antibodies have been mobilised and have moved up the sampler. This is important to indicate that if there is no line in the first window, a negative result is correct.

Figure 2.23 Antibodies are used to detect the hormone hCG in urine of pregnant women.

Kidney failure

Kidneys may fail for a variety of reasons, such as blood loss in an accident, high blood pressure, diabetes, overuse of some drugs (e.g. aspirin) and some infections. If kidney failure happens, then this is likely to prove fatal within a short period of time as urea, water, salts and toxins are retained and not excreted. Less blood is filtered so the glomerular filtration rate decreases. Below 60 cm^3 min^{-1} is a dangerous level for glomerular filtration rate. There is a build-up of toxins in the blood and the concentrations of ions and charged compounds are not maintained.

The build-up of potassium ions in the blood is associated with abdominal cramping, tiredness, muscle weakness and paralysis. If the concentration rises too far the impulses from the sinoatrial node in the heart may slow down, leading to arrhythmia and cardiac arrest. Sodium plays a major role in fluid balance, neuromuscular function and acid/base balance. The kidneys either conserve or excrete sodium depending on the body's needs. If a person's kidneys are not able to excrete sodium, he or she can become disorientated and have muscle twitches, increased blood pressure and general weakness.

In some cases the failure occurs suddenly and lasts for a short time; in others it is a long-term condition. There are two ways in which kidney failure is treated

- dialysis, in which toxins, metabolic wastes and excess substances are removed by diffusion through dialysis membrane

- kidney transplant.

Haemodialysis and peritoneal dialysis

Dialysis means separating small and large molecules using a partially permeable membrane. People receiving dialysis may use haemodialysis, as shown in Figures 2.24 and 2.25, or peritoneal dialysis. Haemodialysis involves regular treatment in hospital or at home.

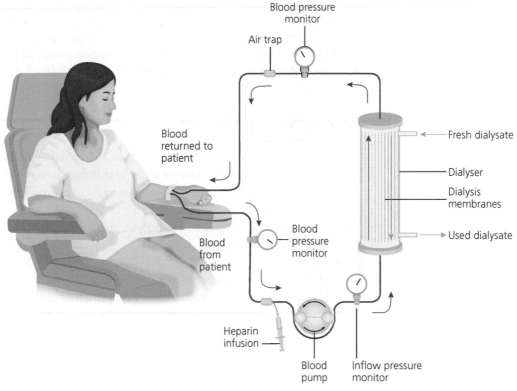

Figure 2.24 Haemodialysis.

Figure 2.25 Haemodialysis. Blood in the haemodialyser flows in the opposite direction to the flow of the dialysis fluid. Why is this?

In the haemodialyser partially permeable dialysis membranes separate blood from the dialysis fluid (dialysate). Blood flows through tubes of dialysis membrane and the tubes are surrounded by dialysate. The dialysate contains the substances that are required in the blood, such as glucose and sodium ions and these are in the correct concentrations. The dialysate does not contain any urea, so urea diffuses from the blood down its concentration gradient into the dialysate, which then goes to waste. Dialysis fluid and blood flow in opposite directions so there is a concentration gradient along the whole length of the dialyser. Each time a unit of blood passes through the dialysis machine it loses some of its urea. After about 3 hours almost all the urea in the blood is removed. During the treatment, heparin is added to the blood to prevent clotting.

The other form of dialysis is peritoneal dialysis. Dialysate is placed through a catheter into the abdominal cavity and urea, other wastes and substances excess to requirements diffuse from the blood across the lining of the abdomen (the peritoneum) into the dialysate, which is replaced after a while.

Dialysis places severe restrictions on people's lives. They may have to make regular trips to a hospital or clinic to receive treatment. Their diet is controlled carefully so that they do not produce too much urea and so that they do not ingest too much salt. A solution to these problems is to carry out a kidney transplant. Only one kidney is required, but the problem is finding a donor. The blood groups of donor and recipient must be compatible. It is less crucial to tissue match kidneys from living donors, but matches are important if the kidney comes from a dead donor. There are risks associated with the kidney transplant operation, as with all operations. In addition, kidney transplant patients need to take drugs to prevent their immune system rejecting the kidney, which will often be of a different tissue type.

Test yourself

24 Draw a flow chart diagram, similar to Figure 2.19, to show what happens after someone drinks a large volume of water so that there is excess in the body.
25 Explain why the release of ADH is an example of neurosecretion.
26 What are the causes of kidney failure?
27 a) Suggest the composition of dialysis fluid.
   b) Explain your answer.
28 Suggest
   a) the problems that occur with haemodialysis
   b) the risks associated with kidney transplantation.

Exam practice questions

1 Which region of a nephron is responsible for determining the volume of urine produced?
   A   Collecting duct
   B   Glomerulus
   C   Loop of Henle
   D   Proximal convoluted tubule   (1)

2 Which is the effect of anti-diuretic hormone on the kidneys?
   A   Decrease in the rate of formation of glomerular filtrate
   B   Decrease in the absorption of water in the proximal convoluted tubules
   C   Increase in the permeability of collecting ducts
   D   Increase in the opening of aquaporins in loops of Henle   (1)

3 The table shows some waste products of metabolism, the site of their production and the metabolic processes that produce them.

Waste product of metabolism	Site of production in the body	Metabolic process
Carbon dioxide	All respiring tissues in the body	
Ammonia		
Urea		
Bile pigments		Breakdown of haem from haemoglobin

   a) Copy and complete the table.   (4)
   b) Ammonia is a waste substance but is converted to urea in the body. Explain why this is necessary.   (2)
   c) Excretion also involves the removal of substances that are required by the body but are in excess. Name two of these substances.   (2)
   d) Outline the roles of the kidneys in homeostasis.   (4)

4 The table shows the hydrostatic pressure and oncotic pressure of the blood plasma in glomerular capillaries and in the filtrate in the Bowman's capsule.

	Hydrostatic pressure/kPa	Oncotic pressure/kPa
Blood plasma in glomerular capillaries	6.0	−3.3
Filtrate in Bowman's capsule	1.3	−0.3

Which is the net filtration pressure within the glomerulus?
   A   1.7 kPa
   B   2.7 kPa
   C   3.7 kPa
   D   4.7 kPa   (1)

5 There are many causes of acute kidney failure. One of these is rhabdomyolysis in which muscle tissue breaks down with the release of the contents of muscle fibres into the blood. Myoglobin is a globular protein with a relative molecular mass (RMM) of 17 800. It is very common in muscle tissue where it acts as a store of oxygen. People with rhabdomyelitis need to have haemodialysis.
   a) Explain why myoglobin does not appear in the urine of healthy people, but does in the case of people with rhabdomyolysis.   (3)
   b) Explain why renal dialysis is necessary to treat people with acute kidney failure.   (4)
   c) i)   Explain how haemodialysis regulates the composition of blood plasma.   (5)
       ii) During haemodialysis blood passes in the opposite direction to the dialysis fluid. Explain why.   (2)

6 The glomerular filtration rate (GFR) is the volume of fluid filtered from the glomerular capillaries per unit time. One method that has been used to determine the GFR involves injecting a known volume of a solution of inulin into a vein. Inulin is a small polysaccharide with a relative molecular mass of 5500. The rate at which inulin is filtered determines the rate at which it appears in the urine.
   a) i)   State three features of any substance, such as inulin, that make it suitable to determine the GFR by this method.   (3)
       ii) Explain why inulin cannot be given orally in this investigation.   (2)

A healthy person was given an injection of an inulin solution. The urine produced was collected over 3 hours and the concentration of inulin in the blood was determined at intervals. The table shows the results.

Mean concentration of inulin in blood/ mg cm⁻³	Volume of urine collected in 3 hours / cm³	Concentration of inulin in urine/g 100 cm⁻³
0.70	315	5.00

The GFR, in $cm^3\ min^{-1}$, is calculated using the following formula:

$$GFR = \frac{\text{concentration of inulin in the urine in mg cm}^{-3} \times \text{urine formation rate in cm}^3\ min^{-1}}{\text{concentration of inulin in the blood in mg cm}^{-3}}$$

b) Calculate

   i)   the concentration of inulin in the urine in $mg\ cm^{-3}$ (1)

   ii)   the GFR in $cm^3\ min^{-1}$ (2)

c) Explain why it is important to take into account the volume of urine in the calculation.   (2)

d) Suggest three factors acting at the glomerulus that could influence the GFR.   (3)

7 The table shows the concentrations of six substances in three body fluids: blood plasma, glomerular filtrate and urine.

Substance	Concentration/g 100 cm⁻³			
	Blood plasma	Glomerular filtrate	Urine	Increase
Protein	7–9	0	0	–
Glucose	0.1	0.1	0.0	–
Urea	0.03	0.03	2.00	
Ammonia	0.0001	0.0001	0.0400	
Sodium ions	0.32	0.32	0.30–0.35	×1
Water	90–93	97–99	96	–

a) The concentrations of urea and ammonia are higher in the urine than in the blood plasma. Calculate the factor by which the concentration of each substance increases.   (2)

b) Explain the concentrations of protein, glucose and urea in the three body fluids shown in the table.   (6)

c) Explain why water is present in urine.   (3)

d) Explain why the control of the water potential of the blood is an example of negative feedback.   (4)

Stretch and challenge

8 The table shows information about six species of mammal. The unit used for the concentration of the urine is a non-SI used for osmotically active solutions. Urine is a mixture of different solutes including urea and sodium chloride.

Species of mammal	1 Percentage of nephrons with long loops	2 Relative thickness of the medulla	3 Maximum concentration of urine /milliosmoles	4 Maximum ratio between concentration of blood plasma and urine
Beaver	0	1.3	520	2
Pig	3	1.6	1100	4
Human	14	3.0	1400	4
Cat	100	4.8	3000	10
Desert rat	27	8.5	5500	18
Spinifex hopping mouse	100	11.0	9000	25

Plot four scattergraphs of the information in the table

- column 1 against column 3
- column 1 against column 4
- column 2 against column 3
- column 2 against column 4.

You may find it easier and quicker to use a spreadsheet program to plot your graphs rather than drawing them on graph paper.

What conclusions can you make from the four scattergraphs?

Chapter 3

Neuronal communication

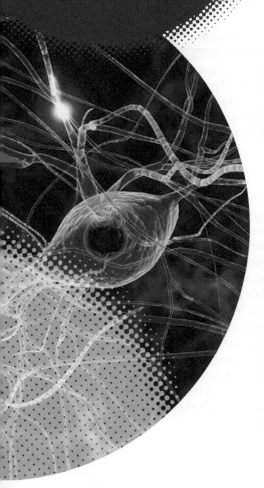

Introduction

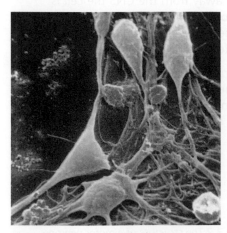

Figure 3.1 Neurones in the brain.

The nervous system is the most complex system in the human body, made up of nerve cells, or **neurones**, which are among the most unusual and specialised cells that we have. These are living cells which can generate and transmit electrical impulses, organised in such a way as to give us senses, coordinated movement, memories, logic and intelligence.

Here are some amazing facts about the human nervous system:

- There are about 13 500 000 neurones in the spinal cord, but that's nothing compared with the brain, which contains about 100 billion cells.

- If it was possible to line up all the neurones in the body, they would stretch for about 960 km.

- Nerve cells connect to one another by synapses. Every nerve cell in the brain connects with between 1000 and 20 000 others (Figure 3.1).

- Neurones can transmit electrical impulses at speeds of up to 100 m per second.
- Mature nerve cells are so complex that they can no longer divide by mitosis. This is why nerve damage is very difficult or sometimes impossible to repair.

The coordination of different organs and systems in different parts of the body requires a communication system. For example, if you are exercising, your muscles need an increased supply of glucose and oxygen for respiration, yet the muscles themselves cannot provide this without help from the blood and respiratory gas exchange systems. The lack of oxygen and nutrients or the increase in carbon dioxide in the exercising tissues must be detected and then a 'message' needs to be sent to the heart and the muscles of the thorax to boost circulation and breathing.

Communication

Multicellular organisms are complex systems and have an optimal state that they need to maintain by the process of homeostasis. Sometimes, circumstances change and a new optimum state is required, but the original optimum state will be restored once normal conditions return. Maintaining a steady optimal state and adapting to new conditions need extensive communication between different parts of the body. Certain organs need to send signals to others, either directly or indirectly, and the 'target' organs have to be able to detect the signal and respond appropriately. It is also important that *only* the target cells respond. The process by which one cell or group of cells communicates with others, either distant or in close proximity, is called **cell signalling**, and both nerves and hormones provide such signals. A living organism has to act as an integrated whole, even though there are specialised organs carrying out their different functions in widely separated parts of the body.

> **Tip**
>
> There is an introduction to the principles of cell signalling on page 86 of *OCR A level Biology 1 Student's Book*.

Nerve cell structure

Nerves are bundles of nerve cells (or **neurones**), each of which is highly specialised to conduct electrical impulses. We shall see later in this chapter that each neurone conducts impulses in only one direction. If this is towards the central nervous system (CNS) the cell is referred to as a **sensory neurone**. If the impulse travels away from the CNS the cell is a **motor neurone**. Sensory and motor neurones have different structures, but with many features in common.

The structures of sensory and motor neurones are shown in Figure 3.2. Some differences between the two types are shown in Table 3.1.

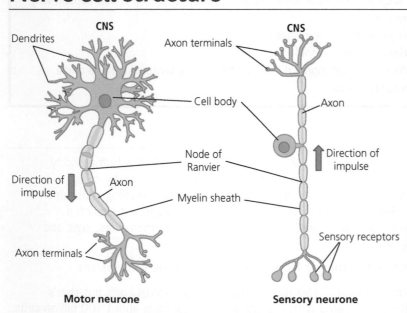

Figure 3.2 The structure of a motor neurone and a sensory neurone.

Table 3.1 Some differences between sensory and motor neurones.

Motor neurone	Sensory neurone
Cell body is at one end of the cell	Cell body is not at the end of the cell
Cell body has dendrites	Cell body does not have dendrites
Impulse travels away from the CNS to an effector	Impulse travels towards the CNS from a sensory receptor

The neurone **cell body** contains the nucleus. The cell body of a motor neurone has processes that connect the cell to other neurones in the CNS.

The axon, which carries the nerve impulse, is surrounded by a **myelin sheath**. This is produced by cells called **Schwann cells**, which wrap themselves around the axon (see Figure 3.2). The myelin is composed of lipid and protein, and has electrical insulating properties. There are gaps in the myelin sheath, called **nodes of Ranvier**, down the length of the axon. Here the surface membrane of the neurone is in direct contact with the extracellular fluid and is uninsulated. The nodes of Ranvier have an important role in nerve transmission, as we shall see below.

When the axon of a motor neurone reaches the effector organ (a muscle or a gland) it splits to form a number of **axon terminals**, which transmit impulses to the effector to bring about a response. The axon of a sensory neurone also has axon terminals, but these connect to other neurones in the CNS or directly to a motor neurone in some reflex arcs. **Sensory receptors** detect the stimulus and the sensory neurones transmit impulses to the CNS.

Other types of neurone

In addition to sensory and motor neurones, there are some other types. **Relay neurones** (also called **interneurones**) are cells with short axons that transmit signals from one neurone to another in the CNS. The sensory and motor neurones described above are myelinated – their axons have a myelin sheath – but some neurones in the nervous system are **non-myelinated** and have bare axons. We shall see below that the myelin sheath helps to increase the speed of the impulse, but when an impulse only has to travel a very short distance (as is often the case in the CNS) myelination is of no benefit.

> **Test yourself**
>
> 1 Why do only the target cells respond to a nerve signal?
> 2 State three structural differences between a sensory neurone and a motor neurone.
> 3 Which cells produce the myelin sheath around an axon?
> 4 What two types of organ form effectors in the mammalian body?

> **Key term**
>
> Effector An organ that becomes active in response to a nerve impulse.

Sensory receptors

> **Key term**
>
> Transducer Something that changes one form of energy into another.

People commonly refer to humans as having five 'senses', but in fact we have many more than that. A sensory receptor will convert some form of energy into a nerve impulse which is then sent to the CNS via a sensory neurone. Receptors therefore function as transducers. Figure 3.3 gives some examples of the sensory receptors scattered around the human body.

39

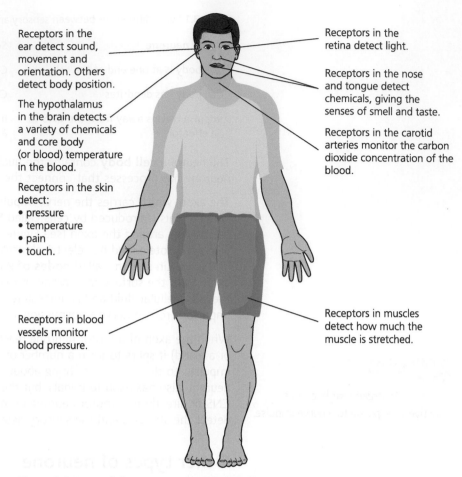

Receptors in the ear detect sound, movement and orientation. Others detect body position.

The hypothalamus in the brain detects a variety of chemicals and core body (or blood) temperature in the blood.

Receptors in the skin detect:
• pressure
• temperature
• pain
• touch.

Receptors in blood vessels monitor blood pressure.

Receptors in the retina detect light.

Receptors in the nose and tongue detect chemicals, giving the senses of smell and taste.

Receptors in the carotid arteries monitor the carbon dioxide concentration of the blood.

Receptors in muscles detect how much the muscle is stretched.

Figure 3.3 Some of the sensory receptors found in the human body.

Sensory receptors are of different types, depending on the nature of the stimulus they detect.

● Photoreceptors detect light.

● Chemoreceptors detect chemicals.

● Mechanoreceptors detect mechanical strain or stretching (the ear detects sound by this method: there is no specific type of receptor for sound).

● Proprioceptors detect body position.

● Baroreceptors detect blood pressure.

● Osmoreceptors detect concentration of body fluids.

● Nociceptors detect damage giving the sensation of pain.

An example of a sensory receptor

Pressure is detected in the skin by sensory receptors known as Pacinian corpuscles (see Figure 3.4). Pressure distorts the lamellae (which are made of fibrous connective tissue) and is transferred to the naked axon ending of a sensory neurone. This distortion causes ion channels to open, which initiates a nerve impulse in the neurone. The frequency at which nerve impulses are sent down the neurone is related to the amount of pressure, so the brain knows not only that the pressure exists, but also its intensity. The number of Pacinian corpuscles stimulated also provides information so that you can tell (for example) both the size and relative weight of something put on your hand without looking. If you actually lift an object, extra information becomes available from the stretch receptors in your arm muscles.

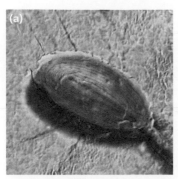

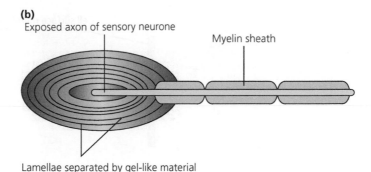

(b)
Exposed axon of sensory neurone

Myelin sheath

Lamellae separated by gel-like material

Figure 3.4 (a) Coloured scanning electron micrograph of a Pacinian corpuscle (x 35) and (b) diagram of a longitudinal section to show the parts.

Activity

Hearing thresholds

The hearing threshold is the lowest volume (in decibels, or dB) that can be detected by the sensory receptors in the ear. This value varies with the frequency of the sound. Scientists investigated how hearing thresholds varied with age. They tested 20 men at five different ages using sound frequencies from 1000 to 6000 Hz. They also tested 20 women aged 60. The results are shown in Figure 3.5.

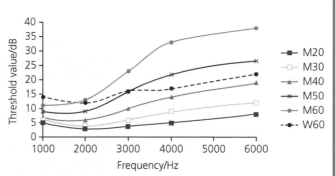

Figure 3.5 The threshold values for hearing at different sound frequencies for men (M) of different ages and for women (W) aged 60.

Questions

1 In each group, the age and sex of the subjects was controlled. Suggest other variables that would need to be controlled when each group was tested.
2 A sample of 20 was used for each age group. Suggest what information is needed to decide if this sample is big enough.
3 From the graph, describe how hearing sensitivity varies with age in males.

4 Compare the hearing sensitivities of males and females aged 60.
5 If you were designing a smoke alarm for the home suggest, with reasons, the most suitable sound frequency to use.

The nerve impulse

Key term

Electrical potential A form of potential energy resulting from behaviour of charged particles, such as their unequal distribution across a membrane.

The resting potential

Nerve cells work by generating and transmitting an electrical impulse. This is a highly unusual and specialised function, and neurones are adapted in many ways in order to carry it out. The nerve impulse results from ions passing through the cell membrane, and even when no impulse is being transmitted there is activity going on inside the neurone. The 'inactive' neurone has an ionic imbalance on either side of its membrane, and this creates what is called a **resting potential** (which is a form of electrical potential).

Just as with concentration gradients, any imbalance of charges across a membrane is usually removed because ions move along electrochemical gradients. If, for example, a positive charge builds up for some reason, the positive ions will tend to repel each other and will also attract negative ions, so any gradient diminishes or disappears (see Figure 3.6).

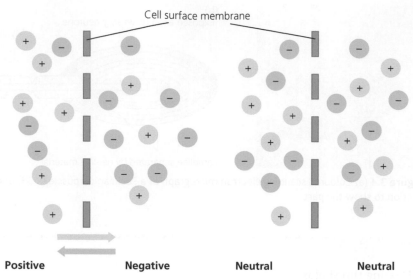

Cell surface membrane

| Positive | Negative | Neutral | Neutral |

The positive and negative ions are repelled by like charge and attracted by the opposite charge.

As a result, the charges on each side of the membrane balance out.

Figure 3.6 The natural balancing of charges on either side of a membrane which occurs in cells *other than* nerve cells.

Nerve cells maintain a potential difference across their cell membranes. The resting membrane potential of a neurone is around −70 mV. This potential is present because the outside of the membrane is positively charged compared to the inside. The difference in charge must be maintained in some way, against the natural tendency for the charges to even out. The resting potential results from the distribution of **sodium and potassium ions**. The membrane contains **channel proteins** that can allow these ions to diffuse through, and carrier proteins that function as **sodium-potassium pumps**. The way in which the resting potential is established is shown in Figure 3.7.

The sodium-potassium pump pumps 3 sodium ions out and 2 potassium ions in.

Potassium ions leak back out again because potassium channels are open.

Sodium ions cannot get back in because the sodium channels are closed.

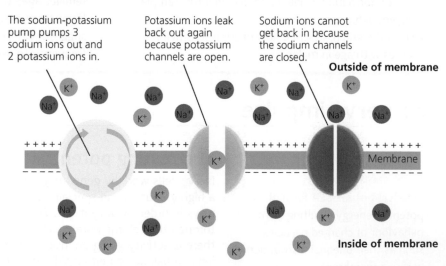

Figure 3.7 The establishment of the neurone membrane resting potential.

The sodium-potassium pump actively transports sodium ions out and potassium ions in using ATP, but does not do this equally: three sodium ions are pumped out for every two potassium ions pumped in. This has a number of effects:

- As a result of the pumps' actions there are more positive ions on the outside, so there is an electrochemical gradient.

- There are now more sodium ions on the outside, so there is a concentration gradient of sodium ions.

- There are more potassium ions on the inside, so there is a concentration gradient of potassium ions.

Sodium ions cannot get back through the membrane at this time because the sodium channels are closed. They therefore cannot follow their concentration gradient.

Although the potassium channels are also closed, they are very 'leaky', so potassium ions can follow their concentration gradient and move back out. This will be restricted to some extent by the electrochemical gradient (the positive potassium ions are moving towards a more positive environment), but the concentration gradient has a much greater effect.

The overall effect of this activity is to establish an equilibrium with more positive ions on the outside of the membrane than the inside, and a clear concentration gradient of sodium ions (more on the outside, fewer on the inside). This resting potential is maintained until a stimulus disrupts it.

Scientists consider that the *main* cause of the resting potential is the fact that the membrane is much more permeable to potassium ions than to sodium ions. The role of the sodium-potassium pump is mostly to create the potassium concentration gradient, although the pump does also affect the membrane potential to some extent.

The action potential

An impulse starts in a neurone when the resting potential is converted into an **action potential**. The events in a myelinated neurone are described below, and are also illustrated in Figure 3.8. The stimulus can be an electrical impulse from another neurone or a chemical change around the membrane. The stimulation causes the voltage-gated sodium ion channels to start opening. Both the concentration gradient of sodium ions and the electrochemical gradient favour the entry of sodium ions into the cell, which was prevented because the voltage-gated sodium ion channels were closed. Once opened, sodium ions diffuse into the neurone. The influx of sodium ions raises the membrane potential from its resting level of −70 mV, but does not immediately start an impulse.

If the stimulus raises the membrane potential to around −55 mV, this causes the wholesale opening of the sodium channels, and sodium ions flood into the neurone. This is an example of **positive feedback**, as the opening of some sodium channels causes a change (a rise in the membrane potential) which then causes more channels to open and the potential to rise further. As the opening of the sodium channels is controlled by voltage, they are referred to as **voltage-gated channels**. The value of approximately −55 mV that causes the channels to open is called the **threshold potential**. No impulse will be created if a stimulus does not reach the threshold potential.

> **Tip**
>
> You will be expected to remember the ways in which substances are transported across membranes from *OCR A level Biology 1 Student's Book* Chapter 5. The sodium-potassium pump uses ATP and so is an example of active transport. The movement of potassium ions through channel proteins is passive, an example of facilitated diffusion.

Depolarisation

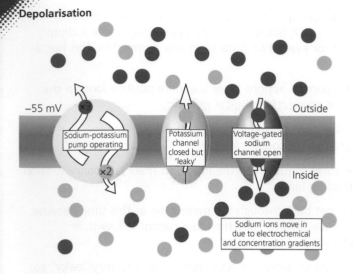

−55 mV

Sodium-potassium pump operating

Potassium channel closed but 'leaky'

Voltage-gated sodium channel open

Outside

Inside

Sodium ions move in due to electrochemical and concentration gradients

Beginning of repolarisation

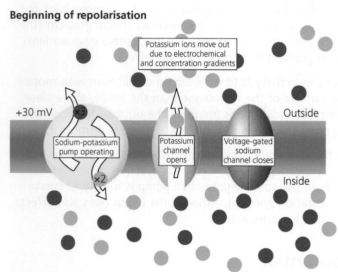

Potassium ions move out due to electrochemical and concentration gradients

+30 mV

Sodium-potassium pump operating

Potassium channel opens

Voltage-gated sodium channel closes

Outside

Inside

End of hyperpolarisation

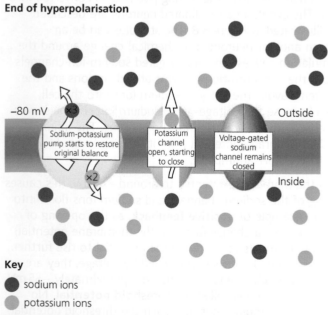

−80 mV

Sodium-potassium pump starts to restore original balance

Potassium channel open, starting to close

Voltage-gated sodium channel remains closed

Outside

Inside

Key

● sodium ions

● potassium ions

Figure 3.8 Events during depolarisation, repolarisation and hyperpolarisation.

Once the threshold potential is reached, so much sodium comes into the cell that the membrane potential is reversed, with the inside of the membrane becoming more positive than the outside. This is called **depolarisation**. The reversal of the potential creates the electrical impulse that will travel down the neurone.

Once the potential reaches about +30 mV the sodium channels close and voltage-gated potassium channels open. There are more potassium ions inside the membrane than on the outside, and the inside of the membrane is also now more positive as a result of the entry of sodium ions. Both of these factors will result in potassium ions moving out. This movement starts to bring the potential difference down again, a process called **repolarisation**.

The exit of potassium ions will continue for some time because the voltage-gated potassium channels do not close until the potential difference reaches around −80 mV; that is, lower than the normal resting potential. This phase is referred to as **hyperpolarisation**.

Although the potential difference is now similar to that of the resting potential, the ionic balance is different. There is more sodium inside the neurone and more potassium outside. The resting state is restored by the sodium-potassium pump and the leakage of potassium ions out of the axon.

The action potential in one part of the neurone acts as stimulus to generate another action potential in the adjacent membrane, so the action potential passes down the axon.

If the changing membrane potential is plotted against time, then a graph like that shown in Figure 3.9 is produced. The **refractory period** is a span of time during which a second stimulus will be unable to cause an impulse. Once the neurone has restored its resting potential and the normal balance of sodium and potassium ions has been restored, the neurone will be able to conduct another impulse.

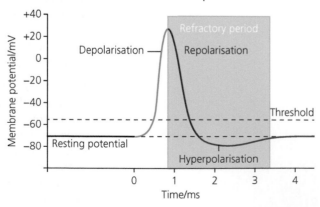

Figure 3.9 Graph of membrane potential against time during the production of an action potential.

Table 3.2 summarises the events that occur during the generation of an action potential.

Table 3.2 Events occurring during the generation of an action potential.

Stage	Inside membrane	Outside membrane	Potential	Sodium-potassium pump working	Voltage-gated sodium channels	Voltage-gated potassium channels
Resting	High K⁺, low Na⁺	Low K⁺, high Na⁺	$-70\,mV$	✔	Closed	Closed but leaking
Depolarisation	High K⁺, high Na⁺	Low K⁺, decreasing Na⁺	Rises from -70 to $+30\,mV$	✔	Open	Closed but leaking
Repolarisation	Decreasing K⁺, high Na⁺	Increasing K⁺, relatively low Na⁺	Decreases from $+30$ to $-70\,mV$	✔	Closing or closed	Open
Hyperpolarisation	Decreasing K⁺, Na⁺ starting to decrease	Low K⁺, Na⁺ increasing	-70 to $-80\,mV$	✔	Closed	Closing

Tip

Take care when answering questions that you do not mix up neurones (nerve cells) with nerves. Individual neurones and whole nerves do not have identical properties. Besides the all-or-nothing principle, neurones can only conduct an impulse in one direction, whereas some nerves conduct impulses both ways because they are mixed nerves containing motor and sensory neurones.

The all-or-nothing principle

An individual neurone cannot produce impulses of different magnitudes. If the stimulus reaches the threshold value, then the neurone will conduct an impulse. If the stimulus does not reach the threshold, then the neurone will not conduct an impulse. This is known as the **all-or-nothing principle**. Bigger stimuli do not produce bigger impulses, but they do cause the neurone to send impulses more frequently, and in this way information about the size of the stimulus is transmitted.

The strength of a stimulus may also affect how many neurones in a nerve reach their threshold potential, with a stronger stimulus causing more neurones to fire. So the all-or-nothing principle does not apply to whole nerves, just to neurones.

Activity

Investigating the sodium-potassium pump
The 'sodium efflux' is a term used to describe the sodium ions pumped out of a neurone. Scientists investigated the sodium efflux in a giant axon of a cuttlefish during treatment with dinitrophenol (DNP). DNP is an uncoupling agent which prevents the production of ATP in respiration. Their apparatus is shown in Figure 3.10.

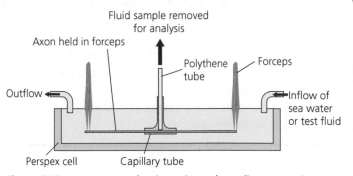

Figure 3.10 Apparatus used to investigate the sodium-potassium pump in a giant axon of a cuttlefish.

Method

1 The axon was isolated and placed in artificial sea water at 18 °C. The artificial sea water contained ions of ^{24}Na, a radioactive isotope of sodium which can be detected.

2 The axon was stimulated electrically (this is represented by 0 minutes on figure 3.11).

3 The artificial sea water was then replaced with sea water which this time did not contain the radioactive isotope

4 Samples of fluid that flowed past the axon were then collected at intervals and the concentration of $^{24}Na^+$ measured. This indicated the sodium ions pumped out by the sodium-potassium pump.

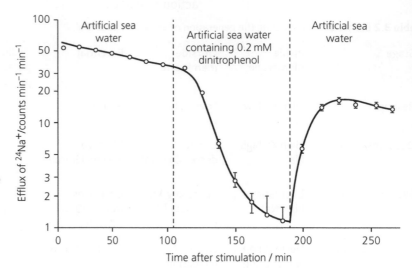

Figure 3.11 Results of the experiment investigating the sodium-potassium pump in cuttlefish giant axon. Note that the y-axis has a logarithmic scale.

5 After 105 minutes, 0.2 mM DNP was added to the sea water for a period of 85 minutes.

6 After 190 minutes, the artificial sea water was again replaced with water that did not contain DNP.

The results are shown in the graph in Figure 3.11.

Questions

1 Suggest a reason for carrying out the experiment at 18 °C.

2 Explain the necessity of stimulating the axon in step 2.

3 Explain the results shown in Figure 3.11 between 0 and 105 minutes.

4 What conclusion can you draw about the mechanism of pumping sodium ions in isolated axons from these results?

5 Using the information in the graph, how much confidence do you think the scientists would have in their results? Explain your answer.

Test yourself

5 What general name is given to sensory receptors which detect some sort of motion (movement or pressure)?

6 State two differences between the behaviour of the sodium ion and potassium ion channels in the membrane of a neurone.

7 Suggest a reason for the presence of many mitochondria in neurones.

8 Explain the cause of hyperpolarisation in a neurone.

9 How does a neurone transmit information about the intensity of a stimulus?

10 Explain why the refractory period would make it impossible for an axon to conduct an impulse in two directions at once.

The myelin sheath and saltatory conduction

The neurones shown in Figure 3.2 have axons which are covered in a myelin sheath, produced by Schwann cells wrapped around the axon. Most neurones do not have a myelin sheath, however. **Non-myelinated**

neurones can be found in the grey matter of the CNS. Non-myelinated neurones tend to be found where the axon is short and does not have to transmit an impulse very far.

What is the function of the myelin sheath and why does it appear to be particularly important if the nerve impulse has to travel long distances? The myelin provides electrical insulation for the axon, due to its lipid content, but in fact it is the *gaps* in this insulation, the nodes of Ranvier, which are really significant. The action potential causes a 'local current' in the axon, which can depolarise the next section of membrane (only in the forward direction, because of the refractory period in the membrane behind the impulse). This is also helped by the attraction of oppositely charged ions on the inside and outside of the membrane of the adjacent sections. In a non-myelinated axons, the impulse travels all the way along the axon, depolarising the membrane as it goes. In a myelinated axon the insulating effect of the myelin means that depolarisation is largely prevented in the covered membrane, and the action potential 'jumps' from one node to the next. This is called **saltatory conduction**, and it results in much faster transmission of the action potential. It is a little difficult to quantify this, because speed is also increased with the diameter of the axon, and myelinated axons are generally thicker than non-myelinated ones, but it has been estimated that myelin speeds up the conduction of impulses by about 20 times.

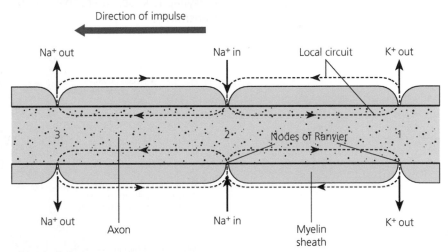

Figure 3.12 Saltatory conduction. The membrane is exposed at the nodes of Ranvier, causing the action potential to jump from node to node, speeding up the rate of transmission.

Synapses

To coordinate activities in the body, nerve cells need to communicate with each other, and to do so they have to connect. The connections are called **synapses**, and they play a key role in the operation of the nervous system. At a synapse the impulse has to travel across a small gap. To do this, the transmitting neurone releases a chemical called a **neurotransmitter**, which initiates an action potential in the receiving neurone. The structure and mode of action of a synapse are shown in Figure 3.13. The events of synaptic transmission can be summarised as follows.

1 The action potential arrives at the **presynaptic knob**, which is separated from the **postsynaptic membrane** by a tiny space, **the synaptic cleft**.

2 The action potential stimulates the opening of **voltage-gated calcium ion channels** in the presynaptic knob.

3 The concentration of calcium ions is higher outside the neurone, so when the channels open the calcium ions diffuse into the neurone. The presynaptic knob has membrane-bound **vesicles** which contain a neurotransmitter, and the influx of calcium ions causes these vesicles to move towards the presynaptic membrane and then fuse with it. Exactly how calcium ions cause this fusion is still the subject of research.

4 The neurotransmitter is released into the synaptic cleft and diffuses down a concentration gradient towards the **postsynaptic membrane**.

5 The postsynaptic membrane has **receptors** which bind the neurotransmitter.

6 The neurotransmitter acts as a chemical stimulus to the postsynaptic neurone, causing ligand-gated sodium ion channels to open and a new action potential to start (if the threshold potential is reached).

If the neurotransmitter were to stay in the synaptic cleft, a further impulse would be caused (after the refractory period), even if another action potential did not arrive down the presynaptic neurone. To prevent this in **cholinergic synapses** (see below) the neurotransmitter is broken down by an enzyme, **cholinesterase**, and the products are absorbed back into the presynaptic knob, where they can be used to make new supplies of the transmitter.

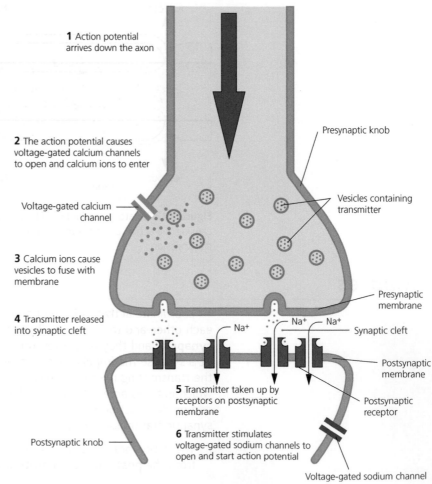

1 Action potential arrives down the axon

Presynaptic knob

2 The action potential causes voltage-gated calcium channels to open and calcium ions to enter

Vesicles containing transmitter

Voltage-gated calcium channel

3 Calcium ions cause vesicles to fuse with membrane

Presynaptic membrane

4 Transmitter released into synaptic cleft

Na⁺   Na⁺   Na⁺   Synaptic cleft

Postsynaptic membrane

5 Transmitter taken up by receptors on postsynaptic membrane

Postsynaptic receptor

Postsynaptic knob

6 Transmitter stimulates voltage-gated sodium channels to open and start action potential

Voltage-gated sodium channel

Figure 3.13 The events at a synapse.

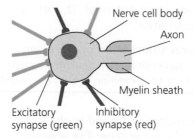

Nerve cell body

Axon

Myelin sheath

Excitatory synapse (green)   Inhibitory synapse (red)

No impulses along any synapse. The neurone is at resting potential.

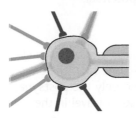

Stimulus from two excitatory synapses is sufficient to reach threshold potential. Action potential initiated in the nerve cell.

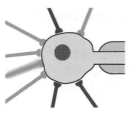

Stimulus from one excitatory synapse increases membrane potential but is insufficient to reach threshold potential.

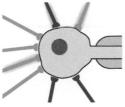

Stimulus from an inhibitory synapse decreases membrane potential so that the threshold potential is no longer reached.

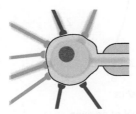

Stimulus from a further excitatory synapse increases membrane potential gain to reach the threshold potential and start an action potential again, despite the inhibitory stimulus.

Figure 3.14 Principle of control by excitatory and inhibitory synapses.

The neurotransmitter

Many chemicals can act as neurotransmitters, but one of the most common is **acetylcholine**. Synapses which release acetylcholine are called **cholinergic synapses**. Some synapses, as described above, cause depolarisation of the postsynaptic neurone and therefore make an action potential more likely. These are called excitatory synapses. As stated, the effect produced is a result of an interaction between the transmitter and the receptor. Some neurotransmitters always seem to have one type of effect. For example, glutamic acid (the most common neurotransmitter in the CNS) and acetylcholine are excitatory and GABA is inhibitory.

The role of synapses in the nervous system

Synapses are the means by which one nerve cell communicates with another. The fact that a single neurone may have between 1000 and 10000 synapses means that there is an incredible level of sophistication in the ability of the nerve cells to communicate.

Excitatory and inhibitory synapses

We have seen that an individual synapse may be excitatory (in other words, make an action potential more likely to form in the postsynaptic neurone) or inhibitory, making an action potential more difficult to achieve. Inhibitory synapses cause chloride channels to open in the postsynaptic neurone rather than sodium channels, and as a result the membrane potential becomes more negative (as chloride ions have a negative charge) and so a bigger stimulus is required if the neurone membrane is to reach its threshold potential. The fact that synapses can be either excitatory or inhibitory allows the body more subtle control of its actions.

We can use a simple example. Breathing is a reflex and is normally done automatically, but there are occasions (e.g. when diving underwater) when you may want to hold your breath. Breathing results from the body detecting high blood levels of carbon dioxide and sending impulses to the diaphragm and the intercostal muscles between the ribs, which brings about breathing movements. These stimuli are excitatory in nature. The neurones that connect to the muscles will be connected to a large number of synapses in the CNS, some excitatory and some inhibitory. If you want to hold your breath, you send signals via the inhibitory synapses. This lowers the membrane potential of the 'breathing neurones' and makes an impulse less likely. Therefore, the normal signal which results from low blood carbon dioxide is not strong enough to reach the threshold potential. As carbon dioxide builds up in the blood, a greater number of excitatory synapses are activated and the impulses become more frequent. Eventually, this overcomes the effect of the inhibitory synapses and action potentials are generated in the relevant neurones. For this reason, there are limits to how long you can keep holding your breath. However, the presence of both excitatory and inhibitory synapses allows a degree of control and the ability to adapt to circumstances, which is invaluable.

The role of excitatory and inhibitory synapses is shown in Figure 3.14, although remember that this is an over-simplification, and that neurones have thousands of synapses.

Summation

Excitatory and inhibitory synapses illustrate another aspect of control by synapses, that of **summation**. The more impulses a neurone receives via its synapses, the more likely it is to generate an action potential. The effects of the synapses add together, and this is summation.

There are two types of summation. **Temporal summation** occurs when a single synapse is activated several times in succession. The neurotransmitter builds up to a concentration which is sufficient to create an action potential in the postsynaptic neurone.

Spatial summation occurs when several synapses (attached to the same postsynaptic neurone) transmit impulses at the same time. The effects of the neurotransmitters from these different synapses add together to produce a strong enough signal to create an action potential.

Synapses and one-way transmission

The structure of a synapse ensures that a nerve impulse can only travel one way along a particular neurone pathway. As only one side (the presynaptic side) has vesicles of neurotransmitter, and only the postsynaptic side has receptors, it is impossible for a nerve impulse to travel in the opposite direction.

Test yourself

11 Why do non-myelinated neurones conduct impulses more slowly than myelinated ones of the same diameter?
12 What type of chemical in myelin gives it its insulating properties?
13 Explain the need for cholinesterase enzyme at cholinergic synapses.
14 Explain how the structure of synapses ensures that impulses only travel in one direction along a series of neurones.
15 Explain the difference between temporal summation and spatial summation.

Activity

The transmission of impulses in the nervous system is dependent on the movement of ions across the cell surface membranes of neurones. These membranes have carrier proteins and channel proteins for the movement of these ions.

1 Explain how carrier proteins and channel proteins in cell surface membranes are involved in establishing the resting potential of a neurone.

The potential difference across the cell surface membrane of an axon was recorded. Figure 3.15 shows what was recorded when an impulse travelled along the length of an axon.

2 Explain what happens to change the membrane potential
   a) between X and Y
   b) between Y and Z.
3 a) Name the neurotransmitter released at cholinergic synapses.

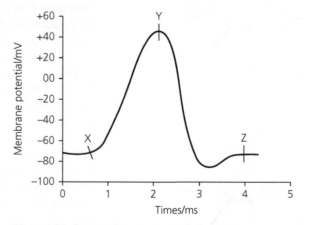

Figure 3.15 The membrane potential of an axon.

   b) Describe how the release of neurotransmitter molecules at a cholinergic synapse leads to an impulse in a relay neurone.

Example

The velocity of a nerve impulse is affected by the diameter of the axon and the thickness of the myelin sheath. A mathematical model was used to predict these effects. The results are shown in the graphs in Figure 3.16.

1 According to the predictions, which has the greater effect on impulse velocity: axon diameter or myelin thickness? Explain your answer.

Earthworms have two giant myelinated neurones (but not in the same way as in vertebrates) running the length of their bodies: the lateral giant neurone (LGN) and the medial giant neurone (MGN). The MGN is 0.07 mm in diameter and the LGN is slightly smaller, with a diameter of 0.05 mm. An experiment was done to determine the velocity of the impulses in these two neurones. The experiment was set up as shown in Figure 3.17.

The two neurones carry impulses in opposite directions. To test each neurone, the worm was touched some distance away from the electrodes. This created an impulse and this could be detected by the electrodes. The time taken for the impulse to travel between electrode A and electrode B (for the LGN) or in the opposite direction (for the MGN) was recorded.

The mathematical model predicted that, for a non-myelinated neurone:

$$\text{Velocity (m s}^{-1}) = \sqrt{\text{diameter (mm)}}$$

The results are shown in Table 3.3.

2 Calculate the velocities of the impulses in the two neurones.
3 The impulses travelled faster than predicted because the neurones were myelinated, but the speed increase in the MGN was greater than in the LGN. Suggest a reason for this.
4 Explain why the myelin sheath increases the velocity of the impulse.

Answers

1 Myelin thickness has more effect than axon diameter, as velocity increases more for any given increase in thickness.
2 Velocity = distance (m)/time (s).
2 cm = 0.02 m, 1 ms = 0.001 s, so 60 ms = 0.06 s and 30 ms = 0.03 s.
Velocity of LGN = 0.02/0.06 = 0.33 m s⁻¹. Velocity of MGN = 0.02/0.03 = 0.67 m s⁻¹.
3 We are told that the thickness of the myelin sheath has an effect on velocity. It is likely that the MGN has a thicker myelin sheath.

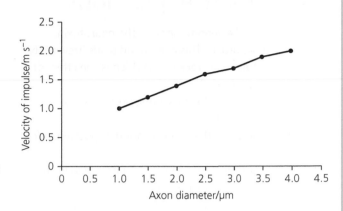
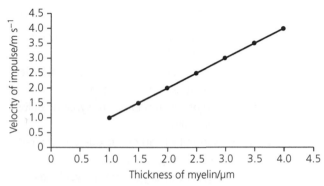
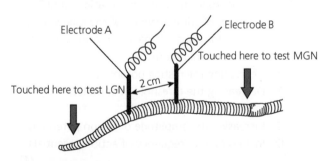

Figure 3.16 Nerve impulse velocity in relation to axon diameter and myelin thickness.

Figure 3.17 Measuring impulse velocity in earthworm giant myelinated neurones.

Table 3.3 Results of nerve impulse speed experiment in earthworms.

Neurone	Time taken to travel between A and B/ms	Velocity/m s⁻¹	Predicted velocity (m s⁻¹) in an unmyelinated neurone
LGN	60		0.22
MGN	30		0.26

4 This is due to saltatory conduction. The myelin sheath insulates the axon, so that the action potential 'jumps' from one node of Ranvier to the next, which is quicker than travelling the same distance through the axon.

Exam practice questions

1 Some snake venoms contain the neurotoxin, α-bungarotoxin. This compound binds irreversibly to acetylcholine receptors. Which is the effect of α-bungarotoxin?

   A It prevents the enzymatic breakdown of acetylcholine.

   B It prevents the depolarisation of postsynaptic membranes.

   C It prevents the fusion of synaptic vesicles to cell surface membranes.

   D It prevents the repolarisation of presynaptic membranes. *(1)*

2 Which of the following is brought about by the process of facilitated diffusion?

   A The efflux of sodium ions at a node of Ranvier on a myelinated motor neurone.

   B The inflow of calcium ions into synaptic bulbs of sensory neurones.

   C The movement of carbon dioxide out of a red blood cell in a lung capillary.

   D The movement of oxygen from alveolar air to a red blood cell. *(1)*

3 Sensory neurones transmit information about the intensity of stimuli. Which of the following is responsible for transmitting information about a high level of stimulation?

   A Decreasing the resting potential

   B Decreasing the threshold of stimulation by receptors

   C Increasing the amplitude of action potentials

   D Increasing the frequency of action potentials *(1)*

4 Four stimuli of increasing intensity (P, Q, R and S) were applied using electrodes to a motor neurone at the base of the axon near the cell body. This region of the neurone is known as the axon hillock. The responses of the neurone to these stimuli were detected at the far end of the axon with recording electrodes. The four responses were displayed on a screen as shown in the diagram.

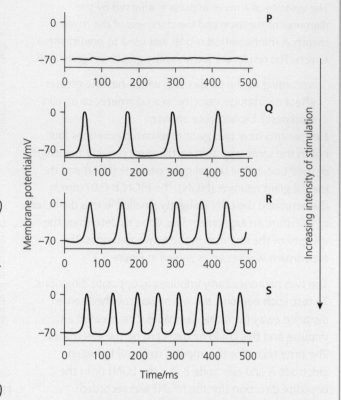

a) Describe and explain the results displayed on the screen. *(6)*

b) The investigation was repeated but this time a small section in the middle of the axon was treated with potassium cyanide, a respiratory poison. The axon continued to be stimulated and about 1000 impulses were recorded over several minutes, but eventually no more were detected. Suggest an explanation for this result. *(4)*

c) Experimental work on neurones shows that it is possible for an isolated neurone to send impulses in either direction. Explain why impulses only travel away from the CNS in motor neurones and not in the opposite direction. *(3)*

5 Myelinated neurones are found in all vertebrates and in a few species of invertebrate. The table shows the diameter and rate of transmission of impulses of myelinated and non-myelinated neurones in four different animals.

Animal	Myelinated or non-myelinated	Diameter/ μm	Rate of transmission of impulses/m s⁻¹
Crab	Non-myelinated	30	5
Squid	Non-myelinated	500	25
Frog	Myelinated	12	30
Frog	Myelinated	14	35
Cat	Non-myelinated	15	2
Cat	Myelinated	20	100

a) The table shows that two features influence the rate of transmission of impulses. Explain how each of these factors influences the rate of transmission. *(4)*

b) Suggest the likely roles of the two different types of neurone in cats, as shown in the table. *(2)*

c) Explain how action potentials are propagated along a myelinated neurone. *(8)*

6 Some sensory receptors are specialised cells that stimulate sensory neurones to send impulses to the CNS. Other receptors are simply the nerve endings of sensory neurones. Sensory receptors of both types are described as transducers.

a) Explain why sensory receptors are described as transducers. *(2)*

b) Use examples to discuss the roles of sensory receptors in mammals. *(6)*

c) Under normal circumstances the response to a painful stimulus on the hand is a rapid movement of the hand away from the source of the stimulus. However, it is possible to override this response and endure the pain. Explain the role of synapses in controlling these two responses to the painful stimulus. *(4)*

Stretch and challenge

7 Chemoreceptors in the taste buds detect five different stimuli:

salt, sweet, bitter, sour, umami (savoury)

The supply of sensory neurones to these chemoreceptors is complex. Some sensory neurones have dendrites on several different chemoreceptors. The figure shows the responses of three different sensory neurones (A, B and C) to increasing concentrations of four of the taste stimuli. The number of impulses recorded from each neurone in 5 seconds is plotted against the concentrations of the four different solutions that were used: salt, sucrose (sweet), quinine (bitter) and hydrochloric acid (sour).

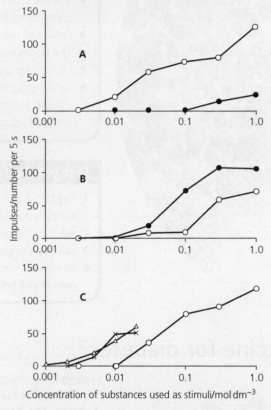

Concentration of substances used as stimuli/mol dm⁻³

● Salt ○ Sucrose △ Quinine (bitter) × Hydrochloric acid (sour)

Comment on the results obtained from the three different sensory neurones from the taste buds.

Chapter 4

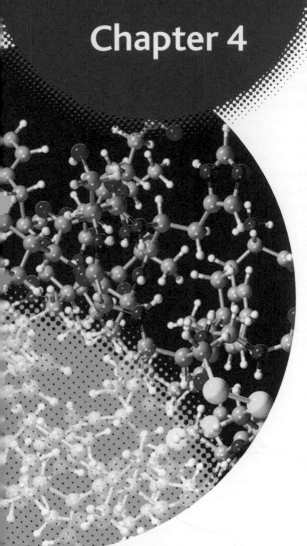

Hormonal communication

Prior knowledge

- Hormones travel around the body in the blood.
- Some animal hormones are proteins.
- Responses controlled by hormones are usually slower than responses controlled by the nervous system.
- The pancreas produces the hormone insulin, which reduces the concentration of blood glucose.
- Failure of the pancreas to produce insulin leads to type 1 diabetes.
- The hormone adrenaline can increase the heart rate.
- Hormones also control the menstrual cycle, fertility and growth.
- Target cells have receptors in their membranes if the hormones are water soluble. Receptors are inside the cytoplasm or nucleus if the hormones are lipid soluble and can cross the phospholipid bilayer.

Test yourself on prior knowledge

1 Suggest a reason why hormonal responses are slower than nervous responses.
2 What is the usual treatment for type 1 diabetes?
3 How do people with type 1 diabetes have to modify their diet?
4 To which chemical group do the hormone receptors in the cell surface membrane belong?

A vaccine for diabetes?

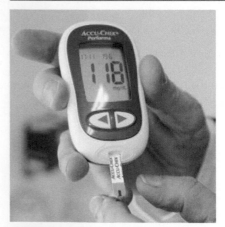

Figure 4.1 People with diabetes can monitor their blood glucose concentrations using a biosensor.

Hormones are vital components in the control of body functions. One important disease that results from a hormone imbalance is type 1 diabetes. The hormone insulin helps to control the concentration of blood glucose, but in people with type 1 diabetes the cells that produce the hormone, the beta cells (often written as β cells) in the pancreas are destroyed by the body's own immune system. A recent scientific trial suggests that it may be possible to 'retrain' the immune system to stop attacking the beta cells using a vaccine. The vaccine targets the T lymphocytes responsible for destroying the beta cells in the islets of Langerhans yet leaves the rest of the immune system functioning normally. The work is at an early stage, and only one trial has been done, with 80 people. However, in that group beta cell function was shown to be better than in patients treated with insulin alone. More trials need to be done and the long-term effects studied, but there is now hope that patients with diabetes may be able to replace multiple daily injections with a vaccine that would last for some months.

Long-term control by hormones

We have seen in the last chapter how the nervous system coordinates actions in the body. However, as a means of control nerves have limitations. They can only control the organ to which they are attached, and they have difficulty in maintaining actions for a long period of time. Although a continuous series of impulses could, in theory, be sent down a nerve, this would be a huge drain on ATP supplies in the neurone and so is not feasible in practice. Long-term control depends largely on hormones, and in this context 'long-term' may mean minutes rather than the seconds over which nervous control is often maintained (although some hormones, like growth hormone, control processes for a period of years).

The methods by which hormones transmit signals are very different from the transmission of impulses along neurones. The nature of the signal is different – being chemical rather than electrical – and the method of transfer is too, via the blood system rather than via nerves. Blood travels to all parts of the body, so a hormone in the blood can reach and affect all organs and tissues. This means that there has to be a mechanism which results in only specific target cells responding to a particular hormone. Cell membranes have hormone-specific receptors and these are different in different cells, so that any one cell can respond to some hormones and not others. Hormones can be produced and act over a long period of time. The production of hormones is carried out by endocrine glands, which secrete the hormone directly into the blood (i.e. they do not have ducts).

In this chapter we will study the actions of two hormones: adrenaline and insulin. Adrenaline, produced by the adrenal glands, is often referred to as the 'fight-or-flight' hormone as it allows the body to respond effectively to dangerous situations. Insulin is important in the control of blood sugar concentrations and its importance is seen when (in diabetes) the body produces insufficient or no insulin (Figure 4.1). Diabetes can be fatal if left untreated.

> **Key term**
>
> Endocrine gland A gland that secretes its product directly into the bloodstream rather than via a duct.

Mode of action of hormones

> **Key term**
>
> Steroids Lipid-based hormones which are related to the four-ring structure of cholesterol and are distinguished by their functional groups or side groups.

> **Tip**
>
> Look back at Figure 2.24 in *OCR A level Biology 1 Student's Book* to remind yourself about the structure of cholesterol.

Hormones are chemical messengers that are secreted by endocrine glands and travel in the blood, having an effect on one or more **target tissues**. Many are proteins (e.g. insulin, growth hormone) but some are steroids (e.g. testosterone, oestradiol). Adrenaline is a catecholamine. Many hormones are peptides, e.g. anti-diuretic hormone, or ADH.

Travelling in the blood, hormones can make contact with any tissue in the body. The cells that need to respond to a given hormone therefore must be different from those that do not: they possess receptor molecules on their surfaces that detect a specific hormone and instigate a response. Other cells, without those receptors, will be unaffected by the hormone. We will now look in detail at the way in which hormones work, using adrenaline as an example.

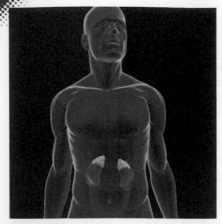

Figure 4.2 Position of the adrenal glands.

Tip

Adrenaline and adrenalin are alternative forms of spelling. Either will be accepted as correct. In the USA, adrenaline is referred to as epinephrine. This sounds similar to EpiPen, which is a type of automatic adrenaline injector. In exams, use the British English forms of the word.

The adrenal glands

The two **adrenal glands** are situated above the kidneys (see Figure 4.2). The adrenal glands consist of two areas: a central **medulla** and an outer **cortex**. The cortex produces steroid hormones, mainly aldosterone and cortisol, which have a variety of functions but are mainly concerned with the control of concentrations of mineral ions, carbohydrates and proteins. The medulla produces **adrenaline**. This hormone is produced at times of stress or excitement. It affects many organs and helps the body respond to emergency situations.

Detection by target cells: first and second messengers

When a hormone affects a cell, there are two messengers involved. The **first messenger** is the hormone itself, which brings the 'message' to the cell from an endocrine gland in a different location from the target tissue. The hormone does not actually enter the cell, it binds to a **receptor** on its surface. A **second messenger** inside the cell then brings about some effect. The process is illustrated in Figure 4.3 for adrenaline, and the main steps are as follows.

1 The adrenaline binds to a specific receptor in the membrane, the **adrenergic receptor**.

2 This binding activates a G-protein complex inside the cell, which contains guanosine diphosphate (or GDP).

3 The GDP is converted to guanosine triphosphate (or GTP).

4 The GTP activates a membrane enzyme **adenylyl cyclase**, which converts ATP to **cyclic adenosine monophosphate** (cyclic AMP, or cAMP).

5 The cAMP starts a cascade of reactions that result in the desired effect within that cell.

The adrenaline does not enter the cell, and it is the cyclic AMP (the second messenger) that causes changes inside the cell.

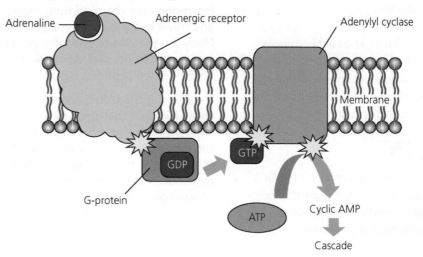

Figure 4.3 Mode of action of adrenaline.

The pancreas

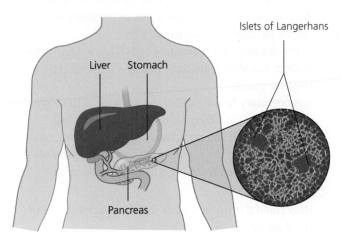

Islets of Langerhans

Liver Stomach

Pancreas

Figure.4.4 Location and microscopic structure of the pancreas.

The **pancreas** is an organ found in the abdomen of mammals, and it functions as both an endocrine and an exocrine gland. Endocrine glands secrete hormones directly into the blood, whereas exocrine glands secrete substances through a duct. The exocrine function in the pancreas is the production of **pancreatic juice** that is delivered to the small intestine where the enzymes in the juice act to digest food. Within the organ, the exocrine and endocrine functions are performed by different tissues, which can be clearly seen in microscope sections (see Figure 4.4). Most of the pancreatic cells are responsible for the secretion of digestive enzymes, but throughout the organ there are small groups of cells called the **islets of Langerhans** that are responsible for hormone production. The islets contain two types of cell, the **alpha cells**, which secrete glucagon, and the **beta cells**, which secrete insulin. The alpha cells are larger but less numerous than the beta cells.

Activity

Drawing sections of the pancreas

Sections of pancreas tissue are usually differentially stained to show up the islets of Langerhans (the endocrine tissue) and the exocrine tissue in different colours. Figure 4.5 shows one such section and some drawings of it done by a student (a low-power plan and high-power drawings of the endocrine and exocrine tissue).

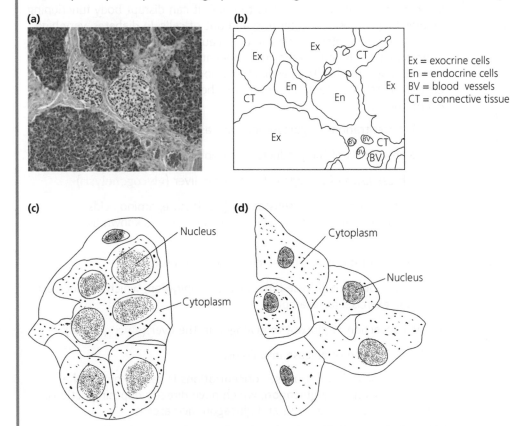

Ex = exocrine cells
En = endocrine cells
BV = blood vessels
CT = connective tissue

Figure 4.5 Section of pancreas tissue and student drawings. (a) photographed at x 200.

Note the following features of the drawings.

Low power

- the low-power plan shows the layout of the tissues, but includes no details of cells
- normally, the convention is that labels should be outside the drawing with labelling lines, but with so many blocks of similar tissues this would result in an untidy mess of labelling lines; so, in this case the best solution is to label on the drawing and use a key
- the lines are clear and do not cross, and there are no breaks.

High power

- only a small number of cells are drawn in each case
- here, the normal labelling convention described above is used
- when drawing the endocrine cells (Figure 4.5(c)), the student has only drawn what can actually be seen: for

instance, in the endocrine cells there is a block of four nuclei and therefore there must be four cells there; no divisions between these cells is shown, because they cannot be clearly seen on the section

- in the drawing of the exocrine cells (Figure 4.5(d)), the student has shown divisions because in those cells they are visible
- it is necessary to distinguish the cytoplasm from 'gaps' in the section: in biological drawings any shading necessary should be done by using dots, as here, not by blocking in or using colours; if there is no possibility of confusion then there is no need to shade at all.

Questions

1 How could the student work out the magnification of the drawings?
2 The white areas on the section are gaps between cells. Suggest why these gaps may not be there in the living tissue.

Control of blood glucose concentration

Key terms

Glycogenolysis The breakdown of glycogen to glucose

Gluconeogenesis The synthesis of glucose from molecules that are not carbohydrates, such as amino acids, lactate, pyruvate, glycerol and fatty acids.

Glycogenesis The conversion of glucose to glycogen

Tip

The terms glycogenesis, glycogenolysis and gluconeogenesis all sound similar and can easily be confused. The names do help. 'Lysis' means breaking down (so glycogenolysis is the breakdown of glycogen) and 'genesis' means creation (so glycogenesis is the creation of glycogen). 'Neo' means new, and so gluconeogenesis is the creation of new glucose. In an extended answer, use these terms only if you are confident of using the right one; otherwise, just describe the process concerned. Also, be very careful not to misspell glucagon and glycogen.

Glucose is important to the body as a source of energy in respiration. It is important that all tissues have a supply of glucose to carry out their functions. However, glucose also has an effect on water potential and if the concentration of glucose gets too high it can disrupt body functioning by affecting osmosis of water into and out of cells. In diabetes, very high concentrations of blood sugar can be a cause of death. It is therefore important that the body keeps the concentration of glucose within a certain range. In a healthy person, the blood glucose concentration is not kept absolutely constant, but never goes below $4\,\text{mmol}\,\text{dm}^{-3}$ or above $8\,\text{mmol}\,\text{dm}^{-3}$.

Blood concentrations of glucose increase as a result of

- the absorption of the products of carbohydrate digestion in the gut
- the breakdown of glycogen stores in the liver (glycogenolysis)
- the conversion of other substances (e.g. lactate, amino acids, glycerol and fatty acids) to glucose: this process is known as gluconeogenesis.

Blood concentrations of glucose decrease as a result of

- the absorption of glucose into cells for respiration, which will increase during exercise, for example
- the conversion of glucose to glycogen in the liver (glycogenesis)
- the conversion of glucose into lipids.

The regulation of blood glucose concentrations is carried out by the hormones **insulin** and **glucagon**, which have directly opposing actions. Insulin reduces blood glucose and glucagon increases it.

Insulin

The beta cells in the islets of Langerhans both detect a rise in blood glucose and also produce insulin to counteract it. The membrane of the beta cell has a potential difference across it at rest. The membrane has ion channels for both potassium and calcium. At rest the potassium channels are open and the calcium channels are closed. Potassium diffuses out of the cell down a concentration gradient and this causes the inside of the cell to be more negative than the outside. When blood glucose concentrations are high, glucose diffuses into the beta cells by facilitated diffusion. Respiration of this glucose produces ATP, and high concentrations of ATP cause the potassium ion channels to close. This causes a change in the membrane potential which results in the voltage-gated calcium channels opening. The influx of calcium causes vesicles containing insulin to move and fuse with the cell surface membrane, so releasing the hormone. The process is shown in Figure 4.6.

> **Key term**
>
> Facilitated diffusion A special case of diffusion involving a carrier protein molecule in the membrane.

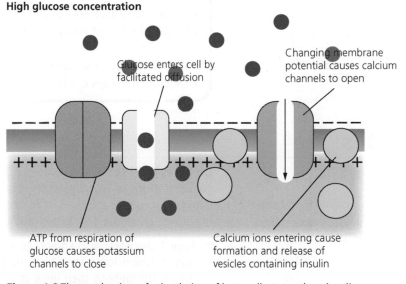

Beta cell at rest

Potassium ions create a relative positive charge outside

Glucose carrier protein

Potassium channel open – potassium ions leave

Calcium channel closed

High glucose concentration

Glucose enters cell by facilitated diffusion

Changing membrane potential causes calcium channels to open

ATP from respiration of glucose causes potassium channels to close

Calcium ions entering cause formation and release of vesicles containing insulin

Figure 4.6 The mechanism of stimulation of beta cells to produce insulin.

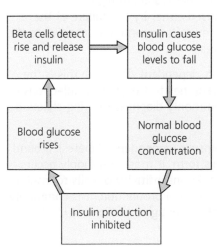

Beta cells detect rise and release insulin

Insulin causes blood glucose levels to fall

Blood glucose rises

Normal blood glucose concentration

Insulin production inhibited

Figure 4.7 The insulin negative feedback mechanism.

Insulin is secreted into the blood and binds to specific receptors on the cell surface membranes of muscle cells and liver cells. It increases their permeability to glucose, which is taken into the cells by facilitated diffusion. Insulin also activates enzymes that can convert glucose into glycogen, a process called glycogenesis. The main store of glycogen is in the liver, although other cells, especially muscle cells, also contain it. Insulin also stimulates an increase in the rate of respiration, particularly in muscle cells, and promotes the conversion of glucose into proteins and lipids.

Insulin secretion brings about a reduction in blood glucose. The lower blood glucose concentration means that the beta cells are no longer stimulated to secrete insulin. This mechanism prevents the concentration of blood glucose dropping too low (see Figure 4.7). This is an example of negative feedback, as the production of insulin leads to a series of events which result in less insulin being produced.

Glucagon

Glucagon is produced by the alpha cells of the pancreas. The alpha cells of the pancreas detect and respond to low glucose concentrations and they make glucagon. The actions of glucagon are antagonistic to those of insulin. Glucagon promotes the conversion of glycogen into glucose and also the formation of glucose from non-carbohydrate molecules.

In the same way as insulin, the concentrations of glucagon in the blood are controlled by a negative feedback mechanism (see Figure 4.8).

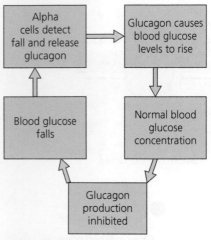

Figure 4.8 The glucagon negative feedback mechanism.

> **Tip**
>
> Try combining the glucagon and insulin negative feedback mechanisms on the same diagram.

Test yourself

1 Name the area of the adrenal glands that produces adrenaline.
2 Explain the terms *first messenger* and *second messenger* in relation to hormones.
3 Name the structures in the pancreas that function as endocrine glands.
4 Explain the role of voltage-gated calcium channels in the secretion of insulin.
5 Define the term *glycogenesis*.
6 Describe the ways in which glucagon raises blood glucose concentration.

Diabetes

Diabetes mellitus is a chronic condition that results in high concentrations of blood glucose. It is caused either by the failure of the pancreas to produce insulin or by the body cells becoming insensitive to the hormone. There are two types of diabetes.

- Type 1 diabetes (also known as insulin-dependent diabetes): this is the less common form, for which the person has to take regular injections of insulin throughout their lives. It is commonly diagnosed during childhood or early adulthood.

- Type 2 diabetes (also known as non-insulin-dependent diabetes): around 90% of people with diabetes have this form. It most commonly occurs in middle age and usually results from an insensitivity of cells to insulin, although there may be a deficiency in insulin production. It is commonly treated by a low-sugar diet and tablets, although injections of insulin may be given.

Causes of diabetes

Type 1

In type 1 diabetes little or no insulin is produced. It is caused by the body's immune system attacking the beta cells and destroying them. The inability to produce insulin in response to raised blood glucose concentration after eating results in the blood glucose reaching very high concentration, a condition known as hyperglycaemia. The excessive concentration of glucose in the blood means that the kidneys cannot reabsorb all the glucose as they usually do, and glucose appears in the urine. Urination becomes much more frequent and because glucose cannot be absorbed into cells for respiration the person feels excessively tired and very hungry. As carbohydrates in the diet can now provide no energy for the body there is considerable weight loss. If untreated, type 1 diabetes is fatal.

The exact cause of type 1 diabetes is uncertain. There may be a genetic component in some cases, and it is thought that the disease might result from some sort of viral infection to which the body's immune system overreacts and destroys the beta cells of the pancreas.

Type 2

In type 2 diabetes the cells of the body become insensitive to insulin and the concentrations of insulin may also be reduced. Type 2 diabetes is caused by a combination of genetic and lifestyle factors, and there are known risk factors including being overweight, high blood cholesterol, high blood pressure and physical inactivity. The genetic aspect is indicated by the fact that certain ethnic groups appear to have a much higher risk of developing type 2 diabetes, and that having a relative with type 2 diabetes makes it more likely that a person will develop it.

The symptoms of type 2 diabetes are very similar to those for type 1, but they develop gradually over a period of years whereas the onset of type 1 diabetes is much more rapid.

Activity

Sugary soft drinks and type 2 diabetes

A study was carried out on a possible link between type 2 diabetes and the consumption of various soft drinks, specifically three types: sugar-sweetened soft drinks, artificially sweetened soft drinks and fruit juice. The participants in this study were drawn from the 330234 people taking part in a long-term investigation, which began in 1991, of nutrition and health in Europe. From this large number, researchers selected 11684 people who had developed type 2 diabetes between 1991 and 2007. They also randomly selected a further group of 15374 people who acted as the control group. Some people initially selected for the control group had to be excluded owing to various factors that made them unsuitable.

The participants filled in a questionnaire about their soft drink intake and many aspects of their diet and health. A statistical analysis of the results was used to estimate the risk of developing type 2 diabetes posed by drinking each type of soft drink. The risk was expressed as a hazard ratio (HR). An HR of 1 indicates no

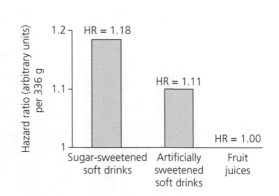

Figure 4.9 The hazard ratio of developing type 2 diabetes associated with consuming soft drinks.

increased risk compared to the control group whereas an HR of 2 indicates that the risk is twice that of the control group. The results are shown in Figure 4.9. The HR is expressed as the hazard ratio resulting from drinking one 336 g portion of the drink (approximately one can).

The HR value of 1.18 for the sugar-sweetened soft drinks was statistically significant (i.e. each sugar-sweetened drink taken increases the risk of type 2 diabetes). However, the HR 1.1 for the artificially sweetened soft drinks was not statistically significant. An HR of 1.0 for fruit juices indicated no increased risk whatsoever.

1 Suggest features of the people excluded from the control group that may have made them unsuitable.

2 The diabetic and control groups were not of equal size. Does this matter? Explain your answer.

3 Suggest a reason why a very large sample size is needed in a survey like this.

4 The scientists reported that drinking sugar-sweetened soft drinks increased the risk of type 2 diabetes, but that drinking artificially sweetened soft drinks and fruit juices did not. Discuss the strength of evidence for their conclusion.

5 The scientists expressed the risk in terms of one 336 g drink. What assumption are they making in doing this? Do you think this assumption is justified?

Producing insulin

For many years, people with diabetes relied on a supply of insulin extracted from the pancreases of pigs and cattle. These types of insulin are effective in the human body, but their molecular make up showed minor differences from human insulin. This meant the insulin dosage had to be slightly higher and it also caused some localised reactions at the injection site.

Nowadays, patients with diabetes are treated with human insulin produced by **genetically modified** (or GM) **bacteria**. This became possible with advances in genetic engineering techniques. The position of the insulin-producing gene in humans is now known, and it can be extracted and placed in a bacterial plasmid. When this plasmid is placed into a bacterium the bacterium divides to produce many others, all of which produce human insulin. GM yeasts are also used to make insulin. The use of this synthetically produced human insulin has a number of advantages.

- The insulin is human insulin, which would have been produced by the person's own body. It does not evoke a significant response by the immune system, and works more efficiently than insulin from a different species.

- The supply of insulin is not dependent on the supply of pancreases from cattle and pigs, and production can be adjusted to suit demand.

- The use of human insulin overcomes objections of certain religious and campaign groups to the use of cattle or pigs or animals in general.

- The production costs of insulin from genetically modified bacteria are lower than extracting the hormone from animal pancreases.

- There is no risk of infection, such as by viruses and prions.

Although type 1 diabetes can be successfully managed, at present there is no cure for the condition. The use of stem cells, which have the potential to develop into any type of human cell, offers the potential of a cure. It is possible that stem cells could be treated so that they develop into pancreatic beta cells, which could then be transplanted into the pancreas of a person with diabetes. They would replace the destroyed cells and allow the person to produce their own insulin. Successful experiments have been carried out using mice, but the research is still at an early stage and it is likely to be some years before this treatment is available for people with diabetes.

Key term

Plasmid Circular double-stranded DNA found in bacteria that is capable of replicating independently of chromosomal DNA.

Tip

You will learn more about genetically modified organisms in Chapter 11 Manipulating genomes.

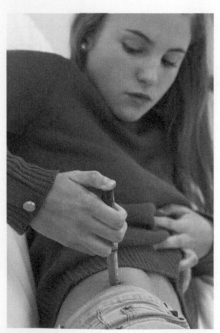

Figure 4.10 People with diabetes now have easy access to a supply of human insulin for injections.

Test yourself

7 Define the term *gluconeogenesis*.
8 Explain why untreated type 1 diabetes can lead to weight loss.
9 What reduces the effectiveness of animal insulin in treating people with diabetes, compared to the use of human insulin?
10 What would be the benefits of stem cell treatment for diabetes compared to the use of insulin?
11 Why would the use of stem cells be unsuitable for many cases of type 2 diabetes?

Example

Glucose tolerance

Glucose tolerance tests are sometimes given to people with suspected type 1 diabetes. The person eats and drinks nothing except water for 12 hours, and then is given a glucose drink (50 g in 200 cm³). Their blood glucose is then monitored every 30 minutes for a total of 3 hours.

The results from a non-diabetic person and two people with diabetes are shown below.

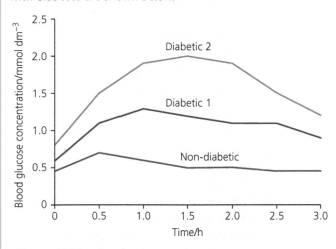

Figure 4.11 Results of a glucose tolerance test.

1 The three test subjects took in no carbohydrates for 12 hours before the test, yet their initial blood sugar concentrations are not the same. Suggest a reason for this.
2 What precautions would be necessary in the 3 hours of the test to ensure a valid comparison?
3 What is the maximum difference between the blood glucose concentrations in the blood of the non-diabetic person and diabetic 2 during the course of this test?

4 Suggest a reason for the difference seen in the responses of the two people with diabetes.
5 What would a diabetic (with type 1 diabetes) normally do to avoid the excessive rise in blood glucose shown here?

> **Tip**
>
> For question 3 you must include units in your answer.

Answers

1 A number of factors could explain this:
- The mass of glycogen stored in the liver and converted to glucose may have been different.
- The amount of physical activity of the three subjects may have varied.
- The concentration of blood glucose at the beginning of the 12 hours might have been different (this could affect the starting concentration in the people with diabetes; it is unlikely to be relevant in the non-diabetic).

2 The amount of physical activity must be the same in the three subjects; there must be no additional intake of food or drink.
3 1.5 mmol dm⁻³ (at 1.5 hours).
4 Diabetic 2 may have fewer surviving pancreatic beta cells than diabetic 1 or diabetic 2 may have a more severe form of the disease.
5 The person would inject insulin.

> **Tip**
>
> In question 4, the two answers are equivalent, but the first is better as it is more specific and detailed.

Exam practice questions

1 Which list includes hormones that all interact with receptors on the surfaces of their target cells?
   A adrenaline, aldosterone, anti-diuretic hormone and cortisol
   B adrenaline, anti-diuretic hormone, glucagon and insulin
   C cortisol, insulin, oestrogen and testosterone
   D aldosterone, glucagon, oestrogen and testosterone *(1)*

2 A homeostatic system was monitored and it was discovered that there was a delay in the transfer of information between the sensors and the effectors. Which of the following shows the most likely effect on the level of the factor controlled in this homeostatic system? *(1)*

A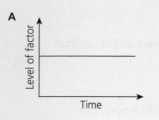

B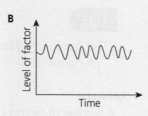

C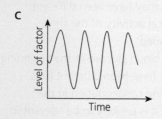

D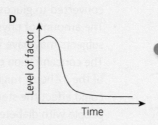

3 Which shows the response of the liver to glucagon and insulin? *(1)*

	Response to glucagon	Response to insulin
A	Increase in glycolysis	Increase in glycogenolysis
B	Increase in glycogenolysis	Increase in gluconeogenesis
C	Decrease in glycogenesis	Increase in glycogenesis
D	Decrease in gluconeogenesis	Decrease in glycolysis

4 a) The table shows information about six different hormones in humans. Copy and complete the table. *(12)*

Hormone	Site of production	Target organ/tissue	Effect on target organ/tissue
Adrenaline		Heart	
Aldosterone		Distal convoluted tubules of kidney	
Anti-diuretic hormone (ADH)			
Cortisol		Liver	Stimulates gluconeogenesis
Glucagon		Liver	
Insulin		Skeletal muscle	

b) The concentration of glucose in the blood normally fluctuates within the range 65–100 mg 100 cm^{-3} (3.6–5.8 mmol dm^{-3}), although it may rise to 140 mg 100 cm^{-3} (7.8 mmol dm^{-3}) after a meal is absorbed. Explain how the concentration of glucose is maintained within this range. *(6)*

5 Some membrane proteins are receptors for cell signalling compounds, such as hormones.
   a) Draw a labelled diagram to show the structure of a cell surface membrane with a hormone receptor. *(4)*
   b) Explain why some hormones have receptors on the cell surface membrane and others have their receptors in the cytoplasm or the nucleus. *(3)*

Glucagon receptors are only found on hepatocytes; there are none on the cells of other tissues, such as skeletal muscle fibres.
   c) Explain why skeletal muscle fibres have no receptors for glucagon. *(2)*

Following activation of hepatocytes by glucagon there is an increase in the concentration of cyclic AMP inside the cytoplasm.
   d) i) Explain how glucagon causes an increase in the concentration of cyclic AMP. *(3)*
      ii) Explain the role of cyclic AMP in hepatocytes following stimulation by glucagon. *(5)*

6 Glucose is a common respiratory substrate. It enters cells by diffusing through glucose transport (GLUT) proteins in cell surface membranes. As soon as glucose enters cells it is phosphorylated by the enzyme glucokinase to form glucose 6-phosphate (G 6-P). This molecule does not pass out of cells.

a) i) Explain why transport proteins are required for glucose. (2)

ii) Suggest why it is not possible for glucose 6-phosphate to pass out of cells. (1)

iii) Explain the advantage of converting glucose to glucose 6-phosphate as soon as it enters cells. (3)

The enzyme glucose-6-phosphatase converts G 6-P to glucose.

b) Explain why liver cells have glucose-6-phosphatase, but muscle cells do not. (2)

Each GLUT protein is composed of a single polypeptide. There are many different types of GLUT protein. The table shows information about four of the GLUT proteins.

Type of GLUT	Main location	Role
GLUT1	All cells	Allows movement of glucose into cells as a substrate for respiration
GLUT2	Liver cells	Allows movement into and out of liver cells
GLUT2	Epithelium of small intestine and proximal convoluted tubule cells in the kidney	Allows glucose to pass from epithelial cells into capillaries
GLUT4	Adipose tissue, skeletal and cardiac muscle cells	Allows movement into cells when cells are stimulated by insulin

c) Suggest and explain the structure of a GLUT protein. (4)

d) When stimulated by insulin GLUT4 proteins appear in the cell surface membranes of skeletal muscle cells much faster than can occur if they are made by protein synthesis. Suggest how molecules of GLUT4 appear in the cell surface membranes of skeletal muscle cells so quickly when the cells are stimulated by insulin. (3)

People who are developing type 2 diabetes produce insulin normally but body cells develop insulin resistance as their cells do not respond to the hormone by taking up glucose from the blood rapidly enough.

e) i) Suggest how body cells develop insulin resistance. (2)

ii) Explain why muscle cells and adipose cells develop insulin resistance more readily than liver cells. (2)

f) The drug metformin may be given as part of the treatment for type 2 diabetes. This drug reduces the rate of gluconeogenesis, which is very high in people with this type of diabetes. One way in which it does this is by inhibiting the activation of cyclic AMP by glucagon. Suggest how metformin inhibits the formation of cAMP in hepatocytes. (3)

g) Another type of glucose transporter is found in the luminal membranes of proximal convoluted tubule cells in the kidney. Explain how the activity of these glucose transporter proteins differs from the activity of GLUT1. (3)

Stretch and challenge

7 Insulin is a hormone that stimulates the conservation of resources, such as glucose and fat, in the body.

a) Describe how insulin is synthesised and secreted by cells in the pancreas.

The cell surface membrane is the site of signal transduction. This occurs when first messengers such as neurotransmitters or hormones arrive at the cell surface and a mechanism transfers the signal into the interior of the cell.

Two mechanisms of signal transduction are:
- ion channels which are activated by neurotransmitters
- second messenger systems.

b) Compare the ways in which these two mechanisms function to transfer signals from the exterior to the interior of cells.

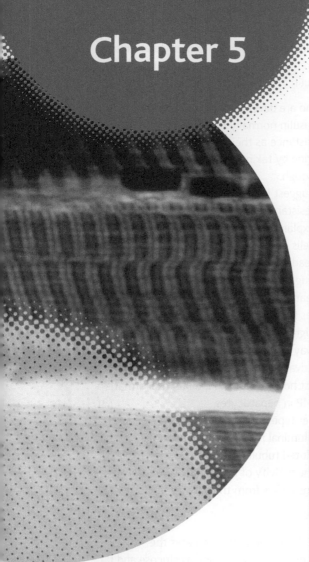

Chapter 5

Animal responses

Prior knowledge

- Animals respond to changes in their environment.
- Such responses are largely under the control of the nervous system.
- The brain and spinal cord make up the central nervous system.
- The nerves branching off the central nervous system form the peripheral nervous system.
- Voluntary responses are under the conscious control of the brain.
- Reflex actions are fast, automatic responses that do not require conscious control.
- Reflex actions usually have a protective function.
- Muscle contraction requires energy from respiration.
- Moveable joints are moved by muscles, which are attached to the bones by tendons.
- Muscles often work in antagonistic pairs, with each member of the pair performing the opposite action to the other.

Test yourself on prior knowledge

1 What are the functions of the nervous system?
2 What name is given to a change in the environment which can be detected by the body?
3 What features define a piece of behaviour as a reflex action?
4 Give three examples of a reflex action.
5 In what chemical form is the 'energy from respiration' provided to muscles?
6 Why do muscles have to work antagonistically, i.e. against one another?

Figure 5.1 Is the speed of a sprinter limited by the speed of muscle contraction?

So far Usain Bolt has recorded a faster time over 100 m than any other sprinter in history. When he recorded a time of 9.58 s in 2009, he travelled at an average speed of about 37 km h^{-1} (which means that his speed must have been even faster than this at some point of the running). However, Bolt has said that he believes that he can run even faster and research seems to back him up. We now know that the power generated by muscles could be enough to generate speeds of 65 km h^{-1}. To achieve such a speed would allow Bolt to run 100 m in an astonishing 5.5 s; but is that really possible? In fact, the speed an athlete can achieve is not just about the power created in the muscles but also depends on the speed at which those muscles can contract (Figure 5.1). The feet of a runner are in contact with the ground for a very short length of time and only during this time can the potential power of the muscles be used to generate forward movement. In this chapter, we will see how muscles are able to produce such extraordinary power and release it in muscle contraction.

Overview

Animals interact with their environment. They need to get certain resources from the environment, and the availability of these resources varies within the environment, so animals have to respond in a way that maximises their chances of getting adequate supplies. They also need to be able to detect threats around them and respond in appropriate ways to ensure their survival. More advanced animals can perform a wide range of movements, which allows them to interact with their environment in complex ways. This has been aided by the evolution of different forms of skeletons with attached muscles and moveable joints.

In order to respond to their environment, animals need systems that are able to detect relevant changes and then produce a coordinated response, involving several parts of their body. As we have seen in the previous two chapters, animals have both nervous and hormonal systems to coordinate their responses.

The human nervous system

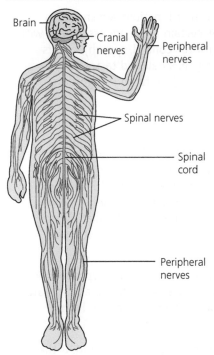

Figure 5.2 The structure of the human nervous system.

The basic arrangement of the human nervous system is common to all mammals. The layout of the nervous system is shown in Figure 5.2. It is organised into two parts, the **central nervous system** (CNS), which consists of the brain and spinal cord, and the **peripheral nervous system** (PNS), which consists of the nerves running to and from the CNS. The peripheral nervous system is further divided into the **somatic nervous system**, which is involved in voluntary responses, and the **autonomic nervous system**, which controls the body's involuntary responses. The distinction between the somatic and autonomic nervous systems is a functional one: the structure of the neurones in the two systems is the same and they run alongside one another, and so are not separated into different areas of the body.

The somatic nervous system

The **somatic nervous system** is concerned with voluntary control of body movements. It consists of three types of nerve. **Sensory nerves** consist entirely of sensory neurones and carry impulses from sense organs to the CNS. **Motor nerves** consist entirely of motor neurones and take signals from the CNS to muscles and glands. The spinal nerves from the spinal cord are **mixed nerves** and contain both sensory and motor neurones.

The autonomic nervous system

The autonomic nervous system is divided into two main parts, the **sympathetic nervous system** and **parasympathetic nervous system**. The autonomic nervous system coordinates involuntary functions, such as heart rate, regulation of blood vessel diameter and peristalsis in the gut. The sympathetic system deals with flight-or-fight responses, whereas the parasympathetic system deals with the 'rest and digest' system.

The brain

The human brain is an incredibly complex organ in terms of both structure and function. Its gross structure is shown in Figure 5.3. The functions of the different areas are listed below.

> **Tip**
>
> Make sure that you understand the distinctions between nerves and neurones. It can be an easy mistake to refer to 'nerves' when you really mean 'neurones' and the other way round.

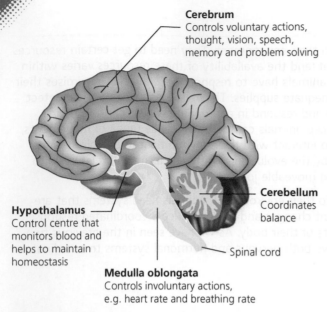

Cerebrum
Controls voluntary actions, thought, vision, speech, memory and problem solving

Hypothalamus
Control centre that monitors blood and helps to maintain homeostasis

Cerebellum
Coordinates balance

Spinal cord

Medulla oblongata
Controls involuntary actions, e.g. heart rate and breathing rate

Figure 5.3 Section of the human brain, showing the general functions of each area.

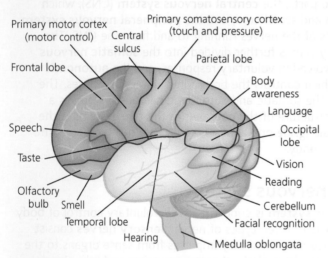

Primary motor cortex (motor control)
Central sulcus
Primary somatosensory cortex (touch and pressure)
Parietal lobe
Frontal lobe
Body awareness
Language
Speech
Occipital lobe
Taste
Vision
Reading
Olfactory bulb   Smell
Cerebellum
Temporal lobe
Facial recognition
Hearing
Medulla oblongata

Figure 5.4 The cerebral lobes and the functions of the cerebrum. The insular lobe is inside the brain and is not visible.

Tip

You need to know the functions of the cerebrum, but will not be expected to name the lobes.

Tip

Look back at Chapter 1 Communication and homeostasis to remind yourself about regulation of the body systems.

The cerebrum

The **cerebrum** is the largest part of the brain in humans (it accounts for about 80% of its mass) and carries out a large variety of functions. It is concerned with conscious activities like vision, hearing, speech, thinking and memory. It consists of five lobes and is divided into two halves, called the **cerebral hemispheres**, joined together by a band of nerve fibres, which make up the **corpus callosum**. The left hemisphere controls the right side of the body and the right one controls the left side. The cerebrum has a thin outer layer called the **cerebral cortex** which is commonly called the grey matter. It consists of neurone cell bodies and is highly folded. This increases its surface area, allowing more nerve cell bodies to be packed into it. The number of connections between nerve cells in the brain increases the complexity of behaviours that can be performed, and the more nerve cells there are, the more connections can be made. Below the cerebral cortex is the white matter, which is composed of myelinated axons.

The cerebral lobes and their functions are shown in Figure 5.4. Notice that certain areas are associated with motor functions (e.g. motor control, speech), others with sensory functions (e.g. smell and taste) and the remainder with association (e.g. facial recognition, reading and language).

The hypothalamus

The **hypothalamus** is an area in the middle of the lower side of the brain, just above (and connected to) the **pituitary gland**. It monitors the blood flowing through it, and releases hormones that are involved in homeostasis, either directly or by stimulating the pituitary gland to release other hormones. Its main functions are as follows.

● Regulation of body temperature: the hypothalamus monitors the temperature of the blood and initiates homeostatic responses when it gets too high or too low.

● Osmoregulation: the concentration of the blood is monitored. If the concentration gets too high the blood becomes more viscous and this initiates the release of anti-diuretic hormone (ADH) from the posterior pituitary gland which acts in the kidneys to increase water retention. A feeling of thirst is also generated by the hypothalamus.

● Regulation of digestive activity: gut secretions and peristalsis are controlled, and a feeling of hunger is created if concentrations of blood nutrients decrease.

● Control of endocrine functions: the hypothalamus releases chemicals that stimulate the release of hormones from the pituitary gland. (see the control of the thyroid gland in Chapter 1)

The pituitary gland

The **pituitary gland** is at the floor of the brain, below the hypothalamus, attached by the pituitary stalk (Figure 5.5). It produces a range of hormones which regulate the body either directly, or by stimulating the release of other hormones at remote locations in the body. It is divided into two areas, called the anterior pituitary and the posterior pituitary. The anterior pituitary produces and releases hormones, while the posterior pituitary does not produce any hormones itself, but stores and releases the hormones ADH and oxytocin that are produced in the hypothalamus.

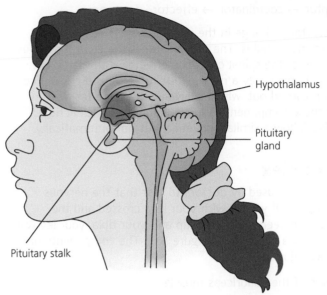

Figure 5.5 Location of the pituitary gland and the hypothalamus.

Hypothalamus

Pituitary gland

Pituitary stalk

The cerebellum

The **cerebellum** lies underneath the cerebrum and it is concerned with motor coordination, including balance. Balance is a complex function, requiring coordination between the eyes, the semicircular canals in the ears and a variety of muscles. Due to the high degree of rapid coordination involved, balance has this whole area of the brain devoted to it. It functions solely at the subconscious level and all of the actions it controls are involuntary.

The medulla oblongata

The **medulla oblongata** (sometimes shortened to the medulla) is located at the base of the brain, where it joins the spinal cord. It contains three centres that control various body functions.

- The cardiac centre regulates the heart rate.
- The vasomotor centre controls blood pressure by regulating the contraction of smooth muscles in the walls of arterioles.
- The respiratory centre controls breathing rate. There are separate inspiratory and expiratory centres within this centre.

> **Key term**
>
> Semicircular canals Sense organs in the inner ear that detect body position and movement.

> **Test yourself**
>
> 1 What structures comprise the central nervous system?
> 2 What are the three types of nerve found in the somatic nervous system?
> 3 Explain the value of the folds in the surfaces of the cerebral hemispheres.
> 4 What part of the brain is responsible for coordinating balance?

Reflex actions

Most of the things we do involve incredibly complex interactions of neurones in many parts of the nervous system. However, **reflex actions** are relatively simple. A reflex action is a behaviour that shows certain characteristics. It is an involuntary response to a given stimulus, is rapid and usually has some protective or survival value. There are a lot of reflex actions that the human body carries out, including

- yawning
- production of saliva
- swallowing
- the withdrawal reflex: pulling a part of the body away from a source of pain
- blinking
- the pupil reflex: the iris muscles constricting in bright light.

Some of these actions can be carried out voluntarily (e.g. blinking), and on such occasions the action would not be classed as a reflex. Others are always automatic (e.g. the pupil reflex).

Any reflex consists of a sequence of components, shown below.

Stimulus → receptor → coordinator → effector → response

The stimulus is a detectable change in the environment. The receptor is an organ that detects the stimulus. The coordinator is the central nervous system, either the brain or the spinal cord. The effector is the structure that carries out the action: usually a muscle, but sometimes a gland. The response is the action carried out, which is always the same for a given stimulus. Note that these components are not unique to reflex actions; voluntary actions also follow a similar sequence, but not automatically.

The knee-jerk reflex

The knee-jerk reflex is often used by doctors to test that the nervous system is working properly. If you sit with your legs crossed and the doctor hits a ligament between your knee cap and your tibia, your leg will involuntarily straighten in a sort of kick (Figure 5.6). The components of the knee-jerk reflex are as follows.

- Stimulus: stretching of the quadriceps muscle

- Receptor: stretch receptors in the quadriceps muscle

- Coordinator: the spinal cord

- Effector: the quadriceps muscle

- Response: contraction of the quadriceps muscle, causing a straightening of the leg.

The nervous pathway of this reflex is shown in Figure 5.7.

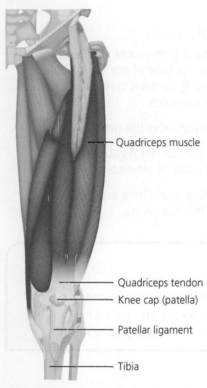

Quadriceps muscle

Quadriceps tendon
Knee cap (patella)
Patellar ligament
Tibia

Figure 5.6 The structures involved in the knee-jerk reflex.

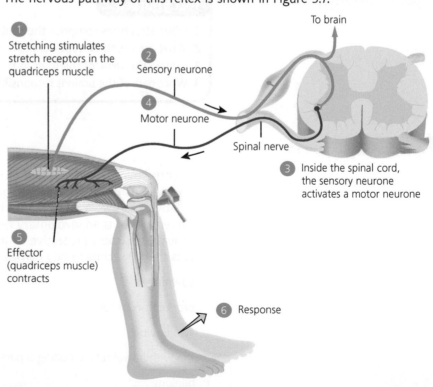

To brain

1 Stretching stimulates stretch receptors in the quadriceps muscle

2 Sensory neurone

4 Motor neurone

Spinal nerve

3 Inside the spinal cord, the sensory neurone activates a motor neurone

5 Effector (quadriceps muscle) contracts

6 Response

Figure 5.7 The nervous pathway of the knee-jerk reflex.

The stretch receptor sends an impulse down the sensory neurone, which connects directly (via a synapse) with a motor neurone in the spinal cord. The impulse transfers to the motor neurone, which takes it to the effector (the quadriceps muscle), which contracts. In many reflexes there is another neurone, the relay neurone, in the spinal cord which functions inside the CNS and conveys impulses between the sensory and motor neurones. In the knee-jerk reflex, however, there is no relay neurone.

Note that impulses would travel along many sensory neurones from stretch receptors in the muscle and then along many motor neurones, each stimulating a group of muscle cells.

This pathway helps to explain why reflex actions are rapid and automatic. Nerve impulses are delayed by synapses, but in this case the signal only has to cross a single synapse. If it went through the brain, it would have to go through many synapses. Connections within the spinal cord will send information about the stimulus to the brain but, by the time it receives and processes the information, the response will have already taken place. The brain therefore has no chance to make a decision, and the response happens automatically.

The protective or survival value of such a reflex is not immediately obvious, but that is because it is induced in an artificial situation for medical testing. The reflex is actually part of our complex balancing system. Pressure on the ligament stimulates stretch receptors in the quadriceps muscle (see Figure 5.6). This also happens if you start to fall, and the kicking action tends to propel you upwards, allowing time for your body to stop you falling.

The blinking reflex

Blinking is a response caused by anything contacting the cornea, or by drying of the cornea. The reflex pathway (shown in Figure 5.8) does go through the brain this time, but not through the decision-making areas, and once again the number of synapses involved is very few.

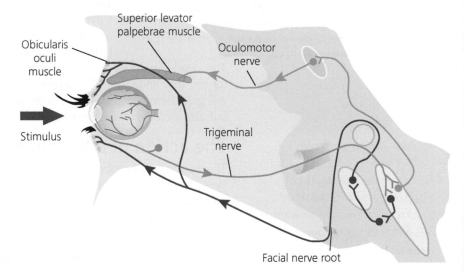

Figure 5.8 The control of the blinking reflex.

The blink reflex is more complex than the knee-jerk reflex, because multiple effectors are needed to close the eyelids. The superior levator palpebrae muscle lowers the upper eyelid, and the orbicularis oculi muscle, which goes around the eye, pulls the eyelids inwards to some extent and helps to close the lids. Any irritation of the cornea sends a signal along the trigeminal (sensory) nerve to the medulla of the brain, where it connects with a number of other neurones which transmit the signal to the effector muscles. Relay neurones are involved in the nervous pathway to the lower eyelid. Other stimuli can also initiate the blinking reflex, such as objects rapidly approaching the eye, or bright light. As both irritating substances and objects travelling towards the eye can cause potential damage, the blinking reflex protects the eye.

Coordination of the nervous and endocrine systems

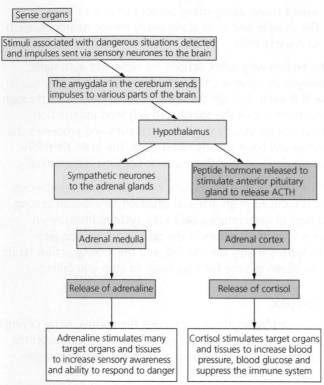

Figure 5.9 Flow chart diagram showing how the release of adrenaline and cortisol from the adrenal glands is controlled. Yellow indicates the roles of the nervous system, green the roles of the endocrine system.

The fight-or-flight response

The two systems involved in coordination of the body – the nervous and endocrine systems – do not work in isolation from one another. They will often work together and complement one another. This can be seen in the fight-or-flight stress response to environmental stimuli. This response is seen in situations where fear or aggression is induced.

The stressful situation in the environment must obviously be detected by the nervous system. The sympathetic nervous system coordinates many of the responses to danger, but its actions are supported by two hormones from the adrenal glands: adrenaline and cortisol. The sequence of events is summarised in Figure 5.9.

You can see that the early part of the response is mediated by the nervous system, but then it is continued by the endocrine system with the hormones adrenaline and cortisol circulating in the blood to affect many different areas of the body. The effects of these hormones focus the body's resources on providing oxygen and sources of energy to the muscles.

The fight-or-flight response has evolved because it has survival value. For example, when seeing a predator, the response allows the potential prey animal to become temporarily stronger, faster and able to react more quickly. This clearly improves its chances of survival. Male animals of many species fight for territory and mates. The aggression induced by the fight-or-flight response will give them a better chance of winning the fight, which has survival value but also increases the chances of passing on their genes to the next generation.

The action of adrenaline

The hormone adrenaline circulates in the blood and affects many different types of cell. This is an example of cell signalling, and only those cells with adrenaline receptors in their membranes will respond. Adrenaline begins to act very quickly and has the following effects on the body:

- increases heart rate and stroke volume (volume of blood per beat)
- increases the concentration of glucose in the blood by stimulating enzymes in liver cells to convert glycogen to glucose
- decreases blood flow to the gut and skin by stimulating vasoconstriction
- increases blood flow to muscles and the brain by stimulating vasodilation
- increases blood pressure because vasoconstriction increases resistance to flow
- increases width of bronchioles by causing smooth muscles to relax so increasing flow of air to alveoli
- stimulates contraction of muscles in the iris of each eye to dilate the pupils.

Adrenaline is broken down very quickly in the body. The concentration of adrenaline halves within 160 to 190 seconds.

Key term

Cell signalling The release of a substance by one cell which transmits information to another cell, either locally or over some distance.

Tip

Action of hormones has been covered in Chapter 4.

Test yourself

5 What is the receptor in the knee-jerk reflex?
6 Why are reflex actions involuntary?
7 Which molecule acts as a second messenger in liver cells which have detected adrenaline?

Nerves and hormones in the control of heart rate

The nervous and endocrine systems also cooperate in controlling the heart rate. Although the heart muscle is myogenic (i.e. the contraction is initiated by the muscle itself), the rate of beating is controlled externally. The various factors which can affect heart rate are shown in Table 5.1.

Table 5.1 Various factors that can affect heart rate.

Nervous system	Effect
Sympathetic nerves	Increases heart rate
Parasympathetic nerve (vagus nerve)	Decreases heart rate
Endocrine system	**Effect**
Adrenaline	Increases heart rate
Noradrenaline	Increases heart rate
Thyroxine (thyroid hormone)	Increases heart rate

Like adrenaline, noradrenaline is a hormone produced by the adrenal glands.

Activity

Exercise and pulse rate

Researchers wanted to investigate the effect of being overweight on heart/pulse rate and exercise in children. One way of checking whether a person is overweight is to calculate their body mass index (BMI). BMI is calculated by dividing your weight in kilograms by your height in metres squared in adults, but in children it is more complicated and age and gender are factored into the calculation. The BMI allows children to be classified as underweight, low weight, healthy weight, overweight, obese or extremely obese.

The study compared 10 children aged 11 who were of healthy weight with 10 who were classed as overweight. Their resting heart rate was monitored for 5 minutes before the exercise, which was the use of an exercise bike (set at a fixed tension) for 5 minutes. Their pulse rate was monitored throughout the exercise using a heart-rate data logger that sent a signal wirelessly to a computer. The time taken for it to return to normal after exercise was measured. The results are shown in Table 5.2.

Table 5.2 Results of measuring heart rate in children while using a bicycle ergometer.

Healthy weight children				Overweight children			
No.	Mean resting heart rate/ bpm	Increase during exercise/bpm	Recovery time/s	No.	Mean resting heart rate/ bpm	Increase during exercise/bpm	Recovery time/s
1	82	46	70	1	87	24	60
2	84	38	55	2	82	59	88
3	79	17	32	3	98	53	82
4	80	20	40	4	88	86	103
5	72	79	91	5	83	38	55
6	82	74	88	6	95	46	79
7	68	65	71	7	79	65	99
8	60	38	56	8	91	33	58
9	82	49	66	9	96	92	135
10	84	31	45	10	89	74	104
Means	77.3	45.7	61.4	Means	88.8	57.0	86.3

Questions

1 The researchers controlled the tension of the bike and the length of the exercise, but the groups were of mixed gender. Do you think that matters? Give reasons for your answer.

2 Certain variables that might affect this investigation cannot be controlled. Suggest what these may be.

3 How could the researchers reduce the impact of such uncontrolled variables?

4 What are the advantages of using a heart rate sensor and data logger for this experiment compared to a stopwatch and counting pulse rate manually?

Example

The researchers decided to analyse the difference between the recovery times of the two groups of children. They chose to use the Student's t test for this analysis.

1 State the null hypotheses for this investigation.
2 Use the formula below to calculate the value for t. Show all your working.

$$t = \frac{|\bar{x}_1 - \bar{x}_2|}{\sqrt{\frac{s_1^2}{n_1} + \frac{s_2^2}{n_2}}}$$

where

$\bar{x}_1$ is the mean of sample 1
$\bar{x}_2$ is the mean of sample 2
n_1 is the number of subjects in sample 1
n_2 is the number of subjects in sample 2

s_1^2 is the variance of sample 1, $\dfrac{\Sigma\left(x_1 - \bar{x}_1\right)^2}{n_1 - 1}$

s_2^2 is the variance of sample 2, $\dfrac{\Sigma\left(x_2 - \bar{x}_2\right)^2}{n_2 - 1}$.

Note the standard deviation is the square root of variance.

3 State the conclusion that you can make based on the value of t that you have calculated. Explain your answer. (You will need to use Table 15.10 on page 313 in order to answer this question.)

Answers

1 The null hypothesis is that there is no difference in the recovery time following exercise between the healthy and overweight groups.

2 Table 5.3 shows how to calculate the variance for each sample.

Table 5.3 Calculating variance.

Healthy children			Overweight children		
Recovery time (s)			Recovery time (s)		
x_1	$(x_1 - \bar{x}_1)$	$(x_1 - \bar{x}_1)^2$	x_2	$(x_2 - \bar{x}_2)$	$(x_2 - \bar{x}_2)^2$
70	8.6	73.96	60	−26.3	691.69
55	−6.4	40.96	88	1.7	2.89
32	−29.4	864.36	82	−4.3	18.49
40	−21.4	457.96	103	16.7	278.89
91	29.6	876.16	55	−31.3	979.69
88	26.6	707.56	79	−7.3	53.29
71	9.6	92.16	99	12.7	161.29
56	−5.4	29.16	58	−28.3	800.89
66	4.6	21.16	135	48.7	2371.69
45	−16.4	268.96	104	17.7	313.29
Totals		3432.4			5672.1

$n_1 = 10, n_2 = 10$

$\bar{x}_1 = 61.40$

$\bar{x}_2 = 86.30$

$$S_1^2 = \frac{\sum(x_1 - \bar{x}_1)^2}{n_1 - 1} = \frac{3432.4}{9} = 381.38$$

$$S_2^2 = \frac{\sum(x_2 - \bar{x}_2)^2}{n_2 - 1} = \frac{5675.1}{9} = 630.23$$

$$t = \frac{\bar{x}_1 - \bar{x}_2}{\sqrt{\frac{S_1^2}{n_1} + \frac{S_2^2}{n_2}}}$$

$$= \frac{61.4 - 86.3}{\sqrt{\frac{381.38}{10} + \frac{630.23}{10}}} = \frac{24.9}{10.06}$$

$$= 2.48$$

> **Tip**
>
> Note that the top line is the *difference* between the means. If it is a negative value, that is ignored.

3 The number of degrees of freedom is $(n_1 + n_2) - 2 = (10 + 10) - 2 = 18$.

At 18 degrees of freedom the critical value at $p = 0.05$ is 2.10. The value for t calculated for this investigation (2.48) is greater than this so there is less than a 5% probability that the results are due to chance. The conclusion is that the null hypothesis is rejected and that there is a significant difference between the means of the two groups of children.

Muscles and movement

The brain and the central nervous system control and coordinate the movement of the body. A huge array of subtle movements can be made, sometimes in direct response to the environment but also to maintain posture. The brain receives a huge amount of information from sense organs, including stretch receptors in the muscles themselves, decides if action is necessary and exactly what sort of action it should be, and then sends signals to a variety of muscles which will coordinate their action to produce the desired movement. The spinal cord often acts as a 'go between' in this process, taking impulses from the sense organs to the brain along sensory neurones and carrying impulses back to muscles via motor neurones.

In humans, as in other vertebrates, movement is dependent on a range of skeletal joints which are moved by muscles. These muscles are arranged in antagonistic pairs, as the joint has to both move and return to its original position. As muscles can only exert force by contracting, an individual muscle can only work in one direction and there must be an antagonistic muscle to move the joint in the opposite direction.

> **Key term**
>
> **Antagonistic** Description of structures or chemicals in the body which perform opposite actions to each other.

Muscle structure

There are three types of muscle: **skeletal** (also called striated or voluntary muscle), **smooth** (also called involuntary) and **cardiac**. The type that moves joints is skeletal muscle, and this has a complex and unusual structure. It is shown in Figure 5.11.

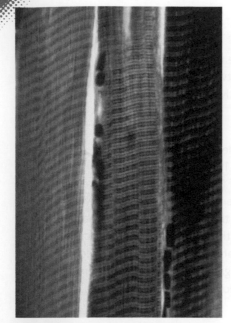

Figure 5.10 A photomicrograph of three muscle fibres teased from skeletal muscle tissue. Compare with Figures 5.11 and 5.12 and identify sarcomeres, Z lines, dark (A) and light (I) bands (x 480).

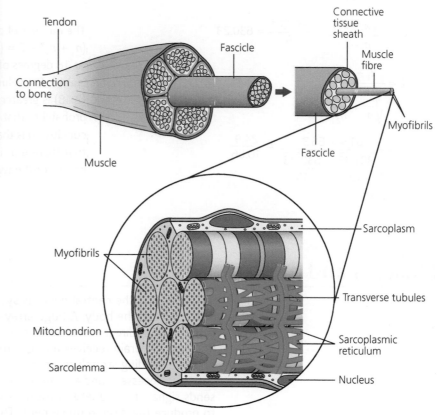

Figure 5.11 Structure of skeletal muscle.

A muscle consists of muscle fibres bundled together into a structure called a fascicle. Many fascicles are bundled together to form the muscle. The muscle fibres are sometimes referred to as muscle cells, although they are multinucleate and have been formed from several cells. The structures within the fibre often have the prefix *sarco-*, so the cell membrane is referred to as the sarcolemma, the endoplasmic reticulum as the **sarcoplasmic reticulum** and the cytoplasm as the sarcoplasm. There are **transverse tubules** (sometimes referred to as T tubules) running through the fibre. These are extensions of the sarcolemma which penetrate the fibre. The fibre contains many **myofibrils**, which are the contractile elements in the muscle. As we shall see below, ATP plays a vital function in muscle contraction, and so the fibre contains many mitochondria.

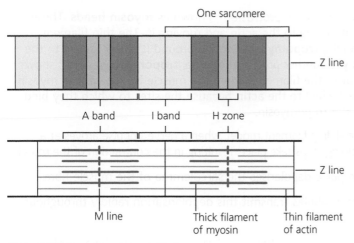

Figure 5.12 Structure of a myofibril.

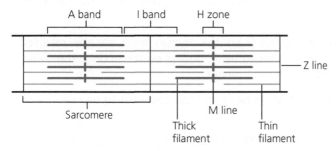

Relaxed

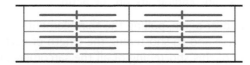

Contracted

Figure 5.13 Arrangement of thick and thin filaments in a relaxed and contracted myofibril.

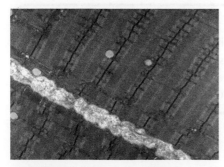

Figure 5.15 A transmission electron micrograph of a longitudinal section of skeletal muscle tissue. Myofibrils, sarcomeres, Z lines and dark and light bands are clearly visible (x 3300).

The sliding filament model of muscle contraction

When a muscle contracts, it is the myofibrils that change in length. The contraction of the myofibrils adds together to contract the whole muscle. In order to understand how muscles contract, it is necessary to know the structure of the myofibrils. This is shown in Figure 5.12. The myofibrils contain bundles of thick and thin **myofilaments** which slide past each other during contraction. Myofilaments are made of proteins. The thick filaments are made mostly of myosin and the thin filaments of actin. Myosin and actin are components of the cytoskeleton of many cells.

The arrangement of the thick and thin filaments gives the myofibrils a striped appearance, which gives rise to the name striated muscle (*striated* means striped). The darker bands represent areas where the filaments overlap. The arrangement of thick and thin filaments repeats regularly along the length of the myofibril. The repeating unit is referred to as a **sarcomere** (see Figure 5.12).

The sliding filament theory of muscle contraction explains how the thin and thick filaments can slide past one another to shorten the myofibrils. The effect of this can be observed in light micrographs of relaxed and contracted myofibrils. The lighter **I band** is noticeably narrower in a contracted myofibril (see Figure 5.13).

The sliding filament theory explains the mechanism by which the thick and thin myofilaments move past one another. In order to understand this, we first need to look more closely at the structures of the thick and thin filaments. This is shown in Figure 5.14.

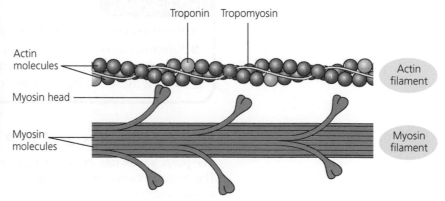

Figure 5.14 Structure of the thin and thick filaments.

The myosin filaments have protrusions, known as **myosin heads**. These are capable of attaching to the actin and moving it. The thin filament has another protein, **tropomyosin**, wound around it and attached to the tropomyosin is yet another protein molecule, **troponin**. Figure 5.12 shows the arrangement of the filaments when the muscle is relaxed. The myosin heads are not attached to the actin because the sites to which they bind are blocked by the tropomyosin.

According to the sliding filament model, when a nerve impulse arrives at a muscle the following steps take place, resulting in the contraction of the muscle.

● The nerve impulse depolarises the sarcolemma of the muscle fibre.

● The transverse tubules transmit this depolarisation rapidly throughout the fibre.

● Calcium ions (Ca^{2+}) enter the myofibril from the sarcoplasmic reticulum.

● These calcium ions bind to the troponin, which then changes shape, dragging the tropomyosin away from the myosin-binding sites of the actin.

● The myosin heads bind to the myosin-binding sites that have now been exposed.

● ADP and inorganic phosphate (P_i) in the myosin heads are now released. This causes the myosin heads to bend, pulling on the actin and moving it towards the centre of the sarcomere.

● Fresh ATP binds to the myosin head and this causes it to be released from the actin.

● The myosin head acts as an **ATPase** enzyme hydrolying the ATP into ADP and inorganic phosphate (P_i). This extends the head once again into a position close to the next myosin-binding site on the actin.

This process is repeated over and over again, pulling the thin filaments towards the centre of the sarcomere and so shortening the length of each sarcomere. The cumulative effect of this shortening of all the sarcomeres is to contract the entire muscle.

The sliding filament theory is summarised in Figure 5.16.

Test yourself

8 Which two proteins form the myofilaments in a myofibril?

9 Which band in the sarcomere gets shorter during muscle contraction?

10 How do calcium ions initiate muscle contraction?

11 What is the role of the transverse tubules in muscle contraction?

Energy supplies and ATP

ATP plays a vital role in muscle contraction, and muscles need a lot of it. The muscle fibres contain many mitochondria and for much of the time aerobic respiration in these organelles provides sufficient ATP. During periods of vigorous exercise, however, the supplies of ATP may run short. Muscles have a variety of ways of coping with this.

● The muscles contain the pigment **myoglobin**, which only releases oxygen when the surrounding concentration is very low. This acts as an emergency store of oxygen for aerobic respiration.

Tip

The sarcolemma and the T tubules conduct impulses in much the same way as the axons of neurones.

(a)

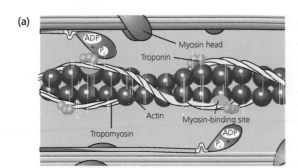

The resting state. The muscle is relaxed.

(b)

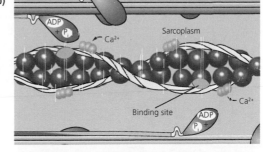

An impulse arrives and calcium ions flood into the myofibril. These ions bind to the troponin, which moves the tropomyosin out of the myosin-binding sites.

(c)

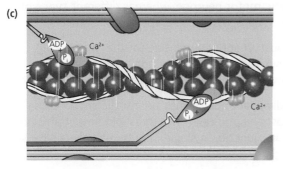

The myosin heads attach to the exposed binding sites.

(d)

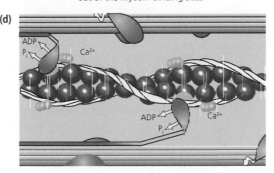

The release of ADP and inorganic phosphate causes the heads to move and pull the actin along.

(e)

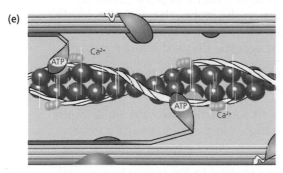

Fresh supplies of ATP enter the myosin heads and this breaks the connection with the binding sites.

(f)

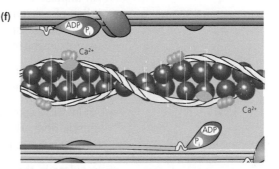

The hydrolysis of ATP to ADP and Pi returns the myosin heads to their starting positions. If calcium ions are still present, each myosin head will then immediately bind to the next myosin-binding site.

Figure 5.16 The stages of contraction in a myofibril. P_i represents inorganic phosphate.

- Muscles are capable of **anaerobic respiration** for short periods. This provides almost all the energy for short bursts of strenuous activity, but only releases far less ATP per molecule of glucose so is very wasteful and not as efficient as aerobic respiration. (See chapter 8.).

- The **ATP-creatine phosphate system** provides extra ATP for a very short period. Creatine phosphate (PCr) is a chemical found in muscle fibres which can provide phosphate to convert ADP into ATP. No lactate is produced in this process so it does not lead to muscle fatigue. Creatine (Cr) is produced. The equation is shown below:

ADP + PCr → ATP + Cr

The creatine phosphate store in muscle is very limited and will run out in a few seconds. The system provides only enough extra ATP for short bursts of activity.

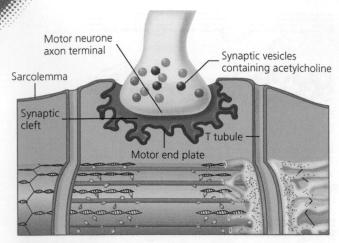

Figure 5.17 A neuromuscular junction.

Labels in figure:
Motor neurone axon terminal
Synaptic vesicles containing acetylcholine
Sarcolemma
Synaptic cleft
T tubule
Motor end plate

The neuromuscular junction

The contraction of skeletal muscle is under nervous control, and results from impulses travelling down motor neurones. At the end of a motor neurone, at the point where it joins the muscle, there is a specialised synapse called a **neuromuscular junction**. This is shown in Figure 5.17.

Neuromuscular junctions are similar in many ways to synapses in the nervous system, but there are certain key differences. These are summarised below.

- Acetylcholine is the neurotransmitter released by a somatic motor neurone, whereas different types of nerve synapse use different neurotransmitters.
- The postsynaptic membrane (of the muscle cell) is folded to increase the surface area, which is not the case in nerve synapses. This allows many more postsynaptic receptors to be present.
- In a nerve synapse an individual impulse may be insufficient to cause an action potential in the postsynaptic cell if the threshold potential is not reached. In a neuromuscular junction a single impulse will always cause a contraction.

The neuromuscular junction functions in the following way.

- An impulse arriving at the end of a motor neurone causes an influx of calcium ions, which results in the release of acetylcholine (the neurotransmitter) from vesicles in the synaptic knob. The neurotransmitter enters the synaptic cleft and diffuses across it.
- The acetylcholine binds to cholinergic receptors on the postsynaptic membrane (which, in muscles, is called the motor end plate).
- An impulse spreads across the sarcolemma over the whole of the muscle fibre. The depolarisation also passes down the transverse tubules throughout the fibre.
- Impulses in the T tubules are coupled to the contraction process. Surrounding T tubules is the sarcoplasmic reticulum (see Figure 5.11). The depolarisation of T tubules is the stimulus for the opening of voltage-gated calcium ion channels throughout the sarcoplasmic reticulum.
- Calcium ions diffuse from the sarcoplasmic reticulum in amongst the myofibrils to start contraction by binding to troponin (see Figure 5.15).
- Acetylcholinesterase enzyme in the synaptic cleft then breaks down the neurotransmitter so that the muscle is not continually stimulated unless further impulses arrive.
- When the sarcolemma, T tubules and sarcoplasmic reticulum are no longer depolarised, calcium ions are pumped back into the sarcoplasmic reticulum and contraction stops.

Note that the events in this account refer to what happens in a single muscle fibre, not the whole muscle. For example, an impulse reaching a motor end plate always causes a contraction in that muscle fibre, but if only very few motor neurones send impulses the number of muscle fibres contracting will be very low and this may not bring about a noticeable contraction in the whole muscle. The strength of muscle contraction will depend on how many fibres within it are contracting.

Muscle types

So far in this chapter we have focused on the actions of voluntary muscles (also called skeletal or striated muscles). These are muscles that can be consciously controlled, even though at times we may be unaware of their action (e.g. when maintaining posture). In a voluntary muscle there may be one or both of two types of muscle fibre. **Fast-twitch muscle fibres** contract rapidly and soon get fatigued. **Slow-twitch muscle fibres** contract less rapidly but fatigue very slowly. Your body does contain purely fast-twitch muscles and purely slow-twitch muscles, each of which consist of only one type of fibre. Many muscles, however, contain *both* fast and slow-twitch fibres. For example, slow-twitch fibres in muscles in the legs will contract for long periods of time to maintain posture, but the fast-twitch fibres in the same muscles can be used to move the legs.

In addition to the voluntary muscles, there are two other types of muscle found in the body. **Involuntary muscle** is used for the unconscious control of many parts of the body. **Cardiac muscle** is a highly specialised form of muscle found only in the heart.

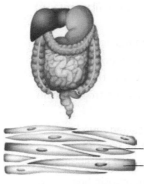

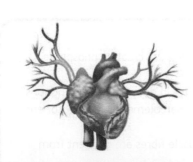

— Nucleus: one per cell; centrally located

— Smooth muscle cell: Elongated and spindle-shaped; much smaller than skeletal muscle cells

Figure 5.18 Involuntary or smooth muscle cells from the wall of the stomach.

Involuntary muscle

Involuntary muscle is also known as **smooth muscle**, as it lacks the striations seen in voluntary muscle, although it does contain filaments of actin and myosin. It is found in the walls of some internal organs (e.g. the gut, bladder, uterus and blood vessels). It is responsible for the control of gut movements and the narrowing of the arteries, among other things. The structure of smooth muscle is much less complex than that of striated muscle. The cells or fibres are smaller, spindle-shaped and have only one nucleus. The structure of smooth muscle cells is shown in Figure 5.18.

Cardiac muscle

Cardiac muscle is found only in the heart. It is a specialised striated muscle that has these unusual properties.

- It is **myogenic**. This means it can contract without any external stimulation from nerves or hormones. It beats on its own at regular intervals, although the speed of the beat can be regulated by the autonomic nervous system.

- The muscle fibres are branched to form a network which extends through the walls of the atria and the ventricles.

  - Cardiac muscle does not tire and can beat continuously throughout the lifetime (more than 2 billion contractions).

  - The fibres are connected to each other by special connections called **intercalated discs**, which allow the transfer of depolarisation from one cell to another.

The continual contraction of cardiac muscle requires a large amount of energy and the cells are packed with mitochondria to produce the necessary ATP. The structure of cardiac muscle is shown in Figure 5.19.

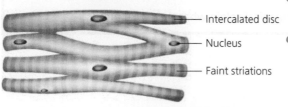

— Intercalated disc

— Nucleus

— Faint striations

Figure 5.19 The structure of cardiac muscle. These muscle cells are uninucleate unlike the multinucleate fibres of skeletal muscle.

Example

Figure 5.21 shows a sarcomere from a myofibril of a muscle fibre. The muscle is relaxed.

1 How would the appearance of the sarcomere change when the muscle contracts?
2 Identify the structures labelled X and Y.
3 Explain the role of each of the following in muscle contraction
   a) calcium ions
   b) tropomyosin
   c) ATP.

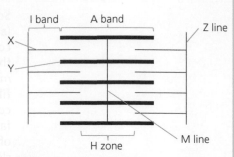

Figure 5.21 A sarcomere from a myofibril of a relaxed muscle fibre.

Answers

1 The H zone and the I band would be narrower.
2 X is a thin filament composed mostly of actin, Y is a thick filament composed mostly of myosin.
3 a) When the muscle is stimulated, calcium ions are released from the sarcoplasmic reticulum. The calcium ions bind to troponin molecules which are attached to tropomyosin molecules on the thin filaments. This causes the troponin to change shape, moving the tropomyosin and exposing the actin binding sites.
   b) Tropomyosin blocks the actin-binding sites in relaxed muscle and so prevents the myosin heads from binding to the actin. During contraction, the tropomyosin moves aside to expose the binding sites.
   c) Once contraction has occurred, ATP binds to the myosin heads and causes them to separate from the actin binding sites. The ATP is then broken down by myosin ATPase and the energy released resets each myosin head, ready to bind to the next actin binding site.

Figure 5.20 A photomicrograph of cardiac muscle cells. The striations across the muscle cells are visible (x 650).

Tip

You will be expected to examine stained sections or photomicrographs of skeletal muscle such as that shown in Figure 5.10.

Test yourself

12 What is the advantage for muscles in using the creatine phosphate system rather than anaerobic respiration?
13 What is the name of the neurotransmitter in neuromuscular junctions?
14 What is the benefit of the folding of the sarcolemma in the motor end plate of a muscle fibre?
15 State one way in which involuntary muscle fibres are different from voluntary ones.
16 Cardiac muscle is *myogenic*. Explain what that term means.

Exam practice questions

1 Which is a difference between skeletal muscle and cardiac muscle?
A Cardiac muscle is striated in appearance.
B Myosin in cardiac muscle is an ATPase.
C Skeletal muscle has multinucleate fibres.
D Skeletal muscle has sarcoplasmic reticulum. *(1)*

2 Which does not supply energy to muscle fibres in skeletal muscle tissue?
A Creatine phosphate
B Hydrolysis of ATP
C Oxidation of glucose
D Oxidation of lactate *(1)*

3 Which leads to a decrease in cardiac output?
A Increase in the rate of return of blood to the heart
B Stimulation by adrenaline
C Stimulation by sympathetic neurones
D Stimulation by parasympathetic neurones *(1)*

4 The following structures are in skeletal muscle tissue.
1 Myofibrils
2 Muscle fibres
3 Sarcomeres
4 Mitochondria
Which is the order of size of these structures, from smallest to largest?
A $1 \rightarrow 2 \rightarrow 3 \rightarrow 4$
B $2 \rightarrow 1 \rightarrow 4 \rightarrow 3$
C $3 \rightarrow 4 \rightarrow 2 \rightarrow 1$
D $4 \rightarrow 3 \rightarrow 1 \rightarrow 2$ *(1)*

5 The diagram shows the arrangement of the neurones that supply cardiac muscle in the heart. These neurones control the heart rate.

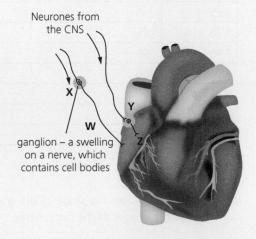

Neurones from the CNS

X
Y
W
Z
ganglion – a swelling on a nerve, which contains cell bodies

a) The neurones shown in the diagram originate from the brain. State the part of the brain that controls the activity of the heart. *(1)*

b) i) The synapses at X, Y and Z are cholinergic. What does this mean? *(1)*

ii) Explain how the effects of the motor neurones supplying the heart differ from the effect of motor neurones that supply skeletal muscle tissue. *(3)*

c) At times of stress, neurone W releases the neurotransmitter noradrenaline, which causes an increase in the heart rate. In some people this natural release of noradrenaline can be harmful. These people may be prescribed beta-blockers to reduce the effect of noradrenaline at synapses in the heart.

i) What are the advantages of increasing the heart rate in times of stress? *(4)*

ii) Suggest how beta-blockers could reduce the effect of noradrenaline. *(3)*

6 Motor neurones of the somatic nervous system terminate at neuromuscular junctions.

a) Describe what happens at the neuromuscular junction to stimulate muscle contraction when an impulse arrives at the end of a motor neurone. *(6)*

A student investigated the effect of stimulating a spinal nerve supplying a leg muscle of a small mammal. The contractions of the muscle were recorded as shown below.

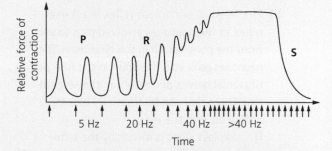

Relative force of contraction

P   R   S

5 Hz   20 Hz   40 Hz   >40 Hz

Time

b) Describe the effect of increasing the frequency of stimulating the motor nerve. *(4)*

c) Outline what happens in the muscle fibres to give the pattern of contractions at P and R. *(6)*

d) Suggest why there are no more contractions at S even though the nerve is still being stimulated. *(2)*

7 a) The diagram shows a vertical section through the human brain.

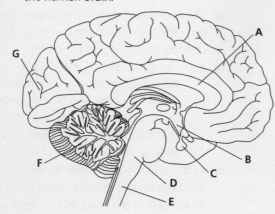

The table shows five functions of the brain. Copy and complete the table giving

- the name of the region that carries out each function
- the location of each region as indicated in the diagram above. (5)

Function	Name of the region of the brain	Letter from the diagram
Control of breathing rate		
Secretion of anti-diuretic hormone into the blood		
Control of core body temperature		
Interpretation of visual stimuli		
Control of balance		

b) The blinking, or corneal, reflex is a cranial reflex as the neurones involved pass to and from the pons Varolii in the hind brain. The neurones pass into the brain in the fifth pair of cranial nerves and return in the seventh pair. This reflex can be tested by blowing a puff of air into the eye.

   i) Explain what is meant by the term *cranial reflex*. (2)

   ii) Describe how the blinking reflex is coordinated in response to a puff of air into the eye. (6)

   iii) Outline the benefits of the blinking reflex. (3)

8 Adrenaline is a hormone that is released during the fight-or-flight response. As part of the response, adrenaline molecules stimulate hepatocytes to convert glycogen to glucose.

The concentration of adrenaline in the blood is very low, but the response by individual cells to adrenaline stimulation is very fast.

a) Adrenaline is water-soluble and does not cross the cell surface membranes of hepatocytes.

   i) Explain how adrenaline stimulates liver cells. (6)

   ii) Suggest how the response by hepatocytes is so fast. (3)

b) Describe the role of the nervous system in the coordination of the fight-or-flight response. (6)

Stretch and challenge

9 An isometric muscle contraction is one in which the length of the muscle remains constant, but the muscle develops tension. In the 1960s, physiologists at the University of London investigated the tension developed during isometric contractions of skeletal muscle with different sarcomere lengths. The diagrams show the arrangement of filaments within the sarcomeres when held at different lengths.

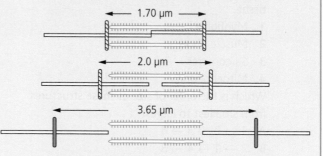

The table shows the force generated by the muscle as a percentage of the maximum force for a variety of sarcomere lengths.

Sarcomere length/μm	Mean tension generated as percentage of maximum
1.25	0
1.70	75
1.90	95
2.00	100
2.25	100
2.30	97.06
3.67	0

a) Draw a graph to show the results of this investigation.

b) Describe and explain the shape of the graph in terms of the images of the sarcomeres.

Chapter 6

Plant responses

Prior knowledge

- Both plants and animals respond to changes in their environment.
- Plants usually respond to their environment by means of growth. This growth is controlled by plant hormones.
- A growth response is called a tropism.
- Roots grow downwards towards the force of gravity (they are positively geotropic) and shoots grow towards the light and away from gravity. They are positively phototropic and negatively geotropic.
- Plant hormones can be used by humans to manipulate the growth of plants artificially.
- The plant hormones responsible for phototropism and geotropism are called auxins.
- In stems, auxins stimulate the elongation of cells, causing growth.

Test yourself on prior knowledge

1 How does the growth response of shoots to light benefit a plant?
2 Plant shoots respond to both light and gravity, but roots respond mostly to gravity. Suggest a reason for this.
3 Explain how light falling on one side of a shoot tip causes it to bend towards the light.
4 Give one example of how people use plant hormones to alter the growth of plants.

Introduction

Figure 6.1 Yields of fruits like red peppers could be improved with treatments involving a plant hormone called gibberellin.

Scientists have discovered a mechanism that has the potential to increase yields of flowers and fruits. For a long time it was thought that the plant hormone gibberellin promoted flowering. However, although high levels of gibberellin inhibit the formation of stem and leaves and promote the formation of flower buds, gibberellin actually inhibits the formation of the flowers themselves. Recently, an enzyme has been discovered which breaks down the gibberellin after the flower buds have started forming. It is now thought that growers could combine treatment with gibberellin with the later application of the enzyme, and so produce more flowers (and, therefore, more fruit). This could potentially increase the supply and decrease the cost of fruit, which is considered a healthy food (Figure 6.1).

Like animals, plants also interact with their environment. Plants require light, carbon dioxide, water and minerals. The availability of these varies within the environment, and plants have to respond in a way that maximises their chances of getting adequate supplies. The responses of plants are much more limited in this respect compared with animals. Animals can move from place to place, whereas plants cannot. Plants are therefore limited to growth responses, usually from a fixed point. Plants do not have sense organs, but they are able to detect relevant changes and then produce coordinated responses.

Types of plant response to the environment

Plants respond to the environment in a number of ways, some of which are more obvious than others. These responses have evolved because they provide some advantage to the organism. Stimuli from the environment can be abiotic or biotic. The term **abiotic** refers to a non-living component, whereas **biotic** refers to the living parts of the environment (i.e. other organisms).

Tropisms

A tropism is a growth response to a stimulus. It can be positive (toward the stimulus) or negative (away from the stimulus). Plants exhibit a number of tropisms, listed below. Light, water and gravity are abiotic stimuli, while touch and chemicals can be either abiotic or biotic.

- Phototropism: a response to light. It ensures the plant gets access to as much light as possible.

- Geotropism: a response to gravity. Shoots show negative geotropism, roots show positive geotropism. This ensures that when a seed germinates the shoot and root grow in the right directions, whatever the orientation of the seed.

- Hydrotropism: a response to moisture. Root tips tend to grow towards damper areas of soil, increasing their access to water.

- Thigmotropism: a response to touch. This is important in climbing plants as it allows them to detect a support and curl around it.

- Chemotropism: some plant tissues have a tropic response to certain chemicals. For example, pollen tubes grow down the flower's stigma towards the ovules due to chemotropism.

Nastic movement: *Mimosa pudica*

A process similar to thigmotropism is seen in the sensitive plant, *Mimosa pudica*. Handling the touch-sensitive leaves of this plant causes its leaflets to fold very rapidly (Figure 6.2). This is caused by rapid uptake of water in cells in the base of each leaflet, with a consequent increase in their volume. In addition, other cells adjacent to the expanding ones lose water and collapse. The response is not related in any way to the direction of the stimulus, and so it cannot be classed as a tropism. Such non-directional responses are known as nastic movements. The movement occurs very rapidly, and is thought to be caused by local bioelectrical signals rather than by plant hormones (which would be too slow).

The adaptive significance of thigmonastic movements to most plants is not well understood. Some evidence suggests that the movements may scare off leaf-eating insects and reduce transpiration when the Sun goes down. Another example with obvious advantages is the movement of the Venus flytrap, where insects are trapped by the movement of the leaves and are then digested to provide minerals for the plant.

Responses to herbivory

Herbivory is the term given to the eating of plants by herbivores. It is equivalent to predation, which is used in relation to carnivores. Many plants produce chemicals that act as a repellent to animals, or may be poisonous. In some cases these chemicals are always present, but in others they are produced as a response to damage or some form of stress. Some examples are given below.

Figure 6.2 Nastic movement in *Mimosa pudica* in response to touch.

Figure 6.3 Any fly that walks over these traps is likely to stimulate the sensitive hairs inside the trap. This may lead to a very rapid nastic movement that closes the trap providing the plant with a useful source of organic nitrogen.

Tannins

Tannins are water-soluble carbon compounds belonging to a group called flavonoids. They are stored in the vacuoles of plant cells and can be fatal to insects, and so they will reduce herbivore populations in the plant's environment. It was thought that the toxic effect was caused by inhibition of the protease enzymes in the insect gut, but recent research has shown that this is not the case, and that the effect is due to the breakdown of the tannins, producing toxic chemicals. Tannins also have a bitter taste, which can put off herbivores.

Alkaloids

Alkaloids are nitrogenous compounds derived from amino acids. Like tannins, they are bitter-tasting and can be toxic. We encounter alkaloids in our everyday lives. Caffeine is an alkaloid produced by tea, cocoa and coffee plants, where its purpose is as a chemical defence (it is toxic to insects and fungi). Nicotine, well known in the tobacco plant but also found in tomatoes, potatoes, aubergines and green peppers, is a potent neurotoxin which can be fatal to insects. Capsaicin is produced in chilli peppers and produces the characteristic burning sensation that many humans actually like but which will deter herbivores.

Pheromones

Pheromones are chemicals that are released by one member of a species and affect the physiology or behaviour of another member of the same species. Ethene (previously called ethylene) is an example, which when released by a plant causes the ripening of fruit in nearby plants, leaf loss and other physiological changes. However, although the mechanism is poorly understood, it also seems to play a key role in plant defences. Oxides of ethene are toxic to insects, but its main role seems to be as a controller of other chemical defences in plants, switching on the genes producing chemicals that deter insects, and so acting as a hormone.

Responses to abiotic stress

Freezing, drought, increased salinity of soil water and the presence of heavy metals are forms of **abiotic stress**. Plants respond to drought by closing their stomata (which restricts water loss) or by losing their leaves altogether. Certain plants produce a chemical in their cells that acts like an antifreeze. It makes it more difficult for ice crystals to form, which is important as ice crystals destroy the cells they form in.

Phototropism in plant shoots

The shoots of plants are **positively phototropic**: they grow towards light. This ensures that the maximum leaf area can be exposed to light for photosynthesis. Thus, if a plant is in a shaded area it will tend to grow out of the shade and into the light. Roots are weakly phototropic, but that response will always be over-ridden by their geotropic response.

The existence of phototropism has been known for a very long time and in the late nineteenth century efforts were made to understand how the process worked. Early experiments were carried out by Charles Darwin, and ideas were gradually developed over the next 40–50 years, as shown in Figure 6.4. Many of the experiments were done on coleoptiles. A coleoptile is a sheath that surrounds the young shoot in grasses.

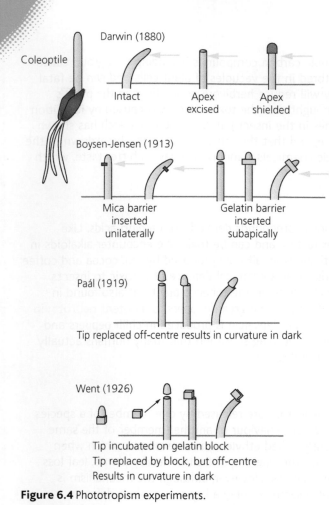

Darwin (1880)

Coleoptile

Intact · Apex excised · Apex shielded

Boysen-Jensen (1913)

Mica barrier inserted unilaterally · Gelatin barrier inserted subapically

Paál (1919)

Tip replaced off-centre results in curvature in dark

Went (1926)

Tip incubated on gelatin block
Tip replaced by block, but off-centre
Results in curvature in dark

Figure 6.4 Phototropism experiments.

Darwin's experiment (1880)

Darwin found that when the tip of a coleoptile was removed the response to unidirectional light did not occur. To discount the possibility that this was due to wounding he repeated the experiment, this time covering the coleoptile tip with an opaque cover, and once again there was no response. This indicated that the tip was responsible for detecting the light.

Boysen-Jensen's experiment (1913)

Boysen-Jensen did two experiments. He found that when a cut tip was replaced onto its coleoptile with a gelatin barrier inserted the phototropic response was restored. This indicated that the stimulus for growth was a chemical, as chemicals would travel through the gelatin barrier. The second experiment used a mica barrier. Mica is impermeable to chemicals. Boysen-Jensen inserted the mica halfway through the coleoptile, either on the side where the light was falling or on the opposite side. When the mica was on the lit side, the coleoptiles bent towards the light, but when placed on the side opposite to the light, the mica prevented the response. These experiments led to two conclusions. The signal which caused the response was a chemical (and therefore a hormone) which was produced in the coleoptile tip and travelled down on the side opposite to the stimulus. Secondly, it meant that the stimulus acted by causing growth on the unlit side rather than inhibiting growth on the lit side.

Paál's experiment (1919)

Paál cut the tips off coleoptiles and replaced them off-centre in the dark. Whichever side the tip was placed on grew more than the other side, causing curvature of the coleoptile. This confirmed that the response was caused by a hormone diffusing through the plant tissue and stimulating growth.

Went's experiment (1926)

Went placed the cut tip of a coleoptile onto a gelatin block, which absorbed the hormone. The gelatin block could create curvature in the dark when placed on the cut coleoptile. By placing different numbers of coleoptile tips on the gelatin, Went could increase the concentration of the hormone. He found that the degree of curvature showed a positive correlation with the number of tips used.

Test yourself

1 What is the difference between a tropism and a nastic response?
2 Explain how causing poisoning to a herbivore can be beneficial to a plant species, even though the plant has already been eaten.
3 What is abiotic stress?
4 Explain how a directional light stimulus causes positive phototropism.

Plant hormones

There are a variety of chemicals in plants that control aspects of their growth and development. As these are chemical messengers, they are referred to as plant hormones (or, sometimes, plant growth regulators). Plant hormones differ from animal hormones in a number of ways. They are not produced in specific organs, although they may be produced in a restricted region of the plant. They are produced by unspecialised cells and their effects on the plant may vary in different circumstances.

The responses coordinated by plant hormones include tropisms, leaf loss, stomatal closure, seed germination, flowering and fruit formation, and ripening.

The phototropic mechanism

Phototropism is now known to be coordinated by the plant hormone **indole-3-acetic acid** (IAA), which is an **auxin**, influencing how genes are expressed. Auxins may 'switch' some genes on and others off. IAA is synthesised in the meristem and passes down the stem to stimulate extension growth. They stimulate proteins in the cell wall known as expansins. These make cell walls more flexible by loosening the bonds between cellulose fibres. This is how shoots grow in length. When the tip is lit from one side, the auxin moves to the far side of the tip away from the light. The cells on this side then elongate more than those on the side nearer to the light, and so the tip curls towards the source of the light. This is shown in Figure 6.5. The exact mechanism by which the auxin is transported across the tip is still the subject of research. However, genetic studies on *Arabidopsis thaliana* (related to the cabbage) have identified special channel proteins that seem to transport auxin across membranes.

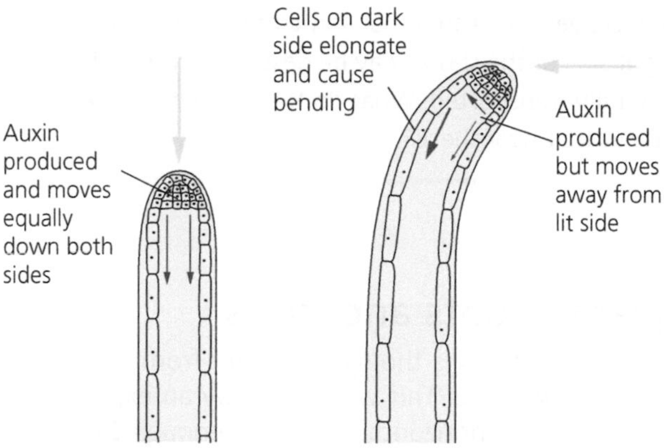

Cells on dark side elongate and cause bending

Auxin produced but moves away from lit side

Auxin produced and moves equally down both sides

Figure 6.5 The mechanism of phototropism.

Activity

Distribution of auxin in shoot tips illuminated on one side

An experiment was carried out to test the hypothesis that when a shoot tip is lit from one side auxin is transported to the dark side.

Shoot tips 7 mm long were cut from oat seedlings (coleoptiles). A razor blade was inserted to separate their lower ends into right and left halves. The tips were placed on agar blocks as shown in Figure 6.6. The end of the tip was removed and an agar block containing auxin radioactively labelled with carbon-14 (^{14}C) was placed on it.

A total of 22 shoot tips were treated in this way and then illuminated by a 150 W bulb at a set distance for 10 seconds. The whole experiment was repeated four times. The side from which the illumination came was chosen randomly in each experiment. The temperature

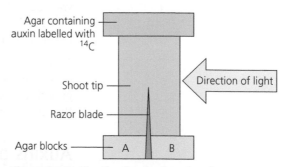

Agar containing auxin labelled with ^{14}C

Shoot tip

Razor blade

Direction of light

Agar blocks — A   B

Figure 6.6 Testing the distribution of auxin.

in the room was maintained between 25 °C and 27 °C. The humidity was not controlled.

The levels of radioactivity in the source block and blocks A and B were then measured (expressed as a percentage of the auxin added). The results are shown in Table 6.1.

▶▶▶

Table 6.1 Results showing the distribution of radioactive auxin. Although the direction of illumination was random, in each case block A represents the side of the tip furthest from the light source.

Experiment no.	Mean percentage of auxin transported	Mean percentage of auxin remaining in source block	Mean percentage of radioactivity in block A	Mean percentage of radioactivity in block B
1	6	94	72	28
2	11	89	78	22
3	11	89	77	23
4	15	85	74	26
Mean	11	89	75	25

Questions

1 Suggest a reason why the temperature of the room was maintained within a narrow range.
2 The humidity was not controlled. Do you think this matters? Explain your answer.
3 Explain why it was necessary to use the razor blade.
4 The experimenters concluded that auxin was transferred from the light side to the dark side of the shoot. Evaluate the strength of evidence for this conclusion.
5 Suggest a reason why such a small percentage of the auxin was transported down the tip.
6 It has been suggested that some of the auxin on the light side of the shoot may be destroyed rather than laterally transported. What do these results suggest about that hypothesis?

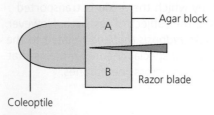

Figure 6.7 Dolk's experiment on geotropism.

Geotropism in plant shoots and roots

The negative geotropism of plant stems is thought to result from an accumulation of auxin on the lower side. This was shown by an experiment by H.E. Dolk. Coleoptiles were grown horizontally for long enough to produce a geotropic response. They were then decapitated and the tips placed laterally on agar blocks, which were divided horizontally by a razor blade (see Figure 6.7).

The auxin concentrations in blocks A and B were measured, and block B was found to contain significantly more auxin than block A. It is thought that gravity modifies the distribution of auxin so that it accumulates on the lower side, thereby increasing the rate of growth of that side.

The mechanism of positive geotropism in roots is thought to be a result of a similar distribution of auxin, to which the response is different. The growth of root cells is inhibited by higher concentrations of auxin, and so the root tip grows downwards.

Auxins and apical dominance

Auxins produced at the growing point at the apex of a plant stem not only cause the stem to grow upwards, they also inhibit the growth of lateral (side) buds. This is referred to as apical dominance. It is normally best for plants to grow upwards towards the light, increasing the amount of energy available for photosynthesis. Sideways growth is not so beneficial, and the action of auxins ensures that growth is preferentially upwards. However, if

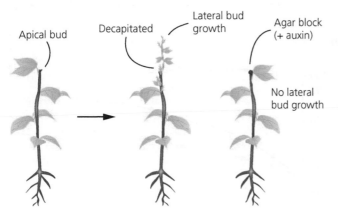

Figure 6.8 Experiment showing the role of auxins in apical dominance.

the growing point at the apex is removed (e.g. eaten by a herbivore) then the removal of the source of auxins results in the lateral buds growing. The lateral shoots will curl towards the light and so continue the growth of the plant upwards.

This has been shown experimentally, as shown in Figure 6.8. Removal of the apical bud allows the lateral buds to grow, but if the cut tip is immediately replaced with an agar block containing auxin then the inhibition of lateral bud growth is restored.

Gibberellins

Gibberellins are another type of plant hormone involved in growth. They are a group of plant hormones which are produced in young leaves and seeds, and also in the root tip. Their functions include the stimulation of germination, stem growth and flowering. Dwarf varieties of plants have been shown to have very low levels of gibberellins and treating them with gibberellins results in them growing to the same height as normal varieties (see Figure 6.9).

Gibberellins work with auxins in a synergistic way to stimulate stem growth. Their effect in combination, or synergism, is greater than their individual effects added together. Like auxins, gibberellins stimulate cell elongation, but also promote cell division.

Key term

Synergism A relationship between two factors where their action together is greater than that of their separate effects added together.

Figure 6.9 The effect of gibberellins on a dwarf variety. The dwarf mutant variety on the left was treated with gibberellin showing that the mutation has occurred to a gene involved in the synthesis of gibberellins.

Tip

You may read about the plant hormone abscisic acid (ABA), which sounds as though it would control abscission. Actually, ABA plays no significant role in the process. When discovered, it was thought that it did regulate abscission (hence its name) but recent research has shown that this is not the case.

Hormones and leaf loss

Deciduous plants lose their leaves when it is very hot and dry and shedding the leaves reduces water loss. This is also true during winter in temperate climates when absorption of water can be difficult from frozen soils. In winter in temperate areas leaves are also shed because photosynthesis is limited due to low temperatures and reduced light.

This leaf loss is controlled by hormones and their action is stimulated in temperate climates by the reduced length of the day in the autumn. Leaf loss is caused by the development of a layer of cells called the **abscission layer** at the bottom of the leaf stalk. This is a layer of parenchyma cells with thin walls, which makes them weak and easily broken. The plant hormone **ethene** stimulates the breakdown of cell walls in the abscission layer and this causes the leaf to break off.

Investigating the effects of gibberellins on plant growth

The effect of applying gibberellins to dwarf pea plants was investigated. Dwarf peas do not produce gibberellins. Two groups of 20 dwarf pea plants were germinated and grown to a height of approximately 30 mm. The exact height of each plant was recorded. The plants in the experimental group were sprayed with a solution in water of gibberellic acid (gibberellins can be absorbed through the leaves); the plants in a control group were sprayed with an equal volume of water. Both groups of plants were then grown for eight days in an identical growth medium (vermiculite). The heights of the plants were measured at intervals throughout that period. The results are shown in Figure 6.10.

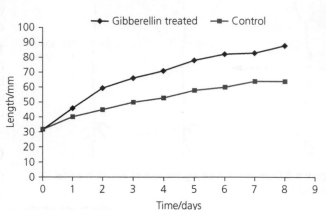

Figure 6.10 The heights of dwarf pea plants treated with gibberellic acid or with water.

1 Explain why it is better to use dwarf pea plants than normal pea plants for this experiment.
2 What are the control variables in this investigation?
3 What extra information would you require to judge whether a sample of 20 was adequate in this experiment?
4 Does this experiment provide evidence that gibberellins stimulate cell elongation? Explain your answer.
5 Assess the strength of evidence for the hypothesis that plants treated with gibberellins grow taller than the control group.
6 Suggest a possible source of inaccuracy in this experiment.

Abscission The shedding of an old or mature organ from a plant. It can apply to leaves, flowers and fruits.

Auxins are also involved in leaf loss. Auxins normally inhibit leaf loss and are produced by young leaves. They make the leaf stalks insensitive to ethene. As the leaf gets older, the concentration of auxins decreases and this allows leaf loss to occur in response to ethene. The process of abscission is summarised in Figure 6.11.

Hormones and stomatal closure

Abscisic acid (ABA) is a plant hormone that has a role in the closure of the stomata in the leaves by the guard cells. It is produced in the roots of plants in response to decreased soil water potential when the soil is drying out. It is then translocated to the leaves where it affects the guard cells. Guard cells lose water and become less turgid, so the stomata close. This is accomplished by the following steps (Figure 6.12).

1 ABA binds to receptors on the cell surface membrane of guard cells.

2 A complex series of events is set in motion which results in the opening of calcium channels, causing Ca^{2+} ions to enter. The pH of the cytoplasm is also raised.

3 These events cause K^+ ions, and also NO_3^- and Cl^- ions, to leave the cell.

4 As a result, the water potential of the cell increases, and water is moved by osmosis into surrounding cells. The resulting loss of turgor causes the stomata to close.

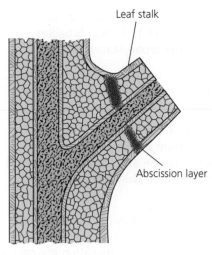

Leaf stalk

Abscission layer

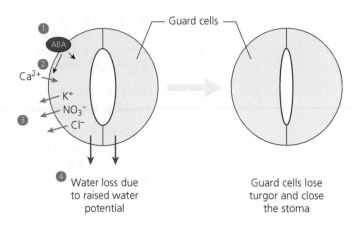

Guard cells

Water loss due to raised water potential

Guard cells lose turgor and close the stoma

Figure 6.12 The mechanism of stomatal closure as a result of stimulation by ABA.

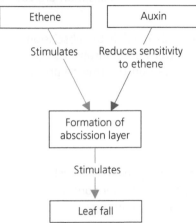

Ethene — Stimulates

Auxin — Reduces sensitivity to ethene

Formation of abscission layer

Stimulates

Leaf fall

Figure 6.11 The abscission layer and the control of leaf fall.

Hormones and seed germination

Seed germination is a complex process which involves mobilisation of food reserves and the promotion of growth. Environmental factors such as temperature and the availability of water are important, but many seeds have mechanisms that allow them to remain 'dormant' even if the environmental conditions might actually favour germination. Such dormancy has to be broken before the seed can germinate. A number of plant hormones have been shown to play some part in germination, but the three most important are gibberellins, abscisic acid and auxins.

Gibberellins stimulate germination and (if necessary) the breaking of dormancy. This has been shown by a number of experiments. Mutant strains of the plant *Arabidopsis* produce no gibberellins but can be induced to germinate if gibberellins are applied. Seeds of the Grand Rapids variety of lettuce normally require light to germinate but will germinate in the dark if they are treated with gibberellins. In cereals, such as wheat and barley, gibberellins released by the embryo travel to the aleurone layer, a layer of cells surrounding the endosperm of the seed. Gibberellins cause the aleurone layer to produce amylase, which breaks down starch stored in the endosperm into sugar, providing energy for the growth of the embryo, and breaking dormancy.

Abscisic acid has the opposite role, helping to maintain dormancy by inhibiting the production of amylase. The balance of gibberellins and abscisic acid is therefore important in determining the start of germination. Auxins are produced early in the germination process to stimulate cell growth, but this effect is dependent on the auxin concentration (higher concentrations will inhibit germination). The actions of the auxins and the abscisic acid are interrelated, and again the end result depends on the balance of the two hormones.

The commercial use of plant hormones

The growth and development of plants is controlled by hormones, so it is clear that anyone growing plants commercially would be interested in their use. Some examples of such uses are given below.

● Auxins can be used as selective weed killers. Although they normally promote growth, high concentrations of auxins cause such rapid growth that the tissues (especially the roots) become distorted, causing damage and the easy entrance of pathogens. **Synthetic auxins** such as 2,4-D are used in concentrations 100 times greater than would normally be found in plants. It is particularly useful for the treatment of fields of cereals and lawns, because grasses are far less sensitive to it than broadleaved weeds, so the weed killer selectively kills the broad-leaved weeds, leaving the grass unharmed.

● At lower doses, auxins are used to stimulate root growth in cuttings. They are available commercially as rooting powders.

● **Ethene** stimulates the ripening of fruit. This can be useful in fruits that are delicate and liable to damage in transport (e.g. bananas and tomatoes). The fruits can be picked when unripe and harder, transported and then artificially ripened in transit or at their destination using ethene.

● Auxins and gibberellins can be used to treat unpollinated flowers, causing them to develop fruit. This is useful in producing seedless fruits (e.g. grapes). Fruits that have been formed without fertilisation are called **parthenocarpic** fruits.

● As well as inhibiting leaf abscission, auxins also inhibit the abscission of fruit. In orchards, fruits dropping from the trees are unusable as they soon rot on the ground. Auxins can be sprayed on the fruit to prevent fruit drop.

Activity

Investigating rooting hormone
Indole-3-butyric acid (IBA) is an auxin that can be used as a rooting powder. High concentrations will stimulate adventitious roots (roots arising from the stem), while lower concentrations stimulate the growth of primary roots (which develop from any part of a plant other than the main root). In most cases, gardeners take stem cuttings and so need adventitious roots to be formed.

A group of students investigated the effect of different concentrations of IBA on root growth. IBA is insoluble in water so the solution was made in alcohol. They made a 5% solution of IBA in alcohol. The investigation was carried out as follows.

1 Serial dilutions of the 5% IBA were made in this way:
   a) 1 cm³ of the 5% solution (solution A) was added to 9 cm³ of alcohol to produce solution B
   b) 1 cm³ of solution B was added to 9 cm³ of alcohol to produce solution C
   c) 1 cm³ of solution C was added to 9 cm³ of alcohol to produce solution D
   d) 1 cm³ of solution D was added to 9 cm³ of alcohol to produce solution E.
2 Stem cuttings were taken from Busy Lizzie, *Impatiens walleriana*.
3 The ends of 10 cuttings were dipped into solution A.

4 The cuttings were embedded in plant compost.
5 Steps 3 and 4 were repeated with solutions B–E.
6 A control group of a further 10 cuttings was set up, and these cuttings were dipped in the alcohol used to make the IBA solution.
7 After a week, during which the plants were watered regularly, the cuttings were removed. The adventitious roots were removed and weighed.

Questions
1 What were the percentage concentrations of IBA in solutions B–E?
2 Due to the large number of cuttings required, two Busy Lizzie plants had to be used. To what extent might the use of different plants affect the validity of the results?
3 What other factors would need to be controlled?
4 Why was the control group dipped in alcohol before planting?
5 The students decided to weigh the roots rather than record the total length of the root growth. Suggest why they made this choice.
6 The control group did grow some roots. How could the students employ statistical methods to assess the minimum concentration of IBA that was having an effect?

5 Explain the value of apical dominance to a plant.
6 Define the term *synergism*.
7 What are the two direct effects of ABA on guard cells?
8 Suggest a reason why a gardener might use a synthetic auxin as a weed killer on a lawn, but not on flower beds.
9 Explain the benefit to fruit growers of using ethene to stimulate the ripening of fruit.

Exam practice questions

1 Which of the following is a response by plants to temporary water stress?
A Increased respiration
B Leaf fall
C Production of tannins
D Stomatal closure (1)

2 Climbing plants have tendrils that grow around supporting objects, such as stems of other plants. Touching a tendril with a pencil encourages it to grow in the direction of the stimulus. Which is a suitable explanation for this?
A Extension growth occurs on the side of the tendril that is stimulated.
B Extension growth occurs on the side of the tendril that is not stimulated.
C More mitosis occurs on the side of the tendril that is stimulated.
D More mitosis occurs on the side of the tendril that is not stimulated. (1)

3 An investigation on the effect of different concentrations of two plant hormones on the growth of stems found that growth was stimulated at some concentrations and inhibited at others. Which was the dependent variable in this investigation?
A The concentrations of the plant hormones
B The daily change in the length of the stems
C The overall growth rate of the plants
D The types of plant hormone investigated (1)

4 Which of the following is not a role of abscisic acid (ABA)?
A Inhibiting seed germination
B Promoting dormancy of buds
C Stimulating leaf fall
D Stimulating stomatal closure (1)

5 Researchers investigated apical dominance by decapitating seedlings of green ash, *Fraxinus pennsylvanica*. Some of the seedlings had 1% IAA dissolved in lanolin applied to the cut end of the stem. After 15 days the researchers counted the number of lateral shoots that had grown and the length of each lateral shoot. The results are shown in the table.

Treatment	Number of seedlings	Number of seedlings with outgrowing lateral shoots	Mean number of outgrowing lateral shoots per seedling (±1 SD)	Mean length of outgrowing lateral shoots/mm (±1 SD)
Intact seedlings	8	0	0	0
Decapitated seedlings	10	10	2.8 (±0.9)	31 (±26)
Decapitated seedlings treated with 1% IAA	9	1	0.2 (±0)	5 (±0)

a) i) Why were intact seedlings included in the investigation? (2)
   ii) Summarise the results. (5)
b) Explain what is meant by *apical dominance*. (3)
c) Suggest the advantage to plants of apical dominance. (3)

d) 'And all the good things which an animal likes have the wrong sort of swallow or too many spikes.' Outline the different ways in which plants protect themselves from herbivory. (8)

6 One of the roles of gibberellin (GA3) is the activation of genes controlling the production of amylase during germination of cereal grains, such as wheat and barley. GA3 is often sprayed on barley grains to encourage the production of maltose during the malting process at the start of brewing. The production of amylase during germination of cereal grains can be investigated by placing germinating grains on starch–agar plates and finding out how much starch has been broken down, as shown in the diagram.

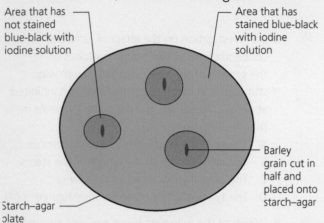

Area that has not stained blue-black with iodine solution

Area that has stained blue-black with iodine solution

Barley grain cut in half and placed onto starch–agar

Starch–agar plate

You are required to consider some of the steps that you would take in an investigation to find the effect of soaking barley seeds in different concentrations of GA3 on the production of amylase.

a) The concentration of GA3 in barley seeds is approximately 3.46×10^{-4} g dm^{-3}, which is equivalent to 1.0 µmol dm^{-3}. Explain how you would make up a range of concentrations of GA3 from a stock solution of 1.0 µmol dm^{-3}. *(5)*

b) State and explain what control experiments you would use. *(4)*

c) State the independent and dependent variables in this investigation. *(2)*

d) Explain why it is important to maintain a constant temperature while the seeds are on the starch–agar. *(2)*

e) Suggest how you would modify this procedure to find out how much maltose is produced by the action of amylase from the germinating seeds. *(6)*

7 Seeds of a wild-type variety of *Arabidopsis thaliana* and a variety deficient in an auxin-transport protein were germinated and grown on agar plates. The plates were positioned vertically so that the roots were growing in a horizontal position as shown in the diagram.

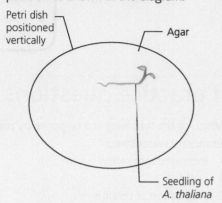

Petri dish positioned vertically

Agar

Seedling of *A. thaliana*

The roots of the seedlings responded by growing downwards. The degree of curvature of 10 seedlings of each variety was measured at intervals over a period of 24 hours. The results are shown in the graph.

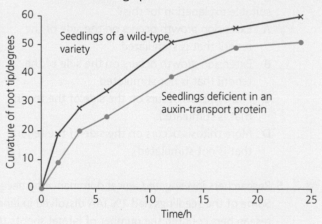

Seedlings of a wild-type variety

Seedlings deficient in an auxin-transport protein

Curvature of root tip/degrees

Time/h

a) i) State the response shown by the two groups of seedlings. *(1)*

   ii) Compare the results of the two varieties of *A. thaliana*. *(3)*

   iii) Explain the advantage of the response shown by the seedlings. *(3)*

b) Explain how the response is coordinated. *(4)*

c) Laser confocal scanning microscopy was used to detect the position of the protein transporter proteins inside cells. They were located in the cell membranes and in the cytosol.

i) Explain how root tissue would be treated to be able to detect the position of a protein within its cells. (2)

ii) Explain the role of auxin transporter proteins in the responses of shoots and roots to gravity. (4)

d) Shoots of *A. thaliana* were exposed to unidirectional light of different wavelengths. The responses in most wavelengths were very weak or non-existent. The greatest response was shown in blue light of wavelength 450 nm. What does this suggest about the mechanism that controls the response of *A. thaliana* seedlings to unidirectional light? (2)

8 Semi-dwarf rice plants were an essential part of the Green Revolution of the latter part of the twentieth century. Cereal crops, such as rice, were improved by selective breeding using varieties that had different genes for dwarfness. Dwarfness in rice is the result of a deficiency of the enzyme GA_{20} oxidase that catalyses three steps in the synthetic pathway that leads to gibberellin (GA3).

a) Explain what is meant by a synthetic pathway. (2)

There are three different mutations that occur in the gene that codes for GA_{20} oxidase. These three mutations all lead to dwarfness. The three mutations are shown in the diagram.

Gene for GA_{20} oxidase

Deletion of 383 base pairs

Substitution: CTC to TTC

Substitution: GAC to CAC

b) Explain how the deletion and substitution mutations shown in the diagram could lead to dwarfness in rice. (6)

c) The Green Revolution did much to alleviate the prospect of hunger in the twentieth century. Explain the advantages of growing dwarf varieties in increasing food production. (3)

Stretch and challenge

9 There are repressor proteins in cells that inhibit the expression of genes for growth in height. Gibberellic acid (GA3) combines with the soluble receptor G1D1 to form a compound that binds to these repressor proteins, marking them for destruction. When the repressor proteins are destroyed the genes for height can be transcribed.

Ornamental plants often grow stems that are too long. Growers have used various chemical treatments to prevent too much growth of these stems. A research team in California investigated the effect of silencing the gene for the soluble receptor G1D1. They used a bacterial vector to infect Petunia plants with a sequence of DNA that prevents transcription of the gene for the receptor, G1D1.

The researchers compared the growth of side shoots of plants with two batches of control plants. Batch A received the same treatment, but with a different sequence of DNA to silence another gene unrelated to the action of GA3. The plants in batch B were untreated. The results are shown in the table (n = the number of plants in each sample).

Analyse these results to find out whether silencing the gene for the receptor is effective or not.

Experiment	Mean length of side shoots (± 1 SD)/mm		
	In plants with the 'silenced' gene for G1D1	In control plants	
		Batch A	Batch B
1	187 ± 82	270 ± 26	287 ± 24
	$n = 15$	$n = 10$	$n = 10$
2	131 ± 83	213 ± 33	224 ± 37
	$n = 30$	$n = 10$	$n = 10$

Chapter 7

Photosynthesis

Prior knowledge
- Living organisms must have a source of energy.
- Energy flows through ecosystems from the sunlight to plants and then to animals and decomposers.
- Autotrophs convert simple inorganic compounds into complex organic compounds. Most use light as a source of energy in photosynthesis.
- Producers such as plants use energy from sunlight to convert to chemical energy. This process is called photosynthesis.
- In photosynthesis, light energy is used with carbon dioxide and water to form glucose with oxygen released as a waste product.
- Heterotrophs obtain their energy from complex organic molecules.

Test yourself on prior knowledge
1 Why do all living organisms need a source of energy?
2 Autotrophic organisms are classified into three kingdoms. State the names of these kingdoms.
3 Distinguish between autotrophic and heterotrophic nutrition.
4 State the energy source and the raw materials that are required for photosynthesis.
5 Name the waste product of photosynthesis and explain why it is regarded as a useful by-product.

Stromatolites are rock-like structures sticking up above the sea floor. They are made of microbial mats formed in limestone by communities of photosynthetic microorganisms such as cyanobacteria and algae. Modern stromatolites are found in just a few locations, such as Shark Bay in Australia (Figure 7.1). Stromatolites are among the oldest known fossils dating back to 3.5 millions years ago. They are thought to be responsible for the first appearance of oxygen in the Earth's atmosphere.

Introduction

Photosynthesis almost certainly evolved in free-living prokaryotes. At some time some of these were taken into eukaryote cells by a process called **endosymbiosis**. These became incorporated into the cells as chloroplasts by a similar process occurring with mitochondria. There is a lot of compelling evidence for this theory:

- chloroplasts are only produced from division of other chloroplasts, which is a separate process from plant cell division

- chloroplasts have their own genome consisting of a circular loop of DNA, much as is present in prokaryotes today

- chloroplasts have their own ribosomes (known as 70S ribosomes) and their own protein synthesis mechanism, which is related to the mechanism found in prokaryotes and which produces some of the proteins used by chloroplasts

Figure 7.1 Stromatolites at Shark Bay in Australia.

- chloroplasts contain similar pigments to photosynthesising cyanobacteria (Figure 7.2).

The similarity of chloroplasts to these prokaryotes suggests that they were once separate organisms that at some time became incorporated into plant cells and ultimately became cell organelles.

Why is photosynthesis important?

In photosynthesis, light energy from the Sun is trapped and converted into chemical energy in organic molecules. In the process, inorganic molecules such as water and carbon dioxide are used and oxygen gas is released.

Photosynthesis is also the means of trapping sunlight energy and converting it into a form that can be used to synthesise food molecules for the plant and for the animals that feed on them. Photosynthesis forms the basis of many food chains providing energy for consumers and decomposers depend. It is the most vital process on Earth since all other life depends on it either directly or indirectly as a source of energy. It is therefore an essential process in most ecosystems and for humans it is an essential part of the food industry, both arable and livestock farming. There are some ecosystems that do not have any photosynthetic organisms. Deep sea vent communities, for example, rely on bacteria and archaeans that use chemical energy rather than light energy to drive the reactions that make complex organic material from simple inorganic compounds.

The interrelationship between photosynthesis and respiration

The process of photosynthesis is vitally important to ensure that the balance of oxygen and carbon dioxide in the atmosphere is maintained at a constant level. Aerobic respiration uses oxygen and releases carbon dioxide while photosynthesis uses carbon dioxide and releases oxygen.

> ### Tips
>
> See Chapter 8 Respiration, to make a full comparison of photosynthesis and respiration.
>
> The balanced chemical equation for photosynthesis that you learned at GCSE will give you the overview of photosynthesis. You will also have learned a similar equation for respiration which will allow you to compare the two processes in simple terms.

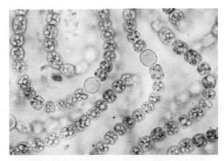

Figure 7.2 Photosynthesising cyanobacteria are related to some of the early prokaryotes (x 1000).

> ### Key terms
>
> **Consumer** A living organism that obtains its nutrients by feeding on other organisms.
>
> **Decomposer** Fungi and bacteria that feeds on dead organic matter. It breaks down dead matter into simple molecules, such as carbon dioxide and ammonia.

The chloroplast

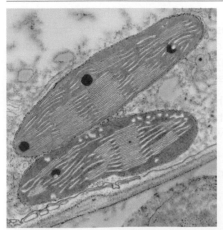

Figure 7.3 A transmission electron micrograph of chloroplast structure (x 6000).

The chloroplast is the cell organelle that is the site of photosynthesis in plant cells and some protoctista (Figure 7.3). It is similar to a mitochondrion in that both types of organelles are thought to have developed from prokaryotes that were taken into eukaryote cells by endosymbiosis. This theory is supported by many structural features.

Features of the chloroplasts

Chloroplasts are small in size, about 2–10 μm, which is larger than mitochondria. Like the mitochondria, they have a double membrane, called the chloroplast envelope (Figure 7.4). The outer membrane of the envelope allows many small ions and small molecules to pass through into the chloroplast. In contrast, the inner membrane relies on transport proteins to allow transport of certain chemicals, thus controlling molecules entering or leaving the chloroplast, moving between the stroma and the cytosol.

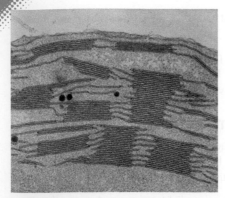

Figure 7.5 Ultrastructure of the granal stacks and lamellae of the chloroplast (× 11 000).

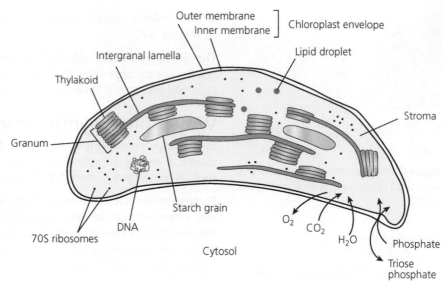

Figure 7.4 The structure of a chloroplast.

Key term

Photosystem A funnel-shaped collection of accessory pigments with a reaction centre containing a complex of proteins and chlorophyll molecules at its base, embedded in the thylakoid membrane. The primary pigment is chlorophyll *a*.

However, this inner membrane is not folded in the way that the mitochondrial inner membrane is folded. Instead chloroplasts have a third internal system of membranes folded into thin interconnected plates called lamellae or **thylakoids**. In places these are stacked up to form piles called **grana** (singular: *granum*) (Figure 7.5). The flattened membranes of the grana carry the pigments needed for photosynthesis in photosystems, and they also have carrier molecules. This is where the light-dependent stage of photosynthesis occurs.

Location of the stages of photosynthesis

- The light-dependent stage occurs on the thylakoid membranes of the grana (Figure 7.6).

- Light is trapped by the reaction centres in the grana.

- The light-independent stage occurs in the stroma.

All the pigments needed for the light-dependent reactions are held in reaction centres called photosystems which capture the light energy and pass it from one pigment to another within the photosystem. The granal stacks increase the surface area of the thylakoids, allowing many photosystems to be present for maximum light absorbance. They also allow many electron carriers and ATP synthase enzymes to enable the light-dependent reactions to continue.

The gel-like medium called the **stroma** forms another region of the chloroplast, which is similar to the cytosol of eukaryotic cells. The stroma is the site of the light-independent stage of photosynthesis. It also carries starch grains and oil droplets as well as the enzymes necessary to catalyse the light-independent reactions, the DNA loops and the small ribosomes (70S).

The granal stacks are surrounded by the stroma so it is easy to transfer the products of the light-dependent stage to the light-independent stage in the stroma. The lamellae between the granal stacks are called intergranal lamellae.

Table 7.1 summarises the ways in which chloroplasts are adapted to photosynthesis.

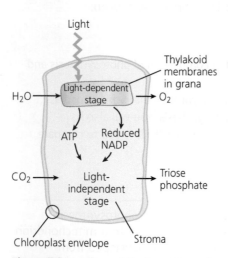

Figure 7.6 Locations of the two stages of photosynthesis.

Table 7.1 Adaptations of chloroplasts to photosynthesis.

Structure	Adaptive feature
Stroma	A cytosol-like gel for carrying enzymes which catalyse the light-independent reactions. The stroma surrounds the grana and membranes so transfer of products from the light-dependent stage into the stroma is rapid.
Grana	The granal stacks have a large surface area to allow many photosystems for maximum light absorbance and high numbers of electron carriers and ATP synthase enzymes. The pigments for the light-dependent reactions are held in photosystems.
Inner membrane of chloroplast envelope	Has transport proteins which control chemicals moving between stroma and cytoplasm.
DNA	Chloroplast DNA codes for some of the proteins and enzymes needed for photosynthesis. Others are coded for by genes in the cell nucleus.
Ribosomes	Translation of proteins coded by chloroplast DNA.

The photosynthetic pigments

Key term

Chlorophyll The main pigment involved in photosynthesis. Chlorophyll *a* is the primary pigment found at the reaction centre of the photosystems. Chlorophyll *b* is one of the accessory pigments. They are both involved in trapping light of certain wavelengths and both reflect green light and so do not absorb light of this wavelength.

Tip

Look back at Chapter 3 of *OCR A level Biology 1 Student's Book* to refresh your memory about haemoglobin structure.

Tip

An action spectrum shows the wavelengths of light used most in photosynthesis. An absorption spectrum is used to measure the wavelengths absorbed by the pigments.

The pigments involved in photosynthesis are found in the chloroplast and are arranged into **light-harvesting complexes** around a reaction centre. This is formed of the primary and accessory pigments held in place by proteins and arranged into funnel-shaped structures of the light-harvesting complex into a **photosystem**. The pigments are the molecules that absorb light of certain wavelengths. Each pigment molecule has its own specific range of wavelengths at which it absorbs light and reflects others, and it is these reflected light wavelengths that give the colour of leaves that we see.

The main pigment is chlorophyll, which is actually a mixture of chlorophyll *a* and chlorophyll *b* pigments. They are similar in structure, having a long hydrocarbon chain (a phytol) and a porphyrin group containing magnesium at the centre. This group is similar in structure to the haem group in haemoglobin except that an atom of magnesium, rather than an iron atom, is found in the centre.

The light-harvesting complexes contain: protein molecules, chlorophyll *a* and molecules of accessory pigments, such as chlorophyll *b*, carotenes and xanthophylls. These light-harvesting complexes absorb light energy and pass this energy to the reaction centre (Figure 7.7) at the base of the photosystems in an area known as the **primary reaction centre**. There are two types of reaction centre, each of which is composed of proteins together with a specific form of chlorophyll a molecule and cofactors:

● photosystem 1 (PSI) usually known as P700

● photosystem 2 (PSII) usually known as P680.

In PSI, the chlorophyll a molecule has maximum absorption of light in the red region of the spectrum at 700 nm and in PSII the chlorophyll a molecule has a maximum absorption at 680 nm.

The proteins in the light-harvesting complexes orientate the pigments precisely in the thylakoids. Accessory pigments such as chlorophyll *b*, carotenes and xanthophylls help collect light from other wavelengths. Absorption spectra for the various pigments are shown in Figure 7.9 along with the action spectrum for photosynthesis. See how the action and absorption spectra coincide showing the light absorbed by the pigments is actually used in photosynthesis.

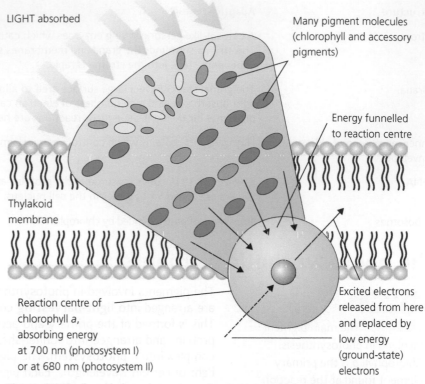

LIGHT absorbed

Many pigment molecules (chlorophyll and accessory pigments)

Energy funnelled to reaction centre

Thylakoid membrane

Reaction centre of chlorophyll *a*, absorbing energy at 700 nm (photosystem I) or at 680 nm (photosystem II)

Excited electrons released from here and replaced by low energy (ground-state) electrons

Figure 7.7 The funnel-shaped structure of a photosystem showing the primary pigment chlorophyll *a* at the base and the accessory pigments clustered around the funnel.

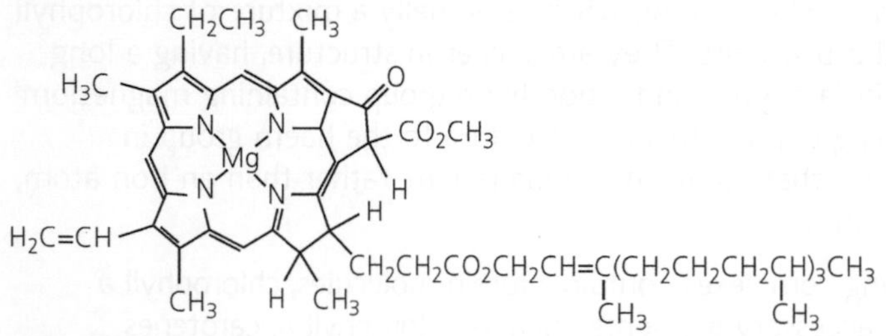

Figure 7.8 Structure of chlorophyll *a*.

The accessory pigments

The accessory pigments consist of chlorophyll *b* and carotenoids. Chlorophyll *b* absorbs light around 500 and 640 nm and reflects blue-green light. Carotene and xanthophyll are the two main carotenoid pigments. They have no porphyrin ring structure although they do have a long hydrocarbon chain. Carotenoids absorb blue light (450–470 nm) and reflect yellow and orange light.

The accessory pigments harvest light at the wavelengths described and pass it to the primary pigment, chlorophyll *a*. In the photosystems they therefore have no direct part in the light-dependent stage of photosynthesis. Their only role is light harvesting. Carotenoid pigments are also used as food dyes.

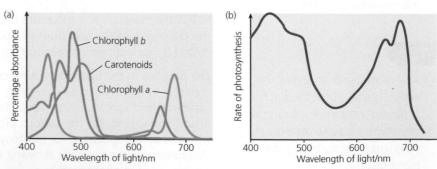

(a)

Chlorophyll *b*

Carotenoids

Chlorophyll *a*

Percentage absorbance

Wavelength of light/nm

(b)

Rate of photosynthesis

Wavelength of light/nm

Figure 7.9 (a) Absorption spectra for photosynthetic pigments and (b) the action spectrum for photosynthesis.

Activity

Using thin-layer chromatography to separate photosynthetic pigments

Thin-layer chromatography (TLC) is a technique that uses a thin layer of adsorbent material spread onto a thin glass, plastic or metal sheet. It can be used to separate photosynthetic pigments on the basis of their physical properties. The glass, plastic or metal sheet is the adsorbent medium that carries and separates the coloured pigments. A solvent is used to dissolve the pigments and move them as the solvent moves. The pigments move with the solvent at different rates depending on how soluble they are in the solvent and their adsorption to the medium, which is related to their polarity.

A pencil line is drawn across the plate just above the line the solvent will reach once the plate is placed into the container. A sample of the chloroplast extract is spotted onto a thin-layer plate onto the centre of the pencil line. The spot is loaded with a large amount of pigment by spotting a small sample, drying it and then adding further samples. Each time the spot must be completely dry each time to make sure the spot becomes highly concentrated and does not simply spread into a large area of pigment.

Once the final spot is added and dried the thin-layer plate is lowered into the chromatography chamber. This should be already saturated with vapour evaporated from the solvent. The plate is carefully supported so that it stands with the spot just above the surface of the solvent. The solvent will rise rapidly through the thin-layer plate taking the pigments with it. The apparatus is left until the solvent is close to the top of the plate. This forms the solvent front. The plate can now be removed and a pencil line used to record the position of the solvent front. The front edge of each pigment spot is also marked with pencil.

A number of pigments should be detected (Figure 7.10):

- carotene: an orange-yellow pigment
- phaeophytin: a grey-brown pigment, a component of the electron transport chain
- xanthophyll: a yellow pigment
- chlorophyll *a*: a yellow-green pigment
- chlorophyll *b*: a blue-green pigment.

The pigments may be identified using R_f values. This is a ratio based on the relative front for each pigment. It will vary according to the solvent used and on the exact type of chromatogram but the order in which the pigments occur will always be the same, so it can be used to identify pigments. It is calculated by using the following equation.

$$R_f = \frac{\text{distance moved by the pigment from the original spot}}{\text{distance moved by the solvent from the original spot}}$$

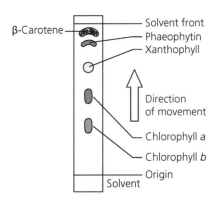

Figure 7.10 A completed chromatogram and the pigments that can be seen.

Questions

1 When setting up the chromatography chamber the solvent is first placed into a container that can be closed with a lid and left for 10 minutes. Suggest why the chamber is set up in this way.
2 Describe the principle used in chromatography.
3 What is the function of the solvent?
4 What limitations in this procedure could affect the pigments identified?
5 Why is it important to the process of photosynthesis for there to be several different pigments rather than a single pigment in chloroplasts?

The light-dependent stage of photosynthesis

Tip

Chapter 8 covers respiration and includes a comparison with photosynthesis.

The light-dependent stage is the stage of photosynthesis that occurs in the photosystems embedded in the membranes of the chloroplasts. Photosystem II is on the inner membranes of the granal stacks, but photosystem I is on the outer surfaces of the grana and the intergranal lamellae. This is where the pigments involved in trapping light energy are found and where light energy is converted into chemical energy. Light is absorbed in the light-dependent stage and provides energy to split water and form ATP and reduced NADP. The process involves transporting electrons through a series of electron carriers similar to the cytochrome electron carriers in respiration. It involves:

- absorption of light by pigments in the light-harvesting complexes of photosystems I and II
- conversion of light energy to chemical energy
- splitting of water (photolysis) to give protons (H^+) and electrons (e^-)
- production of oxygen
- use of energy to pump protons from the stroma into the thylakoid spaces
- using a proton gradient to drive the formation of ATP
- formation of reduced NADP.

ATP and reduced NADP are the products of the light-dependent stage that are used in the light-independent stage of photosynthesis.

Light of many wavelengths is absorbed by all the pigments in the light-harvesting complexes and 'funnelled' to P700 in the reaction centre of PSII. This energy excites electrons in P700 providing high energy levels which is enough for them to leave the chlorophyll molecule. The electrons either return to P700 with the energy released as fluorescence or they are accepted by an electron carrier, in the thylakoid membrane. As the electrons pass from carrier to carrier energy is released in small quantities. As an electron leaves a carrier it is oxidised and the carrier gaining the electron is reduced. The energy released as the electrons pass through the carrier chain is used to actively move protons (hydrogen ions) from the stroma into the thylakoid space. This active movement of the protons is against their concentration gradient so energy is needed.

Once the electrons reach photosystem I they have a lower energy level. Light absorption by PSI re-energises the electrons, raising their energy level so they leave PSI and pass to another electron carrier. The electrons then move to coenzyme NADP, which accepts two of these electrons with two protons from the stroma to become reduced NADP. The enzyme NADP reductase, which is on the outer surface of the thylakoid membrane, catalyses this reduction of NADP. Figure 7.13 shows the flow of electrons in a thylakoid membrane.

Because protons are actively pumped into the thylakoid space they accumulate there creating a higher concentration than in the stroma. This results in the pH inside the thylakoid space becoming lower than the pH in the stroma and creating an electrochemical gradient or proton-motive force. The membrane is impermeable to protons except for the channels through the molecules of ATP synthase. The protons move down their electrochemical gradient and pass through ATP synthase. Part of this remarkable ATP synthase protein spins as the protons pass through it and as it does the active site accepts ADP and a phosphate ion. Energy is transferred to allow a bond to form between the terminal phosphate on ADP and the phosphate ion and so ADP is phosphorylated to form ATP. The movement of protons through ATP synthase is by facilitated diffusion. The movement of protons by facilitated diffusion down their electrochemical gradient with the formation of ATP is chemiosmosis (Figure 7.11). There is more about this process in Chapter 8.

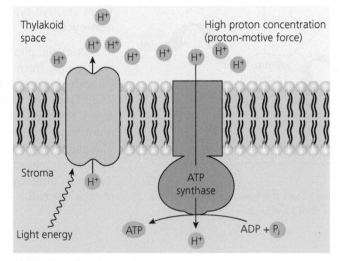

Figure 7.11 Chemiosmosis.

The non-cyclic pathway

The pathway of electrons from PSII to PSI to NADP is a non-cyclic pathway. Electrons leaving PSII move to NADP and so have to be replaced. This occurs by the splitting of water: the reaction centre of PSII catalyses the splitting of water by a process called photolysis, because it only occurs in the presence of light. Water is therefore the source of electrons needed by PSII with PSII acting as an enzyme to catalyse the reaction:

$$2H_2O \rightarrow 4H^+ + 4e^- + O_2$$

This reaction provides the electrons to PSII and the protons for non-cyclic photophosphorylation and the pool of protons in the thylakoid space.

> **Tip**
>
> The Z scheme is actually a graph with the *y*-axis showing the energy levels of the electrons rising as they pass through the different stages of the pathway. The energy levels rise as energy is absorbed from the light, sufficient to raise the electrons' energy level, and they fall as the electrons lose energy to the proton pump.

The light-dependent stage is often depicted using the Z scheme, which shows the route the electrons take as they pass from water to NADP in the thylakoids (Figure 7.12). As the electrons lose energy some of it is used to pump the protons into the thylakoid space.

Photophosphorylation

When ATP is made using light energy the process is known as **photophosphorylation**.

As light strikes a chlorophyll molecule it excites the molecule enough to raise the energy level of two electrons, so that they are released from the chlorophyll molecule and then captured by an electron carrier molecule. The electrons pass along the chain of carriers from one carrier molecule to another, releasing energy as they move along. The chlorophyll molecule is now oxidised.

There are two types of photophosphorylation, **cyclic** and **non-cyclic**.

In non-cyclic photophosphorylation the flow of electrons is from photosystem II to photosystem I, so both photosystems are involved in series (Figure 7.13). Once the electron has passed along the carrier chain, the electron passes on to NADP, which becomes reduced as a result. ATP is formed by chemiosmosis. The electrons lost from photosystem II are replaced by those released by photolysis of water and those lost from photosystem I are replaced by those from photosystem II.

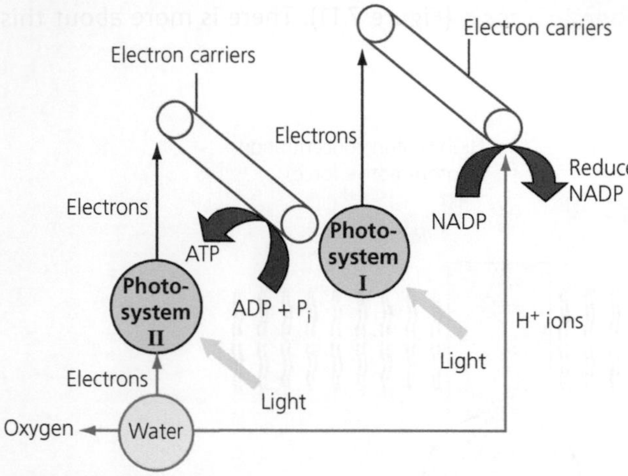

Figure 7.12 The Z scheme shows the flow of electrons in the light-dependent stage of photosynthesis.

> **Key term**
>
> **Photophosphorylation** Formation of ATP using light energy. There are two types: cyclic and non-cyclic.

> **Tip**
>
> The electron carriers are protein-based molecules with an iron group and are similar to the cytochrome carriers found in mitochondria. Chapter 8 has more detail and refers to the type of phosphorylation that takes place in respiration.

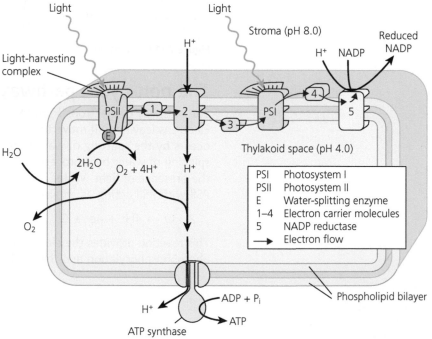

Figure 7.13 Non-cyclic photophosphorylation in the light-dependent stage of photosynthesis.

The flow of electrons in non-cyclic photophosphorylation is not a cyclic process because the electrons flow from photosystem II to photosystem I and then on to reduce NADP. However, another flow of electrons that *is* a cyclic process actually traps more light energy but does not reduce NADP.

Cyclic photophosphorylation involves only photosystem I (Figure 7.14). The electrons pass back to the chlorophyll molecule from which they were originally lost, after passing along the electron carriers. Photolysis of water does not occur and NADP is not reduced but some ATP is generated.

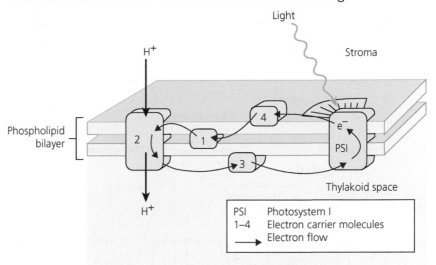

Figure 7.14 Cyclic photophosphorylation in the light-dependent stage of photosynthesis.

Table 7.2 shows a comparison of cyclic and non-cyclic photophosphorylation.

Table 7.2 Comparison of cyclic and non-cyclic photophosphorylation.

Feature	Cyclic photophosphorylation	Non-cyclic photophosphorylation
Photosystems involved	PSI	PSI and PSII
Photolysis of water	Does not occur	Occurs
Electron donor	P700 in PSI	Water
Final electron acceptor	P700 in PSI	NADP
Products	ATP	ATP, reduced NADP and oxygen

Summary

- Light energy is absorbed by the light-harvesting complexes (Figure 7.15).
- Photolysis occurs: water molecules are split to give hydrogen ions and electrons.
- Oxygen is released by photolysis.
- Light energy raises the energy levels of a pair of electrons so they leave the chlorophyll molecules, in photosystem I and photosystem II, and are picked up by the first electron carrier.
- Energy is released as the electrons flow along the electron carriers and is used to pump protons into the thylakoid space. The movement of protons is a form of active transport.

- The protons flow down the proton gradient through ATP synthase, generating ATP in the process called chemiosmosis.
- ATP is formed in both cyclic and non-cyclic photophosphorylation.
- Reduced NADP is formed only in non-cyclic photophosphorylation.

Photosystems I and II work in tandem – receiving light energy, releasing excited electrons and replacing each lost electron by one in the ground state (photosystem II electrons come from split water; photosystem I electrons from photosystem II)

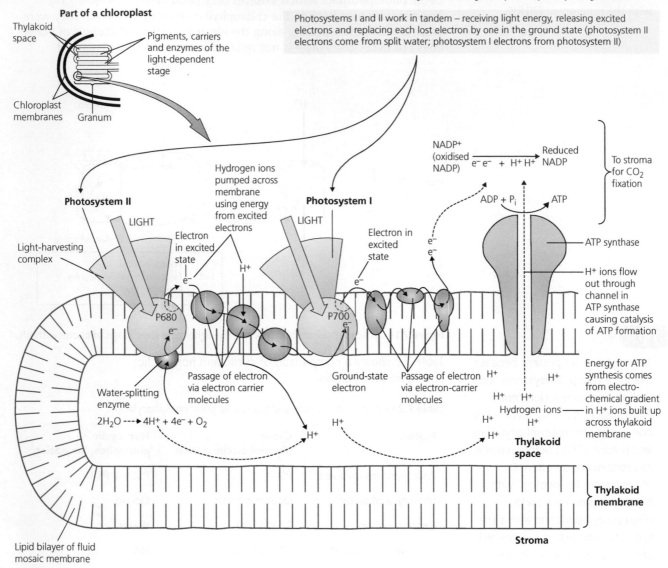

Figure 7.15 A summary of the light-dependent stage of photosynthesis.

The light-independent stage of photosynthesis

This stage occurs in the stroma of the chloroplasts. The products from the light-dependent stage are used in this stage but there is no direct use of light energy. However, if light is no longer available this cycle will only continue for as long as the products of the light-dependent stage (ATP and reduced NADP) are available.

The Calvin cycle

This stage of photosynthesis is also known as the Calvin cycle after Melvin Calvin who led the team that discovered its biochemistry. He used unicellular algae grown with radioactive carbon dioxide in his famous 'lollipop' apparatus.

Carbon dioxide is the carbon source for this cycle, which builds up complex organic molecules from simple inorganic molecules by fixing carbon dioxide (Figure 7.16).

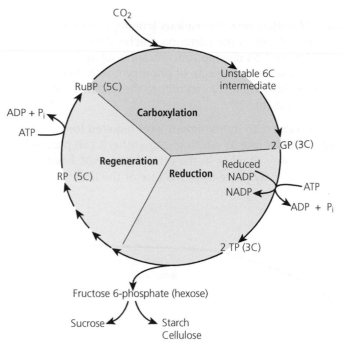

Figure 7.16 The light-independent stage: Calvin cycle.

- The carbon is supplied to the photosynthesising plant by the carbon dioxide gas entering the leaves via the stomatal pores. It dissolves in water and diffuses into the mesophyll cells and then to the chloroplasts in those cells.

- The enzyme Rubisco (which is short for ribulose bisphosphate carboxylase oxygenase) combines carbon dioxide with a five-carbon molecule called **ribulose bisphosphate** (RuBP), both of which are found in the stroma.

- An unstable six-carbon molecule is formed that immediately breaks down into two three-carbon molecules of **glycerate 3-phosphate** (commonly known as GP). This step is known as **carboxylation** because carbon dioxide has become fixed and GP formed.

- Both of the GP molecules are now reduced by the addition of hydrogen from reduced NADP (which now becomes re-oxidised NADP). ATP is converted into ADP during this step (phosphorylation), with the release of energy and phosphate. As a result two molecules of **triose phosphate** (TP) are formed.

- The ATP and reduced NADP are both from the light-dependent stage of photosynthesis.

- Five-sixths of the TP is used to reform ribulose bisphosphate by further phosphorylation with ATP. This keeps the cycle going.

- One-sixth of the TP forms hexose phosphates (phosphorylated sugars), which can be used to form sucrose for plant transport or starch for plant storage.

- Exactly what happens to the remaining TP will depend on the needs of the plant: TP may be exported into the cytosol and used to form other organic molecules (e.g. fatty acids and glycerols for lipid formation, and cellulose for plant cell walls) or TP may remain in the chloroplast where it is used to form amino acids by combining with ammonia or an amine group in the process of amination.

Rubisco evolved in conditions of low oxygen concentration and so it will catalyse reactions with both oxygen and carbon dioxide. However, as it normally encounters high levels of carbon dioxide in the chloroplast stroma it usually fixes carbon dioxide. In hot conditions the concentrations of carbon dioxide decrease and concentrations of oxygen increase and so that oxygen competes with carbon dioxide for the active site of rubisco molecules leading to a decrease in the fixation of carbon dioxide: this is called photorespiration.

Carbon dioxide therefore is the raw material for the light-independent stage of photosynthesis since it is the carbon source that is fixed by combining with RuBP. It is therefore the key to production of all organic molecules required by plants, and by animals feeding on them.

Key term

Rubisco The enzyme involved in fixing carbon dioxide, a process known as carboxylation. It is correctly called ribulose bisphosphate carboxylase oxygenase, but it is commonly referred to as Rubisco.

Factors affecting photosynthesis

For photosynthesis to continue it needs a number of different biochemical molecules. We can look at each one separately but it must be remembered that because photosynthesis is a complex process, consisting of a large number of steps, all the different factors affecting its rate will be operating at the same time.

The factor that is closest to the minimum needed will be the one that limits the rate at which the process proceeds. Therefore the factors are called limiting factors.

The graphs in Figure 7.17 outline how the various limiting factors can affect the rate of photosynthesis in the common orache, *Atriplex patula*. Low light intensity (a), low carbon dioxide concentration (b) and low temperatures (c) can all slow down the rate of photosynthesis. See also how factors interact: high light intensity can reduce the effects of low carbon dioxide concentration (b).

When a factor becomes closest to its minimum level required for photosynthesis it will limit the rate at which photosynthesis can proceed and be the limiting factor. Any changes in the other factors will have no effect since the process can only proceed at the rate of the limiting factor.

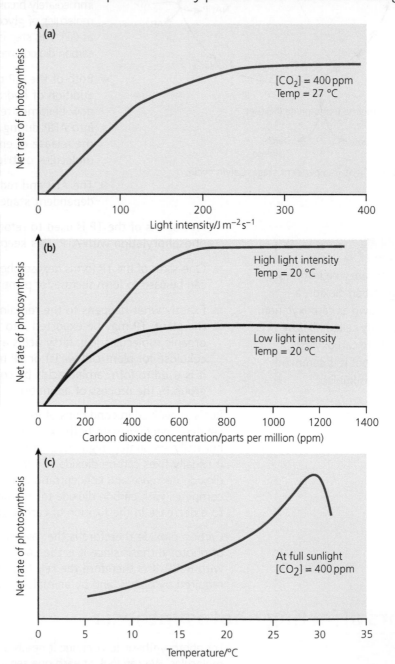

Figure 7.17 The effect of light intensity, carbon dioxide concentration and temperature on the rate of photosynthesis in *Atriplex patula*.

Temperature

Figure 7.17(c) shows how the net rate of photosynthesis in this plant, which grows in North America, rises as temperature approaches 25°C. Here it begins to slow until the peak at 27–30°C (the optimum temperature), provided all other factors are not limiting. There is a brief plateau before a rapid fall in rate due to the enzymes becoming denatured at higher temperatures.

If all the other factors are not limiting the optimum temperature for photosynthesis is 25–30°C depending on the species. Extremely high temperatures will cause the enzymes of the light-independent stage to denature, thus stopping the reactions. At temperatures below the optimum the rate of photosynthesis slows down. Between 0 and 25°C the rate of reaction will double for each 10°C rise in temperature, which is the Q_{10} relationship common to all enzyme-controlled reactions. The light-independent stage is not affected by temperature changes. Temperature will become a limiting factor when it falls below 25°C (unless another factor is closer to its own minimum value).

> ### Tip
>
> The Q_{10} is a temperature coefficient. It is the ratio between the rate of a process at two temperatures that are 10°C apart. See page 70 in *OCR A level Biology 1 Student's Book*

Light intensity

Figure 7.17(a) shows that the net rate of photosynthesis rises as light intensity rises up to approximately $250\,J\,m^{-2}\,s^{-1}$. There is then a plateau when limiting factors other than light intensity now affect the rate. The rate of photosynthesis is closely linked to light intensity at the lower levels of light intensity since it determines the energy available for the light-dependent stage. In many species light intensity becomes a limiting factor when it falls below $250\,J\,m^{-2}\,s^{-1}$.

As the light intensity increases more light energy is trapped in this stage by the chlorophyll and more water is split in photolysis and so more products (ATP and reduced NADP) are formed. In addition, increases in light intensity cause the stomatal pores to open, so gas exchange happens more freely and more carbon dioxide is available.

Carbon dioxide concentration

Carbon dioxide is the raw material for carbon fixation in the light-independent stage. Therefore, a reduced supply of carbon dioxide will limit photosynthesis because there will be less triose phosphate produced in the light-independent stage. In turn this lowers the need for ATP and reduced NADP, which will also slow down the light-dependent stage (although this stage itself does not need carbon dioxide to proceed). This is an example of end-product inhibition of enzymes as the concentrations of ATP and reduced NADP increase as they are not used. The concentration of carbon dioxide in the air is 0.04% (400 parts per million, ppm), which will vary according to where on the Earth the reading is taken. Close to the ground in a crop field on a warm, still day the slow movement of air will result in carbon dioxide concentrations falling by the afternoon and so becoming a limiting factor. In a greenhouse the carbon dioxide concentration will also be low as a result of high growth intensity and plants competing for the available carbon dioxide. Growers who specialise in growing plants in these conditions burn paraffin or similar fuels in greenhouse heaters to raise the carbon dioxide concentrations and the temperature.

Figure 7.17(b) shows that the net rate of photosynthesis rises as carbon dioxide concentration increases to 0.07% or 700 ppm. At this point the plateau in rate indicates that other limiting factors now affect the rate. Carbon dioxide concentration becomes a limiting factor below 0.01% carbon dioxide in the atmosphere.

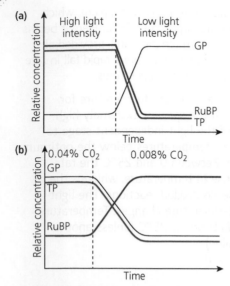

(a)

(b)

Figure 7.18 Effect of reducing carbon dioxide concentration on the relative concentrations of GP, TP and RuBP.

The effect of the three main limiting factors on GP, TP and RuBP

Changes to light intensity, carbon dioxide concentration and temperature influence the concentrations of ribulose bisphosphate (RuBP), glycerate 3-phosphate (GP) and triose phosphate (TP) in chloroplasts. When light intensity decreases, RuBP and TP concentrations decrease but GP levels rise in proportion. This occurs because with little (or no) light available, the light-dependent stage stops and there will be no products (ATP and reduced NADP) for the light-independent stage. As a result, GP accumulates because it cannot be changed to TP and in turn no more RuBP is formed. In time the fixation of carbon dioxide will also cease and no more GP will be formed so its level will plateau. When concentrations of carbon dioxide fall to below 0.01%, GP and TP levels both decrease but RuBP levels will rise, also in proportion to the level of carbon dioxide. This is because RuBP is a carbon dioxide acceptor and without enough carbon dioxide it will remain unfixed. The result will be a fall in both GP and TP (Figure 7.18).

Photosynthesis experiments

In many experiments the rate of photosynthesis is found by measuring the rate at which oxygen is released. Aquatic plants are usually used since it is difficult to collect the oxygen released from terrestrial plants. However, collecting gas evolved only measures the *apparent (net) rate* since it does not take into account any oxygen released from photosynthesis that may be used by respiration. This is especially important in hot weather when the stomatal pores may be closed, limiting the amount of air that can enter the plant. It is also worth considering the situation at night when the plant is not photosynthesising. The actual volume of oxygen given off is therefore difficult to measure and so the actual rate, or *true (gross) rate*, of photosynthesis is also difficult to measure.

Test yourself

7 List the differences between cyclic and non-cyclic photophosphorylation.

8 What is the importance of the photolysis of water?

9 What are the products of the light-dependent stage and what happens to them?

10 Explain the importance of Rubisco.

11 What are the three main stages that occur in the light-independent stage?

12 Describe and explain the effect on the concentrations of triose phosphate, glycerate 3-phosphate and ribulose bisphosphate in chloroplasts when carbon dioxide concentration decreases from 0.04% to 0.008%.

Activity

Factors affecting photosynthetic rate: the effect of light intensity on the pond weed, Cabomba

This investigation determines the effect of light intensity as a limiting factor on the rate of oxygen production as a way of determining the rate of photosynthesis. A student placed a shoot of *Cabomba* under water at the open end of a photosynthometer as shown in Figure 7.19.

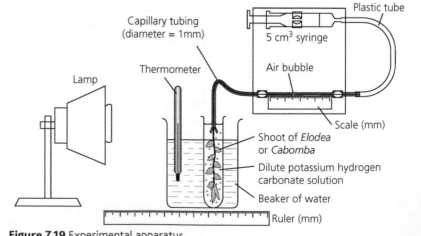

Figure 7.19 Experimental apparatus.

The capillary tubing of the photosynthometer was filled with water and coloured dye added to make it easier to see the oxygen bubble and read the distance travelled. First, the student placed the bench lamp 0.10 m away from the beaker containing the *Cabomba* and allowed it to adjust for 5 minutes. Then the volume of gas produced in 2 minutes was recorded. Two more results were taken for each distance.

The experiment was repeated with the lamp at 0.15 m and again at 0.20, 0.30 and 0.50 m.

Table 7.3 shows the data collected.

Table 7.3 Oxygen collected from *Cabomba* over 2 minutes at different distances from a light source.

Distance/m	Bubble length/mm		
0.10	77	75	73
0.15	51	49	38
0.20	34	32	31
0.30	18	16	17
0.50	8	5	3

Questions

1 Suggest a control to show that all the gas produced comes from the plant while it is photosynthesising.
2 Why did the student perform the experiment three times?
3 What other factor is likely to have the biggest impact on the accuracy of these results? Discuss the validity of these results.
4 Light intensity is proportional to $1/d^2$, where d = distance from the lamp, Use this information to calculate the light intensity for each of the distances in the results table above and draw a graph to show the effect of light intensity on the mean volume of oxygen produced.
5 Describe and explain the results shown by your graph.
6 Suggest another method of measuring light intensity and state the units and unit symbols that would be used.

Example

A student carried out an experiment similar to the one suggested above using *Cabomba*. The student made a decision to measure the gas produced in 5 minutes. The results are shown in Table 7.4.

Tip

See Chapter 18 in *OCR A level Biology 1 Student's Book*, on Maths skills.

Table 7.4 Oxygen collected from *Cabomba* over 5 minutes at different light intensities.

Light intensity/$1/d^2$	Bubble length/mm
62.5	80
15.6	25
3.9	12
0.9	5
0.3	0

The student decided to include another reading with the lamp at a distance of 0.24 m.

1 Use the formula $(1/d^2)$ to determine the light intensity with the lamp at 0.24 m, showing your working.
2 If the diameter of the capillary tube is 1 mm, state how you would calculate the volume of gas produced.
3 It was assumed in this experiment that all the oxygen gas produced by the plant was collected. Suggest why this may not be the case.
4 Another assumption was that all the gas collected was oxygen. However, when analysed it was found to contain only 50% oxygen. The remaining gas was 44% nitrogen and 6% carbon dioxide. Comment on the percentage of carbon dioxide present in the gas and give a reason for this value.

Answers

1 Light intensity with lamp at 0.24 m = $1/0.24^2$ = $1/0.0576$ = 17.36.

2 Use the diameter to calculate the area of the bore using πr^2. Multiply the area of the bore by the length of the bubble in millimetres, h, so the equation for the volume of gas in the bubble will be $\pi r^2 h$. For a light intensity of 0.9 the calculated value is $3.142 \times 0.5^2 \times 5 = 3.9275\,mm^3$. This is $4\,mm^3$ (to 1 sig. fig.).

3 Some oxygen will be used by the plant in respiration, some may dissolve in the water, some may be attached to the leaves and some may simply not be captured by the apparatus.

4 The value for CO_2 is higher than expected; plant respiration produces CO_2; potassium hydrogen carbonate (HCO_3^- ions) has been added to the water as part of the experiment; some will have dissolved in the water and been released into the air above the beaker as the water is warmed by the light source.

Activity

The effect of carbon dioxide concentration on the rate of photosynthesis

To determine the effect of carbon dioxide concentration on the rate of photosynthesis another experiment using the evacuated discs method can be set up. Leaf discs are first cut from thin leaves such as cress. A $5\,cm^3$ syringe is then used to draw up $5\,cm^3$ of $0.2\,mol\,dm^{-3}$ sodium hydrogen carbonate solution and five leaf discs added. The air is then evacuated from the discs allowing the discs to fall to the bottom of the syringe (Figure 7.20). The syringe can be held vertically using modelling clay.

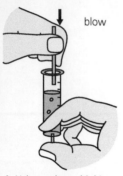  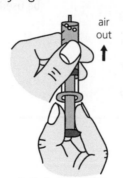 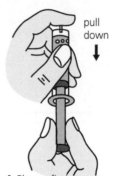 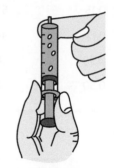

1 Using a clean drinking straw gently blow the leaf discs into the solution in the syringe.

2 Carefully replace the plunger and point the syringe upwards.

3 Push out all of the air.

4 Place a finger over the nozzle, gently pull the plunger down. Many bubbles will appear on the leaf discs.

5 Tap the syringe vigorously so that the air bubbles rise to the top. Repeat steps 3, 4 and 5 until all the discs sink.

Figure 7.20 Procedure for evacuating air from leaf discs.

The time taken for each disc to rise to the surface is recorded with a stopwatch. A student used this procedure to investigate the effect of light intensity on the rate of photosynthesis. The student made the hypothesis that the rate of photosynthesis would be proportional to the light intensity.

Freshly cut leaf discs were used for each light intensity. The time taken for four leaf discs to rise from the base of the syringe was recorded. The table shows the results.

Questions

1 Explain what causes the discs to rise to the top of the syringe.
2 Analyse the student's results and discuss whether the results support the hypothesis or not.
3 Evaluate the procedure as a method to determine the rate of photosynthesis.
4 Plan an investigation using the same method to investigate the effect of carbon dioxide concentration on the rate of photosynthesis. You are supplied with a 1.0M sodium hydrogen carbonate solution that you can use to make solutions of other concentrations.

Light intensity / lux	Mean time taken for leaf discs to rise / min
3.04	14.70
3.15	9.22
3.37	6.79
3.70	4.77
4.26	3.47
5.40	2.79

Exam practice questions

1 What is the role of water in photosynthesis?
- **A** It acts as a source of energy.
- **B** It provides a source of electrons.
- **C** It provides a source of oxygen.
- **D** It provides a suitable medium for electron flow. *(1)*

2 A student investigated the gas exchange of aquatic plants that were in an aquarium tank of water at 17 °C. In the light, the plants released oxygen at a rate of 300 cm³ h⁻¹. In the dark, the plants absorbed oxygen at a rate of 75 cm³ h⁻¹. Which is the best estimate of the gross rate of photosynthesis by the plants at 17 °C?
- **A** 150 cm³ h⁻¹
- **B** 225 cm³ h⁻¹
- **C** 300 cm³ h⁻¹
- **D** 375 cm³ h⁻¹ *(1)*

3 Stomata close in conditions of water stress. In daylight this causes the rate of photosynthesis to decrease. Which combination of factors is responsible for stomatal closure when plants experience water stress? *(1)*

	ABA concentration of leaf	Movement of K⁺	Turgor pressure of guard cells
A	Decreases	Out of guard cells	Increases
B	Decreases	Into guard cells	Decreases
C	Increases	Into guard cells	Increases
D	Increases	Out of guard cells	Decreases

4 The graph shows an action spectrum for a species of aquatic plant.

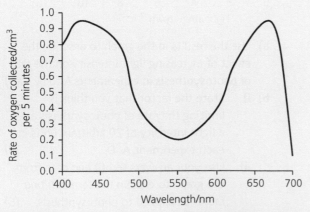

The following statements were made by students about the graph.

1 The maximum absorption of light is at 425 nm and at 675 nm.

2 Plants cannot absorb green light.

3 Wavelengths of light at the far left and far right of the visible spectrum have the highest energy levels.

Which statements are supported by the evidence in the action spectrum?
- **A** 1 only
- **B** 2 only
- **C** 1 and 3
- **D** 2 and 3 *(1)*

5 a) The diagram shows a longitudinal section of a chloroplast as viewed in an electron microscope. The magnification is ×13 000.

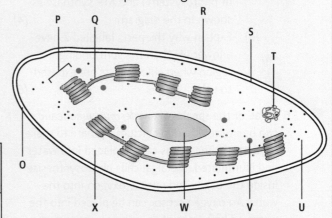

The table shows six processes that occur in chloroplasts. Copy and complete the table, naming the part of the chloroplast that carries out each process and identifying it from the diagram using one of the letters, O–X. *(6)*

Function	Name	Label from diagram
Absorption of light by chlorophyll		
Fixation of carbon dioxide		
Production of oxygen		
Translation of chloroplast genes		
Storage of polysaccharide molecules		
Transcription of chloroplast genes		

b) Calculate the actual size of the chloroplast in micrometres. Write down the formula you will use to make your calculation and show your working. *(3)*

c) Chloroplasts have some similarities with cyanobacteria. State two ways, other than size, in which chloroplasts resemble prokaryotes, such as cyanobacteria. (2)

d) The diagram shows an enlarged view of part of the chloroplast.

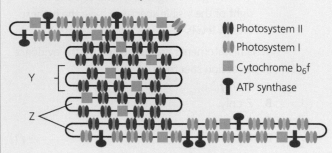

Photosystem II
Photosystem I
Cytochrome b₆f
ATP synthase

i) Name structure Y. (1)

ii) Describe and explain the distribution of photosystem I and ATP synthase as shown in the diagram. (4)

iii) Explain why the parts labelled Z have a lower pH than the surrounding region when chloroplasts are exposed to light. (6)

6 Algae of the species *Scenedesmus quadricauda* can be immobilised in calcium alginate to make algal beads. The beads can be placed into water and illuminated. The algal cells photosynthesise inside the beads and release oxygen into the water. An oxygen sensor can be placed into the water and readings of the oxygen concentration can be logged at intervals as shown in the diagram.

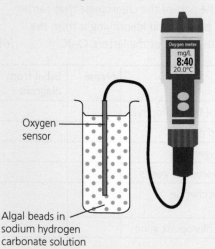

Oxygen meter
mg/L
8:40
20.0°C

Oxygen sensor

Algal beads in sodium hydrogen carbonate solution

A student used the apparatus to investigate the effect of different carbon dioxide concentrations on the rate of photosynthesis of *S. quadricauda*. The student thought that the carbon dioxide concentration would influence the rate of photosynthesis of the algae in the beads.

a) Write a hypothesis that the student could investigate with the apparatus shown in the diagram. (1)

b) State the independent and dependent variables in the investigation. (2)

c) State four variables that the student should control in the investigation with immobilised *S. quadricauda*. (4)

d) The student was supplied with an unlimited supply of algal beads and a 1% solution of sodium hydrogen carbonate. Describe how the student could carry out the investigation to obtain valid results. (6)

e) Explain how the student should process the results obtained from the oxygen meter and present them. (4)

7 A researcher carried out four experiments, A–D, to investigate the effect of light intensity on the rate of photosynthesis of cucumber plants. The plants were maintained at two concentrations of carbon dioxide and at two temperatures. The results are shown in the graph.

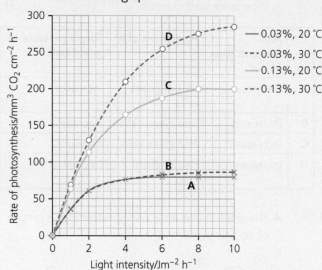

— 0.03%, 20 °C
--- 0.03%, 30 °C
— 0.13%, 20 °C
--- 0.13%, 30 °C

a) Use the results in the graph to describe the effect of increasing light intensity on the rate of photosynthesis in experiments A to D. (5)

b) i) State the factor that you think is limiting the rate of photosynthesis at a light intensity of 20 arbitrary units in each experiment, A–D. (4)

ii) Use your answers to b(i) and data from the graph to explain the term *limiting factor* as applied to photosynthesis. (6)

8 The coenzymes NADP and ATP play important roles in biochemical processes such as photosynthesis.

   a) i) Define the term *coenzyme*. (2)

      ii) Describe the roles of NADP and ATP in photosynthesis. (6)

A suspension culture of the alga *Scenedesmus* was exposed to carbon dioxide with the radioactive isotope $^{14}CO_2$. This was mixed with a suspension of the algae in the dark. The light was turned on and concentrations of glycerate 3-phosphate (GP) and ribulose bisphosphate (RuBP) that were radioactively labelled were determined at intervals of time. After a while, the light was switched off and then on again. In a separate experiment, the concentration of carbon dioxide supplied to the algae was reduced so that none was available.

The graph summarises the effects of the change in light and carbon dioxide on the relative concentrations of radioactively labelled GP and RuBP in the algae.

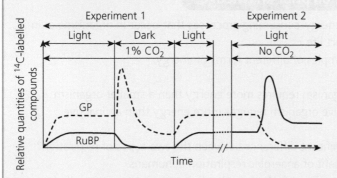

   b) i) Use the information in the graph to explain the effects of the changes that occur to the relative concentrations of RuBP and GP when there is a change from light to dark and when the carbon dioxide concentration decreases from 1% to 0%. (8)

      ii) Describe what happens to triose phosphate (TP) inside chloroplasts after it is formed in the light-independent stage. (6)

     iii) Describe and explain what you think will happen to the relative concentrations of radioactively labelled triose phosphate when the light is switched off and when the carbon dioxide concentration decreases. (4)

Stretch and challenge

9 Two-way thin layer chromatography (TLC) using two different solvents was used to separate and identify chloroplast pigments in marine planktonic algae. The absorbance of seven of these pigments, **A** to **G**, was determined at different wavelengths of light. The table shows the R_f values of these pigments in two solvents and the wavelength(s) of peak absorbance.

Pigment	R_f values		Wavelength of peak absorbance of light/nm
	Solvent 1	Solvent 2	
A	0.83	0.43	430 and 675
B	0.97	0.98	490
C	0.18	0.00	645
D	0.55	0.56	440
E	0.28	0.08	450
F	0.82	0.93	475
G	0.60	0.13	480 and 650

Pigments were extracted from a species of planktonic alga and analysed using two-way TLC. The chromatogram is shown below.

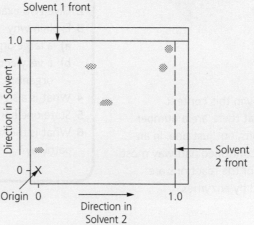

   a) Identify the pigments in this species of alga and explain how you made your identifications.

   b) Use the results to explain the advantage of using two-way TLC rather than one-way.

   c) Suggest why the planktonic algae that were sampled have few pigments with peak absorbance in the range 600 – 700 nm.

Chapter 8

Respiration

Tip

A *pathway* in this context means that there are a number of reactions, not just one. In an enzyme-controlled pathway most, if not all, of the reactions are controlled by enzymes.

The importance of respiration

Respiration is the process needed to support the energy needs of all living organisms and allows life processes to continue. Without energy life processes would proceed very slowly or not at all.

There was no oxygen in the atmosphere before photosynthesis evolved about 3.8 billion years ago. Organisms gained energy from sources other than light. These organisms were probably similar to the prokaryotes in many different environments that derive energy by oxidising various chemical sources to provide 'energised' electrons. Green plants, some protoctists and some prokaryotes use light as a source of energy to drive the formation of ATP and reduced NADP that provide energy to fix carbon dioxide and form complex carbon compounds as you have seen in Chapter 7. The organisms that first used light as a source of energy were most likely similar to the cyanobacteria that are found today in stromatolites (Figure 7.1).

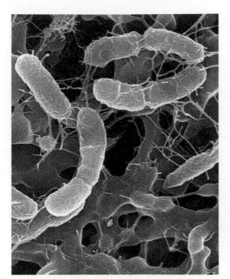

Figure 8.1 A bacterium *Venenivibrio stagnispumantis*, which carries out an ancient form of energy release (x32 000).

Some organisms use chemicals as sources of energy. Figure 8.1 shows several individuals of a species that can use hydrogen (H_2) as a source of energy. These hydrogen-oxidising bacteria have hydrogenase enzymes in the cell membrane and the cytoplasm that breakdown hydrogen to form electrons and protons. In the membranes, electrons pass to electron carriers to provide energy for proton pumping and ATP synthesis. The electrons are used to reduce NADP. So indirectly, hydrogenases provide ATP and reduced NADP to drive carbon fixation in the Calvin cycle in these bacteria. Many other bacteria and archaeans use other sources of chemical energy in the same way. The nitrifying bacteria are an example: read more about them in Chapter 13.

Autotrophic organisms fix carbon as complex organic molecules which they can use as a source of energy and store for use in the dark or when conditions are unfavourable. They also use these organic molecules to build cellular structures. However, they cannot rely on photosynthesis as an energy source all the time and so need the energy that is stored in molecules, such as carbohydrates, proteins and fats, to be released and made available to cells. This is achieved by the chemical process of respiration.

Heterotrophic organisms that cannot fix carbon dioxide rely directly or indirectly on organic compounds made by autotrophs. Heterotrophs gain their supply of energy in the form of complex organic compounds by eating plants or eating other organisms that ultimately derived their energy from autotrophs.

Therefore photosynthesis and respiration are complementary processes with energy and molecular compounds being used and released. Figure 8.2 shows the relationships between them in a green plant.

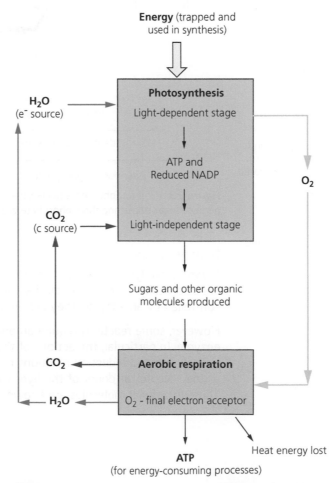

Figure 8.2 The interrelationships between photosynthesis and respiration in a green plant.

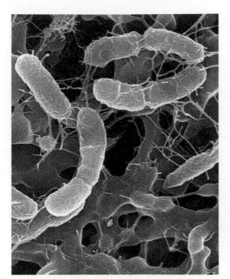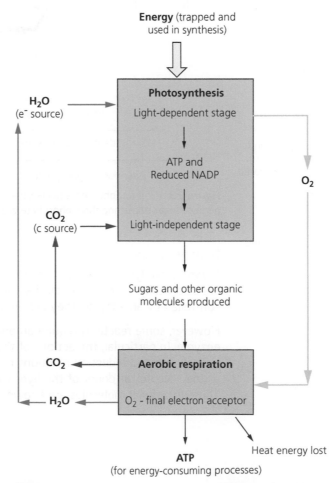

Why is respiration necessary?

Plants, animals and microorganisms respire to transfer energy from carbon compounds to ATP, the 'energy currency' of cells. This 'energy currency' is required for many processes, such as activation of chemicals, active transport, transport of molecules across the cell membrane (endo- and exocytosis), movement, protein synthesis and cell division.

Adenosine triphosphate

Adenosine triphosphate (ATP; Figure 8.3) is a compound found in all living things. Its role is in short-term and immediate energy use as opposed to long-term energy storage in molecules such as carbohydrates and fats. ATP is hydrolysed in a single step to adenosine diphosphate (ADP) by removing one of its inorganic phosphates (P_i) and 30.5 kJ of energy is released for immediate use by cells. This small amount of energy release is extremely efficient as it ensures minimum energy wastage and no damage to the cells, which may occur if large amounts of energy were released quickly. ATP is constantly being reformed and broken down in cells that need energy (Figure 8.4).

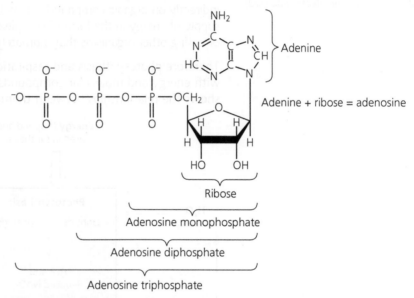

Figure 8.3 ATP is a triphosphate nucleoside composed of the base adenine, the pentose sugar ribose and three phosphate groups.

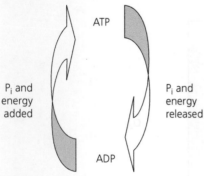

Figure 8.4 The ATP cycle: formation and break down.

Importance of coenzymes in respiration

Enzymes catalyse a wide range of reaction types in all life processes. They are also vitally important in all the stages of respiration since each step is controlled by an enzyme; they are enzyme-catalysed reactions.

However, some reactions require an additional molecule to assist the enzyme. In particular, the actions of the dehydrogenase enzymes involved in oxidation and reduction reactions require an additional molecule to act as the acceptor or donor of the hydrogen atoms removed by the enzyme. These additional molecules are known as **coenzymes**.

Coenzymes are non-protein organic molecules which bind to an enzyme's active site. Some of them assist in enzymic oxidation and reduction reactions. They become reduced when they accept hydrogen atoms, which will later divide into protons and electrons. They become oxidised when the hydrogen is passed on.

NAD (nicotinamide adenine dinucleotide) is a hydrogen carrier molecule that is manufactured in the body (Figure 8.5). The nicotinamide part accepts a pair of hydrogen atoms and becomes reduced as it does so. It is oxidised when it loses the hydrogen atoms. NAD is an important molecule involved in all stages of respiration.

FAD (flavin adenine dinucleotide) is a similar molecule to NAD, but it contains riboflavin with adenosine and two phosphate groups. It also acts as a hydrogen carrier. However, FAD is tightly bound to a dehydrogenase enzyme embedded in the mitochondrial inner membrane. As a result it does not pump hydrogen atoms into the intermembranal space but instead returns them to the matrix.

Coenzyme A (often referred to as CoA) is another organic molecule consisting of adenosine and three phosphate groups together with the amino acid cysteine (Figure 8.6). It carries the 2-carbon acetyl group made from pyruvate in the link reaction to the Krebs cycle. It also carries acetyl groups made from fatty acids and some amino acids to the Krebs cycle, so it is an important entry point for other substrates that may be respired.

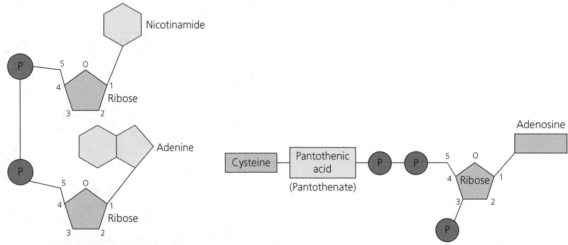

Figure 8.5 Molecular structure of NAD.   **Figure 8.6** The structure of coenzyme A.

The stages of respiration

Respiration is a continuous process involving many reactions organised into four stages (Figure 8.7), as follows.

1 Glycolysis does not need any oxygen and occurs in both aerobic and anaerobic respiration. It takes place in the cytosol, which contains all the enzymes necessary to convert glucose into pyruvate.

2 The link reaction occurs in the matrix of the mitochondrion, where pyruvate is converted to an acetyl group bound to coenzyme A.

3 The Krebs cycle takes place in the matrix of the mitochondrion and uses the acetyl group from acetyl coenzyme A.

4 Oxidative phosphorylation takes place on the inner mitochondrial membranes. In it electrons flow along the electron transport chain, providing energy for the active transport of protons, and ADP is phosphorylated to ATP.

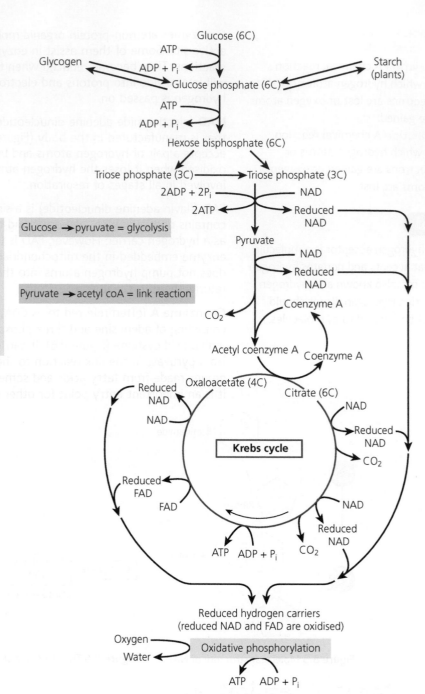

Tip

Hexose bisphosphate in this flow chart is fructose 1,6-bisphosphate. You can use either name. Bis means that the phosphate groups are attached to two different carbon atoms.

Glucose → pyruvate = glycolysis

Pyruvate → acetyl coA = link reaction

Figure 8.7 An overview of all stages of respiration.

Glycolysis

Glycolysis occurs in the cytoplasm of all living cells. It is thought to be an ancient biochemical pathway that satisfied the energy needs of primitive life forms. The pathway is anaerobic and the amount of energy released is small.

It consists of 10 separate reactions, each catalysed by a different enzyme, and can be simplified as the breakdown of one molecule of glucose to two molecules of pyruvate (Figure 8.8). During this process the coenzyme NAD and two phosphates groups from two molecules of ATP are required to allow the pathway to continue.

Phosphorylation

● Phosphorylation consists of a series of steps involving the six-carbon sugar glucose being phosphorylated to hexose 1,6-bisphosphate.

- Phosphorylation occurs first when phosphate is added to glucose to activate it and raise its energy level. This is achieved by transferring a phosphate group from ATP to a glucose molecule on the carbon-6 position to form glucose 6-phosphate, which is converted into fructose 6-phosphate.

- A second phosphate group is transferred from an ATP molecule to the carbon-1 position on the fructose 6-phosphate to form an activated hexose sugar known as fructose 1,6-bisphosphate.

- Two ATP molecules have been used up and the resulting ADP molecules are released, to be reformed into ATP when the energy that is necessary is released in the later stages.

Splitting fructose 1,6-bisphosphate

- Fructose 1,6-bisphosphate is now split into two molecules, each one with a single phosphate attached. These molecules are called triose phosphate as they are three-carbon sugars.

Oxidation of triose phosphate

- Triose phosphate is now oxidised by removal of two hydrogen atoms using dehydrogenase enzymes with coenzyme NAD acting as a hydrogen acceptor. This forms reduced NAD.

- During this step two molecules of reduced NAD are formed per molecule of glucose. During this step, enough energy is available from the oxidation reaction to phosphorylate, both molecules using phosphate ions from the cytosol to form two molecules of triose bisphosphate. This step involves conservation of energy because without it energy would be lost as heat.

Conversion to pyruvate

- Each triose phosphate molecule is then converted to a molecule of three-carbon pyruvate by a number of enzyme-controlled reactions. As a result, enough energy is released to phosphorylate two more ADP molecules to form two ATP molecules by substrate-level phosphorylation.

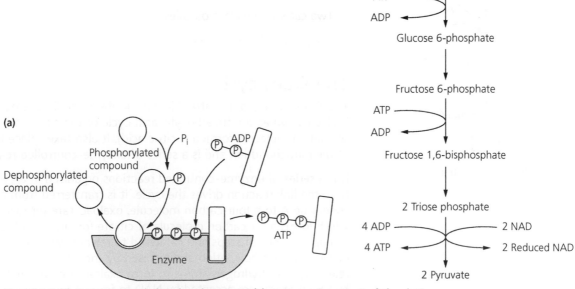

Figure 8.8 (a) Substrate-level phosphorylation and (b) summary flow chart of glycolysis.

Products of glycolysis

For each glucose molecule entering glycolysis, the following molecules are produced.

- Two molecules of reduced NAD. The NAD coenzymes have been reduced by the acceptance of hydrogen atoms. These are taken into the mitochondria by diffusion involving a shunt mechanism (the mitochondrial shunt).

- A net gain of two ATP molecules. Four are actually made but of these two are used up in the initial phosphorylation of glucose.

- Two molecules of pyruvate, which are actively taken into the mitochondria for the link reaction to take place. However, in the absence of oxygen these two molecules are used in fermentation: either to lactate in animal cells or in ethanol fermentation in yeast and some bacteria.

The link reaction

The two pyruvate molecules produced at the end of glycolysis are transported across the mitochondrial membranes into the matrix where the enzymes needed for the link reaction are found. The link reaction takes place in the mitochondria.

Two different reactions occur in the link reaction (Figure 8.9).

1 Decarboxylation, where a carboxyl group is removed from pyruvate using the enzyme complex pyruvate dehydrogenase. The carboxyl group becomes carbon dioxide.

2 Dehydrogenation, where the pyruvate dehydogenase complex removes hydrogen atoms. These are accepted by coenzyme NAD to form reduced NAD.

The final product is a two-carbon acetyl group, which is accepted by coenzyme A to form acetyl coenzyme A (or acetyl CoA), and this carries the acetyl group into the Krebs cycle.

The products of the two molecules of pyruvate entering the link reaction are

- two molecules of reduced NAD

- two carbon dioxide molecules

- two acetyl CoA molecules.

The Krebs cycle

The Krebs cycle (named after Sir Hans Krebs, who discovered the pathway) is also known as the tricarboxylic acid cycle or the citric acid cycle (the first six-carbon molecule formed is citric acid). It also takes place in the matrix of the mitochondrion and is a series of enzyme-controlled reactions.

It is a series of enzyme-controlled reactions organised in a cycle. Acetyl from the link reaction drives the cycle. It is transferred from acetyl coenzyme A to a four-carbon molecule, oxaloacetate (or oxaloacetic acid), which forms a six-carbon molecule of citrate (or citric acid). Citrate is taken through a series of steps and catalysed to release two molecules of carbon dioxide by decarboxylation and oxidised by the process of dehydrogenation, releasing eight hydrogen atoms to form oxaloacetate again. Six of the hydrogen atoms are accepted by NAD to form three molecules of reduced NAD and two are accepted by FAD to form a single molecule of reduced

> **Tip**
>
> Remember, **net gain** is the number of ATP molecules produced in glycolysis less the number needed or used up during the glycolysis.

> **Key term**
>
> **Decarboxylation** Removal of carbon dioxide from a substrate molecule in the link reaction and in the Krebs cycle.

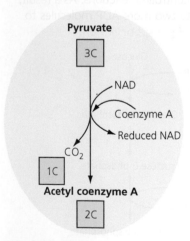

Figure 8.9 The link reaction.

FAD. During these steps enough energy is released to form one molecule of ATP directly by substrate-level phosphorylation for each turn of the Krebs cycle (see Figure 8.10).

Glycolysis results in two molecules of pyruvate. The link reaction produces two acetyl groups for each molecule of glucose. If both of these acetyl groups enter the Krebs cycle then all the reactions shown in Figure 8.10 occur twice for each molecule of glucose respired. In other words there are two 'turns' of the Krebs cycle for each molecule of glucose respired.

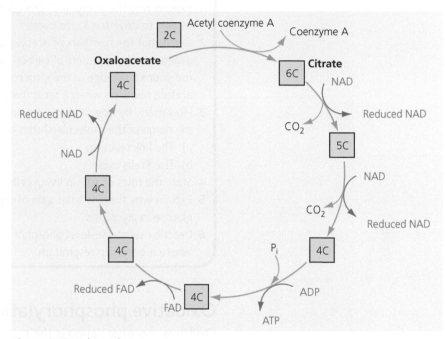

Figure 8.10 Krebs cycle.

Steps in the Krebs cycle

Here's what happens for each turn of the Krebs cycle.

- The two-carbon acetyl group from acetyl CoA combines with four-carbon oxaloacetate to form six-carbon citrate. In the process coenzyme A is released to be re-used again.
- Citrate is now dehydrogenated and decarboxylated to form a five-carbon molecule. This releases a pair of hydrogen atoms which are accepted by NAD (becoming reduced NAD), and one carbon dioxide molecule.
- The five-carbon molecule is dehydrogenated and decarboxylated to form a four-carbon compound, a molecule of reduced NAD and a molecule of carbon dioxide.
- The reorganisation of the four-carbon molecule provides energy to phosphorylate ADP with a phosphate ion from the matrix of mitochondria to form one ATP molecule directly. This is another example of a substrate-level phosphorylation (see Figure 8.8a).
- The next four-carbon compound is dehydrogenated, but this time FAD acts as the hydrogen and electron acceptor to become reduced FAD.
- In the final reaction, a four-carbon molecule is dehydrogenated to reform oxaloacetate and complete the cycle. The two hydrogen atoms are accepted by NAD to form another reduced NAD molecule.

Since one molecule of glucose produces *two* pyruvate molecules, conversion in the link reaction forms two acetyl groups. These acetyl groups both enter the Krebs cycle and create *two* turns of the cycle. The products of two turns of the Krebs cycle are

- six reduced NADs
- two reduced FADs
- four carbon dioxide molecules
- two ATP molecules produced directly.

Oxidative phosphorylation

Oxidative phosphorylation is the process where energy is used to add inorganic phosphate to ADP to form ATP. It occurs on the folded inner membranes of the mitochondria, called cristae. This is where electron carriers are embedded in the membrane itself together with the enzyme, ATP synthase, which projects out from the membrane. The folded structure increases the surface area for the carriers and ATP synthase.

The reduced NAD and FAD from the Krebs cycle, and reduced NAD from the link reaction and possibly glycolysis (if the hydrogen atoms from these steps diffuse into the mitochondria) pass on their hydrogen atoms to the next carrier in the chain and become oxidised so they can be re-used. The hydrogen splits into protons and electrons. The electrons are accepted by NADH-coenzyme Q reductase (also known as Complex I in Figure 8.11) and the protons are passed into the matrix.

The electrons are now passed from one carrier to another, each carrier becoming reduced as it accepts the electrons and oxidised as it passes them on to the next carrier.

Between 1.5 and 2.5 ATP molecules are produced for each pair of electrons passing through the chain as a small amount of energy is released at some of the steps, which is enough to allow a single ATP molecule to be produced each time. The exact amount of energy depends on which hydrogen carrier is involved and whether the hydrogen atoms are already in the mitochondria or whether they need to be moved in from the cytosol.

The final electron acceptor is oxygen, which – together with protons (hydrogen ions) that now rejoin electrons – is reduced to form water (see Figure 8.11).

Key terms

Cristae The highly folded inner membranes of the mitochondria where the electron carriers and enzymes are found.

NADH-coenzyme Q reductase A dehydrogenase enzyme which that catalyses the oxidation of reduced NAD so that NAD can be recycled for reuse in the link reaction and in Krebs cycle.

Tip

FAD is part of the enzyme bound to the membrane and passes electrons accepted by complex III along the electron transport chain. However, the protons do not form part of the proton gradient as they pass to the matrix.

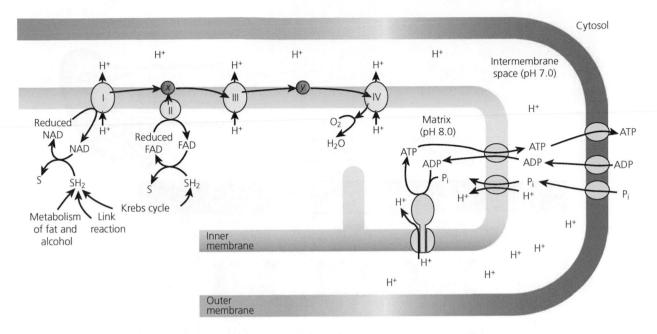

SH$_2$ = intermediate substances in the link reaction, Krebs cycle and metabolism of fat and alcohol. These substances are dehydrogenated to form reduced hydrogen carriers (reduced FAD and reduced NAD), which are oxidised by complexes I (NADH-coenzyme Q reductase) and II

I, II, III and IV are protein complexes

x and *y* are electron carrier molecules

Complexes I, III and IV pump protons; complex II passes electrons from reduced FAD to complex III via electron carrier *x*.
Complex II is not a proton pump

Figure 8.11 Oxidative phosphorylation.

Key term

Chemiosmosis Flow of protons (hydrogen ions) through ATP synthase channels, creating an electrochemical force that is great enough to produce ATP. It occurs in the cristae of the mitochondria, and in the thylakoid membranes of chloroplasts for photosynthesis. It also occurs in the cell surface membranes of prokaryotes.

Tip

You should be able to compare oxidative phosphorylation, photophosphorylation and substrate-level phosphorylation and describe the similarities and the differences between them.

Chemiosmosis

Chemiosmosis is the process of forming ATP from the energy released when electrons pass along the electron transport chain. It occurs on the inner mitochondrial membranes as part of oxidative phosphorylation.

As electrons move along the carrier molecules, they release energy. The reduced coenzymes are oxidised and some of the energy from this process is used to actively pump the protons into the intermembrane space where the protons build up and create a proton and electrochemical gradient. ATP is not used for this pump, as the energy needed comes from the electron flow or electromotive force. The inner membrane is impermeable to ions including protons. They cannot move across the membrane itself but do flow through ion channels in the membrane, down the proton gradient. The ion channels are associated with ATP synthase. As the protons flow through the ion channel of ATP synthase the rotating part of the ATP synthase enzyme is turned and the energy released is used to build one ATP molecule from ADP and P$_i$ (see Figure 8.12).

This is known as chemiosmosis. It is the theory proposed by Peter Mitchell in 1961 and has since been endorsed by other scientists. The process of chemiosmosis is vital for forming ATP in both oxidative phosphorylation in respiration and photophosphorylation in photosynthesis.

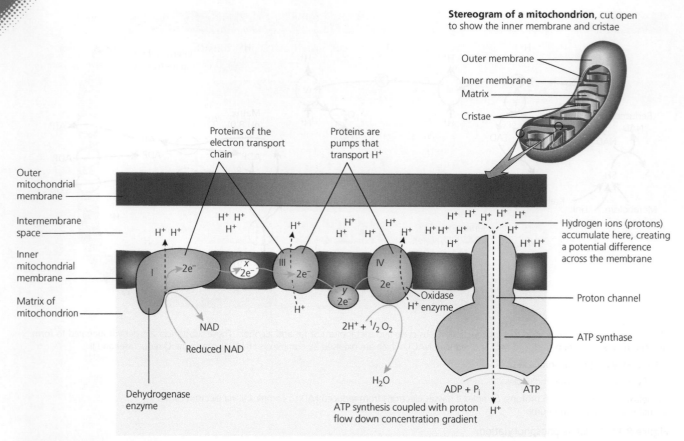

Stereogram of a mitochondrion, cut open to show the inner membrane and cristae

Outer membrane
Inner membrane
Matrix
Cristae

Proteins of the electron transport chain

Proteins are pumps that transport H⁺

Outer mitochondrial membrane

Intermembrane space

Inner mitochondrial membrane

Matrix of mitochondrion

H^+ H^+

H^+ H^+ H^+

I $2e^-$

x $2e^-$

H^+

III $2e^-$

H^+ H^+ H^+

y $2e^-$

H^+

H^+ H^+ H^+

IV $2e^-$

H^+

H^+ H^+ H^+ H^+ H^+

H^+ H^+ H^+

H^+

Oxidase H^+ enzyme

H^+ H^+

Hydrogen ions (protons) accumulate here, creating a potential difference across the membrane

Proton channel

ATP synthase

NAD

Reduced NAD

Dehydrogenase enzyme

$2H^+ + \frac{1}{2} O_2$

H_2O

$ADP + P_i$

H^+

ATP

ATP synthesis coupled with proton flow down concentration gradient

Figure 8.12 The proton pumps, proton channel and ATP synthase involved in chemiosmosis. This diagram shows complexes I, III and IV. Complex II is not shown.

Table 8.1 gives a summary of the stages in respiration and lists the products formed.

Table 8.1 Summary of the stages in respiration and the products formed.

Stage	Location	Substrate molecules	Product molecules	Number of ATP molecules produced	Type of respiration
Glycolysis	Cytosol	Glucose (6C)	2 Pyruvate (3C), 4 ATP, 2 reduced NAD	2 (net production)	Aerobic and anaerobic
Link reaction	Matrix of mitochondrion	Pyruvate	Acetyl coenzyme A Reduced NAD and CO_2		Aerobic
Krebs cycle (per turn of cycle)	Matrix of mitochondrion	2 acetyl groups from coenzyme A	Oxaloacetate recycled 2 CO_2, 1 ATP, 6 reduced NAD, 2 reduced FAD	2 (for two turns of cycle)	Aerobic
Oxidative phosphorylation	Cristae of mitochondrion	10 reduced NAD and 2 reduced FAD and ADP	H_2O and 26 ATP	26–28	Aerobic
Total				30–32	

Tip

The electrochemical gradient resulting from chemiosmosis creates potential energy in the intermembranal space. This is used to drive the ATP synthase reaction and as a result ATP is built up.

How much energy is produced as ATP?

In glycolysis, four ATP molecules are produced directly, by substrate-level phosphorylation. However, as two ATPs are used up, the net gain is two ATP molecules. In the Krebs cycle, two ATPs from each turn of the cycle are produced directly by substrate-level phosphorylation. In oxidative phosphorylation, 10 molecules of reduced NAD can produce 26 molecules of ATP.

Both reduced NAD and reduced FAD donate electrons for the electron transport chain, but only reduced NAD also donates the hydrogen ions used to build up the proton gradient for chemiosmosis. The hydrogen ions from reduced FAD remain in the matrix but will recombine with the electrons and oxygen to form water.

However, some ATP is used to actively transport pyruvate into the mitochondrial matrix and reduced NAD from glycolysis needs to be transported into the matrix via the **mitochondrial shunt** mechanism. There is a possibility that some H+ will be lost by leaking out. Therefore the potential energy may be reduced, which would further limit ATP production. In reality 30 ATP molecules may not be achieved.

Mitochondria

Mitochondria (singular: mitochondrion) are cell organelles found in the cytoplasm of eukaryotic cells. Each mitochondrion is a rod-shaped structure about 2–5 μm long, although some may be longer, and 0.5–1.0 μm in diameter. The numbers of mitochondria in metabolically active cells such as liver or muscle are much greater than in less active cells such as fat-storage cells. Mammalian red blood cells have none.

Mitochondrial structure

- Mitochondria have an envelope that consists of two phospholipid membranes: a smooth outer membrane and a folded inner membrane (see Figure 8.13 which shows the exchanges between a mitochondrion and the surrounding cytosol).

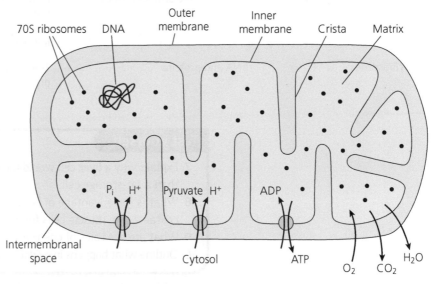

Figure 8.13 Structure of a mitochondrion and the exchanges with the cytosol. The blue circles represent proteins that transport phosphate ions (P$_i$), pyruvate and ADP across the inner membrane into the matrix.

- In the envelope there are channel proteins, or carrier molecules, and enzymes.

- The inner membrane is folded extensively into cristae, which increase the surface area available to hold electron carriers and ATP synthase enzymes. It has a different composition from the outer membrane, which is further evidence in support of the **theory of endosymbiosis**.

- Between the outer and inner membranes there is an intermembranal space that has an important role in energy release.

- The matrix is a gel-like mixture of proteins and lipids found within the inner membrane. It also contains a looped mitochondrial DNA, 70S ribosomes and enzymes.

Table 8.2 outlines the structure and function of mitochondria (see Figures 8.11, 8.12 and 8.13).

Table 8.2 Structure, composition and function of mitochondria.

Structure	Composition	Function
Outer mitochondrial membrane	Phospholipid bilayer and proteins	Permeable to pyruvate, oxygen, carbon dioxide, ATP, ADP but not glucose
Inner mitochondrial membrane – folded into cristae to give a large surface area	Phospholipid bilayer with protein complexes (I to IV) of electron transport chain and ATP synthase	Complexes I, II and IV pump protons from the matrix to the intermembrane space; transport proteins move phosphate ions, pyruvate and ADP from the intermembrane space into the matrix. The movement of phosphate ions and pyruvate is coupled with the inwards movement of protons. ADP is exchanged for ATP; ATP synthase synthesises ATP from ADP and P_i
Intermembrane space	Lower pH than cytosol and matrix as it has a high concentration of protons (hydrogen ions)	Site of higher concentration of protons to drive movement of protons into matrix through channels in ATP synthase
Matrix	Protein-rich region contains DNA loop, ribosomes and many enzyme molecules	Link reaction; Krebs cycle; part of urea cycle (see Chapter 2); protein synthesis
mitochondrial (mt)DNA	Loop of double-stranded DNA	mtDNA codes for 13 of the proteins used in the mitochondrion; genes are transcribed as mRNA; rest of mitochondrial proteins are coded by DNA in the nucleus
70S ribosomes	rRNA and proteins	Translation – assembly of amino acids to form proteins

> **Test yourself**
>
> 7 Explain why a liver cell would have a larger number of mitochondria than a fat-storage cell.
> 8 What is the advantage of larger mitochondria and greater numbers of cristae in the muscle cells for a trained athlete?
> 9 What are the roles of coenzymes in the different stages of respiration?
> 10 Outline what happens in oxidative phosphorylation?

Anaerobic respiration

When oxygen is not available it cannot act as the final acceptor for oxidative phosphorylation and so the electron transport chain stops. This then prevents both the Krebs cycle and the link reaction from proceeding due to the build-up of reduced NAD and FAD. As a result only glycolysis can continue. Even here the reduced NAD must be regenerated or this pathway will also stop.

There are two types of **anaerobic** regeneration of the reduced NAD in eukaryotic cells

- lactate fermentation, which occurs in the muscle cells of animals
- ethanol or alcohol fermentation, which is typically used by fungi, such as yeast, and plants.

No ATP is produced in either of these pathways. However, since glycolysis is the starting point for both pathways there is a net gain of two ATP molecules overall for each glucose molecule respired, by substrate-level phosphorylation.

Lactate fermentation

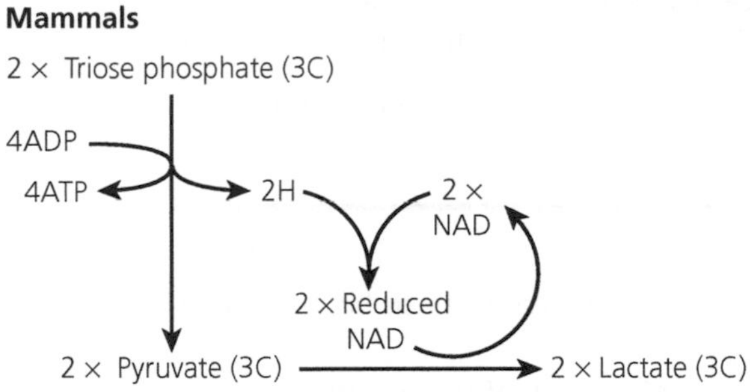

Lactate fermentation occurs in rapidly respiring muscle cells; for example, when a person is exercising intensely. In these conditions the oxygen cannot be delivered quickly enough to supply the oxygen demand of the muscles. ATP can be supplied rapidly for a short period of explosive exercise by anaerobic respiration. However, there will be an oxygen debt that must be repaid after the exercise has ceased before any more rapid exercise can be performed. This is achieved by deep breathing until enough oxygen has been taken in.

- During glycolysis reduced NAD, ATP and pyruvate are produced.
- In order to recycle the reduced NAD, the pyruvate acts as the hydrogen acceptor for the hydrogen from reduced NAD. The NAD is now oxidised and can be re-used to accept more hydrogen atoms.
- Lactate dehydrogenase catalyses the reduction of pyruvate to lactate and the oxidation of NAD.

Once the exercise is over, and as soon as more oxygen is available, the lactate is converted back to pyruvate in the liver. Then the pyruvate rejoins the respiratory cycle by entering the link reaction or it may be converted back to glucose and then on to glycogen. If lactate is allowed to accumulate in the muscle cells the reduced pH prevents the muscles from functioning (Figure 8.14).

Mammals

Figure 8.14 Fate of pyruvate in anaerobic respiration in mammals.

Ethanol fermentation

Ethanol, or alcohol, fermentation will occur in yeast when there is insufficient oxygen for the cells to respire aerobically. In a similar way to lactate fermentation the reduced NAD must be oxidised to allow the pathway to continue.

- During glycolysis reduced NAD, ATP and pyruvate are produced.
- Pyruvate is decarboxylated to form ethanal and carbon dioxide. This reaction is catalysed by pyruvate decarboxylase.

Yeast

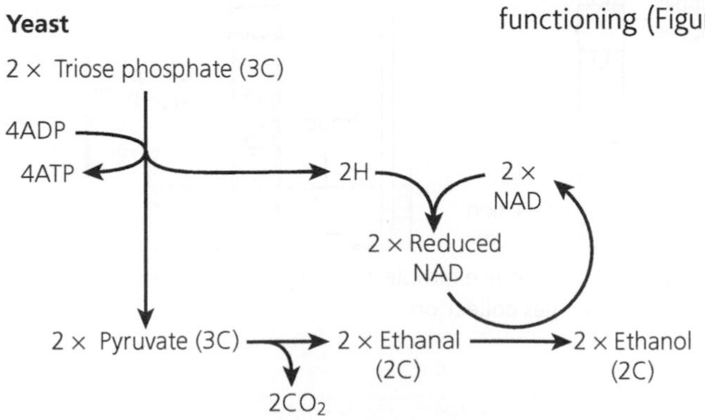

Figure 8.15 Fate of pyruvate in anaerobic respiration in yeast.

- Ethanal acts as a hydrogen acceptor to remove the hydrogen atoms from reduced NAD; NAD is now oxidised and ethanal becomes reduced to ethanol using the enzyme ethanol dehydrogenase (Figure 8.15). The NAD can now accept more hydrogen atoms so glycolysis can continue.

Both forms of anaerobic respiration produce only a net gain of two ATP molecules and are therefore wasteful in terms of energy release. However, they both have an important role in allowing some energy release under difficult conditions. Table 8.3 compares lactate and ethanol fermentation.

Table 8.3 Comparison of lactate and ethanol fermentation.

	Lactate fermentation	Ethanol fermentation
ATP molecules	2	2
Reduced NAD	2	2
CO_2 molecules	0	2
Intermediate molecule	None	Ethanal
Enzymes	Lactate dehydrogenase	Pyruvate decarboxylase, ethanol dehydrogenase
End products	Lactate (lactic acid)	Ethanol (alcohol)

Test yourself

16 Draw a table to compare all the features of aerobic and anaerobic respiration.
17 Comment on the role of NAD in each of these pathways.
18 What are the advantages of anaerobic respiration and what are the main disadvantages?

Activity

Investigating respiration rates in yeast

A student collected the gas produced from a reacting yeast and sugar mixture by assembling a gas-collection apparatus using displacement of water. You will need to know several different methods for measuring gases in this way (see Figure 8.16).

The volumes of gas produced were recorded every 30 seconds for 3 minutes.

A clean set of apparatus was used with a fresh yeast and glucose mixture in the flask. On this occasion the glucose solution was made with boiled and then cooled water. Once mixed with yeast and placed into the apparatus a thin layer of oil was added to the surface before gas collection began. Volumes of gas produced were again recorded every 30 seconds for 3 minutes.

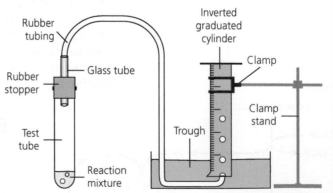

Figure 8.16 Here is one example of the type of apparatus that may be used for gas collection.

The results are shown in Table 8.4.

Table 8.4 Table of results, collecting gas from respiring yeast.

Time/s	Gas collected/cm³	
	Yeast and sugar in aerobic conditions	Yeast and sugar in anaerobic conditions
0	0.00	0.00
30	0.30	0.20
60	0.60	0.25
90	0.90	0.35
120	1.30	0.55
150	1.30	0.55
180	1.10	0.60
Total volume of CO_2 after 180 seconds/cm³	5.50	
Gas produced in 60 seconds/cm³	1.83	

Questions

1 Plot the results as a graph of gas volume against time.
2 How could the results be made more valid? How would this change the type of graph you draw?
3 Copy the table of results and complete the 'Gas produced in 60 seconds/cm³' and the 'Total volume of CO_2 after 180 seconds/cm³' cells for the anaerobic conditions.
4 To be confident of the data-collection procedure a number of variables should be controlled. Suggest three variables that should be controlled and describe how to control each one.
5 To calculate the Q_{10} for this reaction further data must be collected by repeating the experiment with some changes made to the procedure. Look at the equation for Q_{10} below and suggest what changes will be needed in the procedure to collect the data needed.

$$Q_{10} = \frac{\text{Rate of reaction at } (t + 10\,°C)}{\text{Rate of reaction at } t\,°C}$$

6 Suggest a statistical test that could be used if more data were collected and explain how this test would improve the validity of the results.
7 Suggest how the experiment may be modified to investigate the effect of changing respiratory substrates.
8 What is the role of the thin oil layer on the surface of the yeast and glucose mixture? Suggest how the yeast responds to anaerobic conditions.

Respiratory substrates

Table 8.5 The energy values of different respiratory substrates.

Respiratory substrate	Energy/kJ g⁻¹
Carbohydrates	16
Triglycerides	39
Proteins	17

A respiratory substrate is an organic molecule that may be used in respiration to release energy. Different substrates release different amounts of energy when respired (see Table 8.5).

Glucose is the most commonly respired substrate in many human tissues. Indeed, the human brain and red blood cells cannot respire any other substrate, but fat and other molecules may be used by other tissues. In reality the diet of a healthy person includes a variety of food and so a mixture of substrates will be respired at any one time.

The Krebs cycle is driven by the number of acetyl groups entering it and therefore the number of acetyl groups determine the number of hydrogen ions made available for oxidative phosphorylation and the number of ATP molecules formed.

- The greatest amount of ATP is produced by chemiosmosis when protons flow through ATP synthase.

- The more protons available, the more ATP that is produced.

- The more hydrogen atoms in a respiratory substrate, the more ATP can be formed.

- The higher the number of hydrogen atoms taken through the system, the greater the number of oxygen atoms needed to act as the final electron acceptor.

Carbohydrates are the main respiratory substrates. Starch and glycogen can both be broken down into glucose and other carbohydrates can be changed to glucose by isomerisation.

Metabolism of glucose should yield 30 ATPs, which is only 32% efficient. The rest is lost as heat, which is vital to endotherms to maintain a constant body temperature for metabolic processes to occur.

Triglycerides are important respiratory substrates, especially in muscle cells. They are first hydrolysed to fatty acids and glycerol. Glycerol is a three-carbon molecule that can be converted to join the glycolysis pathway but fatty acids are taken into the respiratory process by combining them with coenzyme A. Fatty acids are long hydrocarbon chains with plenty of carbon and hydrogen atoms but little oxygen. They will therefore produce a lot of ATP when the hydrogen atoms are taken into oxidative phosphorylation. Energy from ATP hydrolysis is required to create a fatty acid–coenzyme A complex. This is then actively transported into the mitochondrial matrix where it is converted to acetyl groups to be attached to coenzyme A. This conversion is called the β-oxidation pathway, which yields both reduced NAD and reduced FAD. The acetyl groups formed are then used to drive the Krebs cycle. Since virtually all fatty acids have an even number of carbon atoms and there are few oxygen molecules so that little carbon dioxide is formed, the number in a fatty acid chain can be halved to determine the number of two-carbon acetyl groups formed. In a 16-carbon fatty acid, for example, eight will be formed.

Amino acids are not usually respired. Usually, once a protein is digested into amino acids any amino acid not immediately needed is deaminated in the liver to produce urea, with the remaining portion converted to glycogen or fat. However, metabolism of proteins may occur when an organism is starving or undergoing prolonged exercise. In this case, if used as a respiratory substrate, some amino acids are converted to pyruvate, and then to acetyl groups, whereas other amino acids enter the Krebs cycle directly. The number of hydrogen atoms accepted by NAD per amino acid is slightly more than for a molecule of glucose so a little more energy is released than for carbohydrate metabolism.

Although glucose is the primary respiratory substrate other molecules may be used in its place. An example of this is alcohol. Once chemically converted by a series of enzymes it may enter the Krebs cycle and be metabolised to produce ATP.

Respiratory quotient

Respiratory quotient (RQ) is the ratio of carbon dioxide produced by a respiring organism to oxygen consumed in a given time. As it is a ratio there are no units required.

$$RQ = \frac{CO_2 \text{ released}}{O_2 \text{ uptake per unit time}}$$

The RQ can tell us what respiratory substrate is being metabolised by an organism and what type of respiration, aerobic or anaerobic, is being employed. The RQ values of different respiratory substrates are listed in Table 8.6.

As an example, in an experiment the oxygen uptake of blowfly larvae was found to be 0.5 cm^3 per second and the carbon dioxide expired was 0.4 cm^3 per second. RQ was calculated using the following equation.

$$RQ = \frac{0.4}{0.5}$$

Therefore

$$RQ = 0.8$$

This suggests a mixture of respiratory substrates used but more fat than protein or carbohydrate.

If anaerobic respiration is being used the RQ value will be infinity. A mixture of aerobic and anaerobic respiration gives a value grater than 1, e.g. 4, whereas purely aerobic respiration gives values of 1 or less. Carbohydrate respired in aerobic conditions gives an RQ of 1.

If a pure substrate is being respired aerobically the decimal ratio is clear. However, a mixture of substrates respired aerobically gives a value of approximately 0.85.

Table 8.6 The RQ values of different respiratory substrates.

Respiratory substrate	RQ
Glucose	1.0
Proteins	0.9
Triglycerides	0.7

Activity

Using a respirometer to calculate respiratory rate and RQ

Simple respirometers for determining the respiration rate of yeast can be made by putting yeast suspensions into a 10 cm^3 syringe and attaching a length of glass tubing to the nozzle of the syringe as shown in Figure 8.17.

Some students used this type of respirometer to investigate the effect of different sugars on the rate of respiration in yeast. Yeast suspensions were prepared in solutions of the different sugars and in water. The results are shown in Table 8.7.

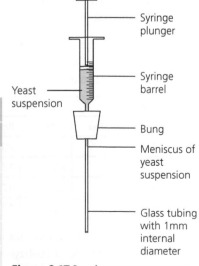

Figure 8.17 Respirometry apparatus.

Table 8.7 Results of an experiment measuring yeast respiration rate.

Type of sugar	Total distance travelled by the meniscus at 15 minute intervals/mm					
	15	30	45	60	75	90
Glucose	0	10	90	180	310	480
Maltose	0	0	10	20	30	50
Sucrose	0	0	10	40	170	350
Lactose	0	10	10	10	10	10
Water	0	10	10	10	10	10

Glucose is a monosaccharide. Maltose, sucrose and lactose are disaccharides.

1 a) State the independent and dependent variables for this investigation.
  b) State three variables that should be controlled in this investigation.
  c) Explain why a yeast suspension in water alone was included.
2 Calculate the volume of carbon dioxide produced at the end of each 15 minute interval for each substrate. State any assumptions that you have made in using the results from the table for calculating the volume of carbon dioxide produced.
3 Use your results from question 2 to plot a graph showing the production of carbon dioxide by yeast.
4 Discuss the conclusions that you can make from the results.

Using a different procedure other students calculated the rate of respiration of yeast in glucose and maltose, as shown in Table 8.8.

Table 8.8 Yeast respiration rate in glucose and maltose.

	Rate of respiration/$cm^3 min^{-1}$		
	Glucose	Maltose	Water
Mean (±SD)	0.86 (±0.085)	1.19 (±0.348)	0.01
Range	0.66–0.95	0.89–2.01	0.01–0.20
Number of replicates (n)	10	10	2

5 Explain the advantage of calculating the standard deviation of these results.
6 Carry out a statistical analysis to discover whether the difference between the results for maltose and glucose is significant. Show your working and state any assumptions you have made.
7 Explain how sensors and data loggers could be used to obtain results on the rate of respiration in yeast.

Exam practice questions

1 In a cell that is respiring aerobically, which process produces most of the carbon dioxide?
  A Glycolysis
  B The Krebs cycle
  C Oxidative phosphorylation
  D The link reaction   (1)

2 Carbohydrates, fats, proteins and alcohol are all respiratory substrates. Which row below shows the correct values?   (1)

	Energy/$kJ g^{-1}$			
	Carbohydrates	Fats	Proteins	Alcohol
A	16	17	29	39
B	39	29	17	17
C	29	39	17	17
D	16	39	17	29

3 Coenzymes fulfil important roles in respiration. What is the role of coenzyme A in respiration?
  A To transfer phosphate groups to hexoses
  B To transfer hydrogen from glycolysis to oxidative phosphorylation
  C To transfer hydrogen from the Krebs cycle to oxidative phosphorylation
  D To transfer two-carbon molecules to oxaloacetate   (1)

4 Potassium cyanide is an inhibitor of the final stage of oxidative phosphorylation. Which activity of neurones is inhibited immediately when exposed to potassium cyanide?
  A Closing of potassium voltage-gated channel proteins
  B Depolarisation
  C Reduction in acetylcholinesterase activity
  D Action of the sodium-potassium pump   (1)

5 Succinate dehydrogenase is an enzyme located in the cristae of mitochondria. The enzyme catalyses this reaction

succinate + FAD → fumarate + reduced FAD

  a) Describe the role of FAD in the reaction.   (4)
  b) Isolated mitochondria provided with succinate as a respiratory substrate will respire. Reaction

mixtures containing mitochondria and different concentrations of succinate were incubated and the rate of respiration determined. Similar reaction mixtures were prepared with the addition of a 1.0 mmol dm⁻³ solution of malonate. The graph shows the results.

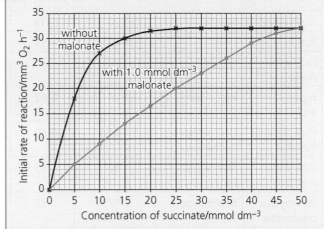

Describe and explain the results shown in the graph. (6)

c) The equation shows the complete oxidation of succinate.

$$C_4H_4O_4 + 3O_2 \rightarrow 2H_2O + 4CO_2$$

i) Calculate the respiratory quotient (RQ) for the complete oxidation of succinate. State the formula and show your working. (3)

ii) Explain why succinate cannot be oxidised in anaerobic conditions. (3)

6 a) List the ways in which ATP is produced in a palisade mesophyll cell. (4)

When seeds first start to germinate they absorb water and swell. This is called imbibition. Then the seed coat cracks open and the radicle (embryonic root) and then the plumule (embryonic shoot) emerge. After germination, the shoot grows into the light, the leaves expand and start to photosynthesise. Until this stage, the seed has relied on stored substances, such as oils and starch, as sources of energy.

A student determined the respiratory quotient (RQ) of germinating soya bean seeds. The student took a large number of seeds, divided them into eight batches and weighed them. All the seeds were soaked and then their seed coats were removed and left to germinate. At intervals each batch was placed into an air-tight chamber as shown in the diagram.

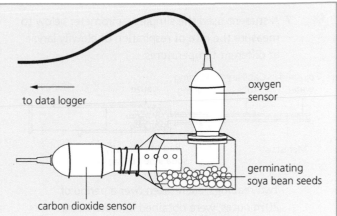

After results were taken in the chamber, each batch was dried to determine the dry mass.

b) Outline the steps that the student followed to determine the RQ of the seeds at each stage of germination using the apparatus above. (6)

c) The student plotted the results as a graph, as shown below. Describe and explain the changes in RQ of these seeds during germination. (5)

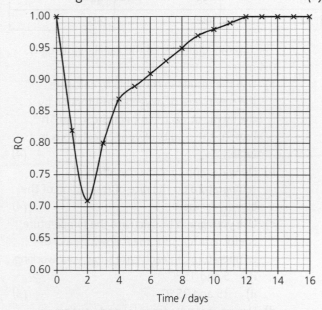

d) The RQ values for seeds immediately after imbibition are above 1.0. Suggest what could have caused these high values. (3)

e) Suggest what extra information can be gained from determining the dry mass of seeds before and after germination. (3)

7 A student used the simple respirometer below to measure the rate of respiration of blowfly larvae at different temperatures.

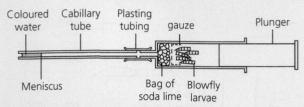

Coloured water · Cabillary tube · Plasting tubing · gauze · Plunger · Meniscus · Bag of soda lime · Blowfly larvae

Two readings, each taken over a period of 20 minutes, were obtained for each of six temperatures. The results are shown in the table.

Temperature/°C	Distance moved by the meniscus in 20 minutes/mm	
	Run 1	Run 2
4	20	17
10	35	29
16	51	46
20	68	58
26	98	94
32	138	146

The internal diameter of the capillary tubing is 0.4 mm.

a) Use the results in the table to calculate the mean rate of oxygen consumption by the larvae at each temperature in $mm^3 h^{-1}$. Present the results of your calculations in a table. *(4)*

b) Draw a graph to show the effect of temperature on the rate of respiration of the blowfly larvae. *(4)*

c) Use your graph to calculate the temperature coefficient (Q_{10}) for the following pairs of temperatures. In each case show your working.
  i) 5 and 15 °C
  ii) 15 and 25 °C *(3)*

d) Use your results from (c) and your graph to describe and explain the effect of temperature on the rate of respiration of the blowfly larvae. *(4)*

e) Evaluate the procedure and data from this investigation. *(5)*

f) Explain how this investigation using the simple respirometer in the diagram above could be modified to determine the RQ of the larvae. Give the practical procedure involved and the measurements that need to be taken. *(6)*

Stretch and challenge

8 The table shows the body masses and the basal metabolic rates (BMRs) of 10 species of mammal.

Species	Common name	Body mass/g	BMR/cm^3 oxygen h^{-1}
Alces alces	Moose	325 000	51 419
Panthera leo	Lion	98 000	16 954
Leopardus pardalis	Ocelot	10 500	3126
Lepus arcticus	Arctic hare	3004	1082
Phyllostomus hastatus	Greater spear-nosed bat	84	388
Dipodomys merriami	Merriam's kangaroo rat	37	43
Peromyscus polionotus	Beach mouse	12	22
Crocidura russula	Greater white-toothed shrew	11	30
Sorex minutus	Pygmy shrew	4	31
Sorex cinereus	Masked shrew	3	31

Enter the data into a spreadsheet program.

a) Use the spreadsheet functions to analyse the data as follows:
  • plot a scattergraph of BMR (*y*-axis) against body mass (*x*-axis)
  • calculate the $\log_{10}$ of each mammal's body mass
  • calculate each BMR as the volume of oxygen used per gram of body tissue
  • plot a scattergraph of BMR per gram of body tissue (*y*-axis) against $\log_{10}$ body mass (*x*-axis).

b) Use your graphs to
  i) explain the advantage of converting the data in the table above before plotting a graph
  ii) comment on the relationship between body mass of these ten mammals and their BMR values.

You can take this analysis further by taking the $\log_{10}$ of the standardised BMR data and plotting against the $\log_{10}$ body mass. What relationship do you find?

9 The endosymbiotic theory was proposed by Lynn Margulis of Boston University in 1967. She noticed similarities between prokaryotic cells and some organelles in eukaryotic cells and suggested that large anaerobic bacteria had, in the past, taken in smaller aerobic bacteria by endocytosis and that these bacteria then developed a mutually beneficial (symbiotic) relationship with the host cell. The aerobic bacteria eventually became the mitochondria that we see in eukaryotic cells. A similar process resulted in ingested photosynthetic bacteria becoming chloroplasts.

a) Make drawings to show the structure of a typical prokaryotic cell and a mitochondrion. Annotate your drawing to show the similarities and differences between the two.

The diagram below shows a photosynthetic bacterium similar to the type that Professor Margulis suggested might be the ancestor of the chloroplasts in eukaryotic cells.

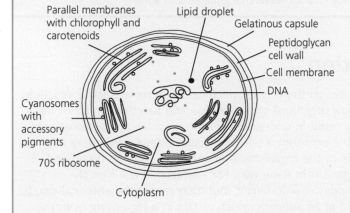

Parallel membranes with chlorophyll and carotenoids
Lipid droplet
Gelatinous capsule
Peptidoglycan cell wall
Cell membrane
DNA
Cyanosomes with accessory pigments
70S ribosome
Cytoplasm

b) Describe the features of the photosynthetic bacterium that support the idea that an organism like it could have been the ancestor of chloroplasts.

c) Make a diagram to show a mitochondrion and a chloroplast from a palisade mesophyll cell. Show, by means of arrows, labels and annotations, the exchanges that occur between
- the chloroplast and the cytosol
- the mitochondrion and cytosol.

Comment on the exchanges that you have indicated.

d) Comment on the statement that mitochondria and chloroplasts each lead a semi-independent existence inside eukaryotic cells.

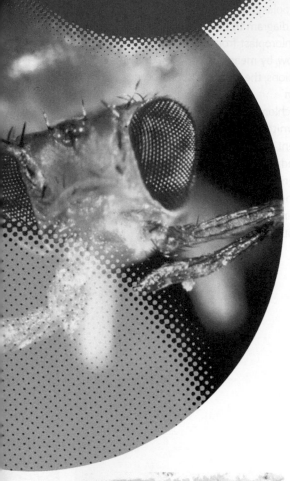

Chapter 9

Cellular control

Test yourself on prior knowledge

1 Name the three components that make up a nucleotide.
2 Explain how the two polynucleotides in DNA are held together.
3 Name the three different types of RNA and give the role that each plays in protein synthesis.
4 What is the relationship between a gene and an allele?

Introduction

This chapter focuses on the way in which cells control metabolic reactions and determine how organisms grow, develop and function. Some of the mechanisms may result in undesirable features, mutations or variations that may cause the organism a selective disadvantage, some may not affect the survival of the organism at all and some may be useful and helpful to the organism in some way. For example, in the fruit fly *Drosophila melanogaster* a dominant mutation causes the antennal disc to form a leg instead of an antenna, which results in a leg arising from the fly's head: clearly a disadvantageous feature under natural conditions. As another example, ladybirds show great variation in both colour and in spot number (Figure 9.1), which has been shown to vary throughout the seasons. These differences are a result of several alleles of the same gene (an example of multi-allelic inheritance). The frequency of each type has been linked to mating preferences.

Figure 9.1 This photograph shows the variation in colour and number of spots among ladybirds.

Key terms

Mutation A change in the genetic material in a cell. It may be a chromosome change or a change in the DNA nucleotide sequence of a gene.

Allele A variant of a gene that is found at the same locus on the chromosome and codes for the same polypeptide as another allele for the same gene. The polypeptides coded by different alleles may differ by one or more amino acids.

Gene mutations

A gene mutation is a relatively small change to a nucleotide sequence in the DNA as a result of mistakes in copying. When a gene is modified or altered, as a result of a change in the sequence of bases, a new form of the gene is formed. This is a new mutation. This alteration *may* cause no change at all in the protein that it produces, because the modified triplet may still code for the same amino acid in the protein. This can only happen because the genetic code is a degenerate code with several triplets coding for the same amino acid. Sometimes the change occurs in a non-coding section of the DNA and so has no effect on the amino acid sequence at all.

When a mutation does cause a change in the amino acid sequence that results in a modified protein structure, the mutation may still not cause an apparent effect. There are a number of reasons for this. For instance, the allele may be recessive and be masked by a dominant allele and so not be expressed. Alternatively, the gene may be one of several that are all involved in the expression of a characteristic, thus reducing the effect of a single mutation. Or, one gene may specifically prevent the expression of others, called the epistasis effect.

There are a number of different types of mutation.

Chromosome mutations can either change the number of chromosomes in a nucleus or the structure of individual chromosomes.

Gene mutations occur when there is any change to a gene caused by changes in the DNA base sequence. There are two main ways in which this can occur: substitution and insertion/deletion.

- Substituting one base pair for a different base pair will result in a change in one triplet only. This is called **substitution**.

- **Insertion** or **deletion** of one or more base pairs results in a complete shift in the code, often called a **frameshift**.

A stutter mutation is where triplets are repeated many times. Huntington's disease is caused by a stutter mutation.

A substitution is where a different base is used in the base sequence in place of the original base, giving a different triplet. This may result in no change at all if the triplet codes for the same amino acid: this is called a **silent mutation**. Or, it may result in a change in the amino acid coded for and used in the polypeptide chain. This does not usually cause a dramatic change in the characteristic coded for although an altered protein may be the result.

For example, in the following triplet sequence the DNA genetic code is for six amino acids from the β globin polypeptide of haemoglobin.

CTG	ACT	CCT	GAG	GAG	AAG
Leu	Thr	Pro	Glu	Glu	Lys

If the code is altered by changing one base as shown below, the amino acid is unchanged (a silent mutation).

CTG	ACT	CCT	GAA	GAG	AAG
Leu	Thr	Pro	Glu	Glu	Lys

as shown on the next page,

However, if another base is changed, as shown on the next page, the amino acid glutamic acid is substituted by valine (Val). This mutant allele is known as Hb^S, people who inherit two copies of this allele develop sickle cell anaemia.

CTG	ACT	CCT	GTG	GAG	AAG
Leu	Thr	Pro	Val	Glu	Lys

A frameshift is caused when one base pair is deleted or added to the DNA base sequence. The result is not a change to a single triplet but a completely different sequence following the changed bases, since the code will now be read as a different sequence of triplets. Below you see what happens when the base G in the first triplet is deleted, i.e. removed completely.

CTA	CTC	CTG	AGG	AGA	AG
Leu	Leu	Leu	Arg	Arg	

All the triplets after the deleted base have now changed. This will affect the polypeptide chain because there has been a shift caused by the deletion. A similar situation occurs with an insertion, when an extra base is added to the sequence.

CTC	GAC	TCC	TGA	GGA	GAA
Leu	Asp	Ser	STOP codon		

As with the deletion, the extra base has caused a frameshift so that all the triplets after the extra base are altered. In this case a stop codon has formed in the middle of the coding strand, which will prevent the addition of further amino acids to the chain, causing it to be cut short.

The effect of gene mutations on protein production and function

If the change in a protein does result in a different characteristic it may be a **neutral mutation**, because the altered characteristic gives no particular advantage or disadvantage to the organism. An example would be many of the physical characteristics of mammals. Cat coat hair length is determined by two alleles at the same gene locus. Short fur is determined by a dominant allele (**L**) and long hair by its recessive allele (**l**). Neither appear to confer an advantage or disadvantage to cats.

A changed characteristic may, however, cause beneficial or harmful effects for the organism.

Beneficial mutations

Early humans living in Africa had dark skin because they produced high concentrations of the pigment melanin. This offered protection from the harmful UV radiation coming from the Sun but still allowed vitamin D to be synthesised because of the high intensity of the sunlight. Pale skin synthesises vitamin D more easily than dark skin. In such an environment any paler-skinned humans would have survived less well due to the threat of skin cancer and burned skin.

Tip

If the number of base pairs deleted or inserted is three or a multiple of three then a frameshift does not occur.

Tip

As the DNA is a double helix it has two strands. One strand is the coding strand which carries the sequence of bases coding for the amino acid sequence. The mRNA molecule is made by completing a strand complementary to the template strand (the second DNA strand) and so it is a copy of the coding strand.

Once humans moved into cooler temperate climates any dark-skinned humans would have been less able to synthesise vitamin D and so paler-skinned individuals would have been at a selective advantage. Lack of vitamin D causes a number of problems such as rickets, which can cause a deformed pelvic girdle; this is dangerous in females during child birth. Vitamin D deficiency also leads to reduced protection against heart disease and cancers. These examples show how a mutation may be beneficial or harmful depending on the environment.

Harmful mutations

Many genetic diseases are the result of gene mutations. Examples include haemophilia and sickle cell anaemia. In 70% of cystic fibrosis sufferers the mutation responsible for the disease is the deletion of three base pairs in the gene coding for the protein CFTR at amino acid position 508, resulting in the loss of the amino acid phenylalanine. People with cystic fibrosis have a number of symptoms but especially lung and pancreatic problems as a result of particularly thickened mucus. Huntington's disorder is caused by a stutter mutation where the triplet code cytosine-adenine-guanine (CAG) is repeated many times. This triplet codes for the amino acid glutamine and the repeat creates a polyglutamine sequence. Once the repeats reach a critical level, usually more than 36, then there is a high probability that the disorder will occur, usually later in life. It causes an increase in the decay of certain neurones in the brain. Some diseases occur because the mutation results in no protein production at all. Phenylketonuria (PKU) is an example of a disease resulting from this type of mutation in the gene coding for the enzyme needed to metabolise the amino acid phenylalanine. The mutation leads to a build up of phenylalanine, causing serious medical problems including mental impairment.

Proto-oncogenes (e.g. *ras* genes) are growth-promoting genes. They code for growth factors or their receptors. They may be regulatory enzymes that can be switched off once the required cell division has been completed, or may restrict progress through the G1 stage of the cell cycle and so prevent progression if a previous step is incorrect or if the DNA is damaged. However, a single mutation in a proto-oncogene may result in the gene becoming an oncogene (e.g. *ras*^{D}), preventing the gene being switched off and leading to unregulated cell division and a tumour.

TP53 is a tumour-suppressor gene, one of another group of genes with a role in cancer development. It encodes the protein p53. It can become mutated by some of the chemicals in cigarette smoke. Some mutations change p53 into an inactive form that prevents it from halting cell division at the G1 stage when damaged DNA or faulty copying has occurred. It is associated with an increased risk of lung cancer and is linked to cervical cancer when modified by the human papilloma virus, HPV.

Neutral mutations

Neutral mutations offer no selective advantage or disadvantage to the individual. Some neutral mutations are the result of a change in one DNA base that does not cause any difference in the amino acid coded for. In other cases the amino acid may be changed but the resulting polypeptide functions in the same way. Alternatively, if function is changed it makes no difference to survival, giving no advantage or disadvantage to the organism.

Tip

Remember that a tumour-suppressor gene is a normal gene that is required to control cell proliferation and prevent excess cell division. The normal allele is dominant and so only one copy is needed to control cell proliferation. Both alleles must be lost, inactivated or mutated for a tumour to develop.

The ability to taste a chemical called PTC (phenylthiocarbamide) is caused by a mutated allele of the *TAS2R38* gene. PTC is a bitter-tasting chemical but since it does not normally occur in our food it is not an advantage to be able to taste it. However, Brussels sprouts contain a bitter compound that is similar to PTC that some people cannot taste and the ability to do so or not is linked to *TAS2R38*. Those who can taste the bitter chemical do not like eating Brussels sprouts. There may have been some advantage in the past of being able to taste such bitter chemicals, as large amounts of a bitter substance may be harmful and many poisons have a bitter taste.

Test yourself

1 State what is meant by a *gene mutation*.
2 Suggest why a mutation may be described as a silent mutation.
3 The mutation causing sickle cell anaemia only changes one amino acid in the β-globin polypeptide. Explain why this suggests that the mutation is not a base deletion and state the type of mutation it does suggest.
4 In 70% of cases of cystic fibrosis the mutation responsible is a deletion of a triplet of base pairs. What effect will this have on the final polypeptide produced?

Regulatory mechanisms

Tip

Transcription and translation are processes involved in protein synthesis. More details of these processes and explanatory diagrams can be found in *OCR A level Biology 1 Student's Book* Chapter 3.

Key term

Operon A section of functional DNA consisting of a number of structural genes under the control of one promoter.

There are regulatory mechanisms that control whether a gene is expressed at different points in development. Some control gene expression at the transcriptional level, some at the post-transcriptional level and some at the post-translational level. Transcription is the process of copying the gene on the DNA strand to form mRNA. The mRNA will then be edited before it is translated from mRNA codons into a polypeptide chain of amino acids. At this stage the polypeptide chain is ready to be processed in the Golgi apparatus.

Each eukaryotic cell has a full complement of chromosomes and therefore a complete copy of the genome. The genes that have been described so far are **structural genes** that code for polypeptides that function as enzymes, membrane carriers, hormones, etc. There are many **regulatory genes** that code for polypeptides and various forms of RNA that control the expression of structural genes at the three levels listed above.

Control at the transcriptional level

Not all genes are transcribed at any one time. Usually only those genes that code for a required protein are actually transcribed unless there is an error in regulation, as in cancerous cells. Associated with each gene there are a number of DNA base sequences, called the promoter region, which control the expression of the gene. These are usually found a short distance away from the gene: about 100 base pairs before the start of the gene. Transcription is initiated when proteins (transcription factors) bind to the promoter region of a gene. In turn this allows RNA polymerase to attach to the promoter so that transcription begins (see Figure 9.3).

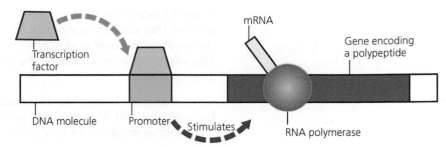

Figure 9.3 The role of a transcription factor and a promoter region in stimulating transcription. The position of a gene encoding a polypeptide is shown in part of a molecule of DNA. In the DNA sequence preceding this gene (referred to as 'upstream' of the gene) is a promoter region. Only if an appropriate transcription factor attaches at this promoter region can RNA polymerase begin to transcribe the gene.

In mammals, the hormone oestrogen is involved in the control of the oestrus cycle and in sperm production. As it is a lipid soluble molecule, it diffuses through the plasma membrane of cells and moves to the nucleus where it binds to an oestrogen receptor. The receptors are transcription factors that can initiate transcription for up to 100 different genes by binding to their promoter regions. When it attaches, the oestrogen changes the shape of the receptor, which then moves away from the protein complex to which it was attached. This allows the receptor to bind to the promoter region for one of its target genes. Now RNA polymerase can bind and begin transcription of that gene (see Figure 9.4).

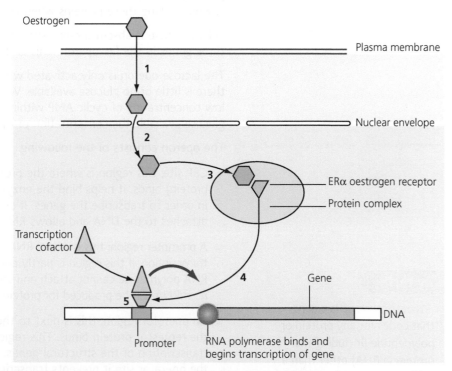

Figure 9.4 Summary of how oestrogen stimulates the transcription of a target gene. (1) Oestrogen diffuses through the plasma membrane of a target cell and then diffuses into its nucleus (2). Here it attaches to an ERα oestrogen receptor that is contained within a protein complex (3). This causes the oestrogen receptor to change its shape and leave the protein complex that inhibits its action (4). The oestrogen receptor can now attach to the promoter region of a target gene (5) where it attracts other cofactors to bind with it. The oestrogen receptor, with combined cofactors, enables RNA polymerase to transcribe its target gene.

In some cases genes may be transcribed and translated at the same time because they code for enzymes involved in the same process, such as the enzymes involved in glycolysis or the Krebs cycle in cell respiration. This ensures efficiency since all the enzymes needed for the process are available.

Genes which are transcribed all the time are known as **constitutive**. However, it would be wasteful to produce enzymes that are not needed and so some enzymes are only produced when they are required.

An example of control at the transcription level is the enzyme β-galactosidase, which hydrolyses lactose in the bacterium *Escherichia coli*. This enzyme is known as an **inducible** enzyme because it will only be produced when lactose is present. Two other proteins are required: lactose permease is needed to allow lactose transport into the cell and another enzyme is needed for the metabolism of lactose. French biologists François Jacob and Jacques Monod described how the regulation of these genes occurs in 1961. They called it the *lac* operon.

The *lac* operon

The *lac* operon is a section of DNA within the DNA loop of some species of bacteria including *Escherichia coli*, consisting of three structural genes and a common promoter. It was the first gene regulatory mechanism to be fully understood. It uses a double mechanism to make sure that the genes in the operon, coding for the proteins needed for lactose metabolism, only produce those proteins when lactose is present. This means energy is not wasted making these proteins when they are not needed.

When lactose is absent a repressor protein stops the transcription of the three genes in the *lac* operon. This is the normal state for the operon.

The lactose operon is only activated when lactose is present and when there is little or no glucose available. When glucose is respired there is a low concentration of cyclic AMP within the cell. When there is a very little glucose available the concentration of cyclic AMP increases.

The operon consists of the following.

- CRP site: this region is where the protein CRP (or cyclic AMP receptor protein) binds. It helps bind the enzyme RNA polymerase to the promoter in order to transcribe the genes. If cyclic AMP (cAMP) is present the CRP attaches to the DNA and allows RNA polymerase to bind.

- A promoter region: this is where RNA polymerase attaches to DNA to start transcription. If this region is partly covered by the repressor protein then RNA polymerase cannot attach and so DNA transcription is prevented and no mRNA can be produced for protein synthesis.

- An operator region: this is next to the structural gene and is where the repressor protein binds. This region acts as a switch by allowing transcription of the structural genes. If the repressor protein is bound to the operator site it prevents transcription and therefore translation of these genes will not occur.

> ### Key term
>
> A structural gene is a gene that codes for any protein or polypeptide (including the protein in RNA) other than a regulatory protein.

- The structural genes: the *lac* operon has three structural genes, next to each other, coding for β-galactosidase, lactose permease and another enzyme. All three are transcribed if the promoter and operator are switched on.

A repressor gene is found on another part of the bacterial DNA, not part of the operon. It codes for a repressor substance: a protein that binds tightly to the operator region of the DNA. In so doing it covers part of the promoter gene in the *lac* operon. It causes the DNA to form a loop, which prevents the RNA polymerase from binding to the promoter and so inhibits the transcription of the three genes. This is the normal state of the operon.

If there is no mRNA produced then no protein synthesis can occur and the three enzymes are not produced.

When lactose is present it acts as an inducer and attaches to the repressor protein and so changes its shape so that it cannot bind to the operator site. As a result the operon is switched on and the genes are transcribed into mRNA and are translated into polypeptides producing the enzymes. The part of the mRNA strand corresponding to each gene has its own ribosomal binding site and so can be independently translated. See Figure 9.5.

(a) High concentration of glucose, low concentration of lactose

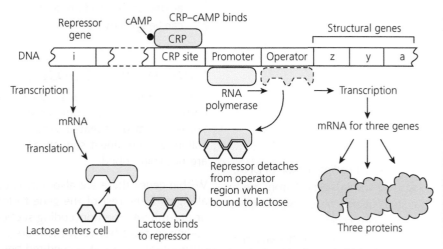

(b) Low concentration of glucose, high concentration of lactose

Figure 9.5 How the *lac* operon functions (a) when lactose is absent in high glucose concentrations and (b) when lactose is present with little glucose.

Example

Lactase is an enzyme needed to digest lactose sugar in milk. It is found in all young mammals. Adults of mammals such as the cat cannot produce this enzyme. They suffer from digestive health problems when drinking milk. Milk for cats that has been treated can be purchased to avoid digestive issues.

1 Suggest why digestive problems would occur when drinking untreated milk and describe how the milk may be treated.
2 Suggest why adult mammals are able to eat yoghurt, which is made from milk, without similar problems.
3 A mutation in a *regulatory gene* allows the lactase protein to be produced in adults with this mutation. From your knowledge of prokaryote gene regulation suggest why this mutation might allow lactase production.

In prokaryotes control of the *lac* operon genes allows uptake and digestion of lactose in the bacterium *E. coli*. The following diagram shows the genes and control regions of the *lac* operon.

Regulatory gene		Promoter	Operator	Structural gene z	Structural gene y	Structural gene a

4 Describe how the three structural genes are switched on when the bacterial food source is changed to lactose.

Answers

1 Digestive problems result because any undigested lactose in the large gut will be fermented by gut bacteria. Treated milk is hydrolysed (with lactase) to remove the lactose.
2 Yoghurt is produced by fermentation of milk by bacteria so the lactose has already been digested.
3 The mutated regulatory gene allows the structural gene to be transcribed and so the enzyme lactase is produced.
4 Lactose binds to the repressor complex and changes the repressor's shape which prevents it from attaching to the operator region. This allows RNA polymerase to bind to the promoter region. The structural genes z, y and a are transcribed to form a single mRNA molecule so that the three proteins can be synthesised.

Control at the post-transcriptional level

In eukaryotic cells there is a lot of non-coding DNA. This DNA does not code for polypeptides, but some may code for functional RNA molecules. Some consist of repeated sequences of bases between the genes, often called tandem repeat sequences or hypervariable sequences. These sections are not transcribed.

Within genes there are also non-coding sections that are transcribed along with the rest of the gene to form a molecule called primary mRNA, or pre-mRNA. The non-coding sections are known as introns and the coding sections between them are called exons. Once transcribed, the primary mRNA is cut and edited before it leaves the nucleus (Figure 9.6). Small nuclear ribonucleic proteins (snRNPs pronounced as 'snurps') are combinations of RNA and proteins. The RNA component of these molecules

Key terms

Intron A non-coding section of DNA, situated within a gene. Introns are edited out after transcription, before translation occurs.

Exon A section of DNA that encodes a polypeptide. Exons are found between non-coding introns.

catalyses cutting and splicing of mRNA. The coding sections, the exons, are now spliced together to make a shorter molecule of mRNA without any introns. This is called **mature mRNA**. This step ensures that only the coding sections of mRNA are used to form proteins by translation. If any part of the introns were included in the mature mRNA the resulting protein would be non-functional. This process ensures only the sequence of bases forming the gene and so coding for the amino acids of a polypeptide is produced. As a result no amino acids are wasted. The introns may in fact be useful. Some introns are further processed after splicing to create **non-coding RNA** molecules which may have a function in gene regulation.

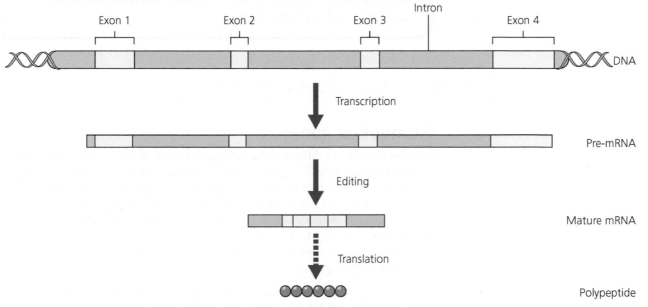

Figure 9.6 During transcription in a eukaryotic cell, the entire base sequence of a gene is copied into a base sequence of mRNA. This produces a molecule of pre-mRNA that contains a copy of the non-coding regions of the DNA (introns) as well as the coding regions (exons). The non-coding regions are edited out before the mRNA leaves the nucleus.

Control at the post-translational level

Once polypeptides have been formed by translation, they are modified in the Golgi apparatus or in the cytosol. Polypeptides may be activated by cyclic AMP (cAMP). This is a derivative of ATP and is formed by the action of an enzyme, adenyl cyclase. Cyclic AMP has a key role in regulation as a second messenger in cells for many processes such as hormone activation, transport into the cells, and activation and regulation of ion channels. Another major role for cAMP is the activation of protein kinases. In eukaryotic cells, cAMP works by activating protein kinase A (PKA), which is an inactive or precursor enzyme. Once it is activated it activates other proteins.

When muscle cells need energy the enzyme glycogen phosphorylase releases glucose from glycogen. The enzyme is activated by cAMP when it attaches to its allosteric site. This activates glycogen phosphorylase by changing the enzyme's shape to expose the active site. The same enzyme is inhibited by ATP and glucose 6-phosphate so only when the cells require energy (as the levels of ATP or glucose are low) will cAMP activate the breakdown of glycogen to glucose.

Test yourself

5 What is the difference between a structural gene and a regulatory gene?
6 a) Explain the control of the *lac* operon when glucose is present without lactose.
   b) Compare this with the control when there is only lactose present.
7 The repressor gene is not part of the *lac* operon. Explain what role it plays in the functioning of the operon.
8 Describe and explain the role of adenyl cyclase in the regulation, control and manufacture of polypeptides.
9 Introns are edited out of a gene sequence after transcription but before translation. Explain why this happens.

Genetic control of body plan development

Homeobox genes

In order for cells to differentiate and specialise for different roles they must be able to regulate which genes are functioning and used. They do this by switching on some genes and switching off other genes. For the process to work correctly in the cell this must be controlled in a specific sequence. The sequence is determined by transcription factors (also called sequence-specific DNA-binding factors). A transcription factor is a protein that binds to specific DNA sequences to control the rate of transcription of that particular gene sequence into mRNA.

Homeobox genes code for transcription factors that regulate transcription by binding to the DNA for specific genes. A homeobox is a sequence of 180 bases coding for 60 amino acids of a part of these proteins. These proteins becomes attached to the DNA at one point and regulate the transcription of other genes, such as those controlling early development in eukaryotic organisms, by turning specific genes on and off in the correct order. Homeobox genes control the early development of animals, plants and fungi, ensuring genes are expressed in the correct order. They help give the basic pattern to the body. They control the segmentation pattern of insects and mammals and the development of wings and limbs. The homeobox genes of *Drosophila melanogaster* are particularly well studied (see Figure 9.7).

Homeobox sequences are all similar because they code for the sequence of amino acids in transcription factors that bind to DNA. These DNA-binding regions must all have the same shape. Any mutations or changes in these sequences lead to organisms that are not viable or are quickly eliminated by natural selection. This is an example of strong negative selection pressure.

Therefore, in many organisms development is genetically regulated by similar homeobox genes, which determine polarity of the whole organism (i.e. the head and tail), the sections of the body and the identity of each section (what each will become and what organs will be present) during early development. Homeobox genes act as master genes controlling which genes will be functioning at each stage of development. If these genes do not function correctly the sequence of development is disrupted and conditions such as antennapedia in *Drosophila* may occur. This is where a pair of legs grow where the insect's antennae should be (Figure 9.8).

Key term

Homeobox A sequence of DNA coding for a protein transcription factor that affects the transcription of DNA into RNA and so controls development.

Figure 9.8 (a) Head of a *Drosophila* mutant fly showing antennapedia (x 55). (b) The head of a normal *Drosophila* (x 70).

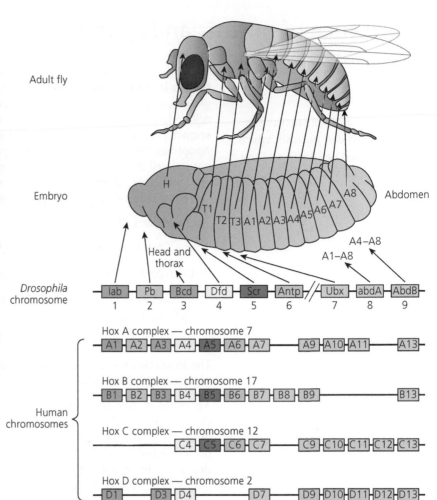

Figure 9.7 Nine homeobox genes in the fruit fly, *Drosophila melanogaster*, control the development of the embryo in sequence: head, thorax and abdomen. The same genes are duplicated and found in four Hox clusters in humans.

Figure 9.9 Roundworms have just one Hox cluster (x 80).

Key term

Hox genes A related group of genes that controls the body plan of an embryo along the axis (head–tail) and later determines the type of structures and where they will be formed. The Hox genes do not actually code for the proteins that make the structures.

Hox genes and their role

The homeobox genes are related to each other in that they are in organised groups known as Hox clusters (these are shown in Figure 9.7). Some simple organisms may have only one or two Hox clusters. For example, roundworms, or nematodes, such as *Caenorhabditis elegans* have just one Hox cluster (see Figure 9.9). Vertebrates have four clusters each of 9–11 genes, which are found on different chromosomes. There is a linear order in the genes in each cluster, which is directly related to the order of the regions of the body that they affect as well as when they are affected. For example, the head–tail axis develops first followed by the type of segmentation in the embryo; that is, when and where structures such as the legs, antennae and wings will develop. As a result any changes in a gene cluster due to mutation generally result in similar changes in the affected regions.

Some of the Hox genes activate genes that initiate apoptosis and so control development by inducing cell death in some areas. In *Drosophila* there is a Hox gene that activates a gene, called *Rpr* (meaning *reaper*), that initiates cell death in the head lobes. This separates the maxillary and mandibular head lobes.

Importance of mitosis and apoptosis in controlling body plan development

Tip

Apoptosis is a useful process and is not the same as necrosis, which is uncontrolled cell death and the subsequent loss of tissue.

Mitosis is the type of cell division in which identical new cells needed for growth, cell replacement or tissue repair are produced. Usually this type of cell division happens a limited number of times and is controlled by proteins known as cyclins and other control measures (discussed later in the chapter). Apoptosis, on the other hand, is programmed cell death, sometimes called natural cell death. In apoptosis older cells that have already undergone approximately 50 mitotic cell divisions, depending on the cell type, are systematically taken through processes leading to cell death.

The processes of apoptosis include:

- the DNA becoming denser and tightly packed
- breakdown of the nuclear envelope and condensation of the chromatin
- biochemical changes and the formation of vesicles containing hydrolytic enzymes
- finally, phagocytosis by phagocytes.

The importance of mitosis and apoptosis is that cells are constantly being replaced and destroyed.

Apoptosis is used to destroy harmful self T lymphocytes during development of the immune system to prevent the cells attacking self body cells. This is important in development because some cells are produced but later may not be needed and so are destroyed as part of the development of the organism. For example, structures such as the fingers and toes initially develop as a single unit and are then separated.

Mitosis is controlled by a number of genes which can be categorised into two groups: those that stimulate cell division are called the proto-oncogenes, while those that reduce cell division are called the tumour-suppressor genes. The tumour-suppressor genes also stimulate apoptosis (the cell death programme) in cells with damaged DNA that cannot be repaired.

Apoptosis, under the influence of tumour-suppressor genes, acts as a protective device for the body because it will destroy and remove any cells that are genetically damaged and could give rise to cancer. (This is in addition to its roles in embryonic development and in maintaining healthy adult tissues.).

Internally, in the cell there are a number of checkpoints during the cell cycle to prevent damaged cells being produced. This control mechanism ensures the cell is ready for the mitosis phase of the cell cycle and repairs any DNA damage. The final checkpoint occurs during metaphase and makes sure the chromosomes and DNA are able to complete mitosis and cell division to produce healthy cells.

This activity is regulated by two groups of protein molecules: cyclins and cyclin-dependent kinases or CDKs. These two groups control the progress of the cell through the cell cycle.

Cyclins are the regulators and the CDKs act as the catalysts once they are activated by the cyclins. When cyclins activate the CDKs they catalyse phosphorylation of certain target proteins, which activates or inactivates them. This moves the cell cycle from one phase to the next. There are different cyclin–CDK combinations, each of which will act on different groups of target proteins. Cyclins are produced at different stages of the cell cycle in response to changes in internal molecular signals.

The genes that regulate and control both the cell cycle and apoptosis are able to respond to a variety of external and internal cell stimuli.

Internal factors

In the cell, apoptosis consists first of the enzymic breakdown of the cell cytoskeleton. The cytoplasm then becomes increasingly dense and tightly packed as the organelles and chromatin condense. At this stage the cell breaks down and forms vesicles that are taken up by the phagocytes to be destroyed without causing any damage to the surrounding cells.

Irreparable genetic damage, RNA decay, internal biochemical changes such as oxidative reactions which lead to cell changes or cellular injury, and the production of cyclin D are all internal factors that affect apoptosis and the cell cycle. These internal factors initiate apoptosis in cells that are experiencing cell stress.

External factors

The process is controlled by a multitude of cell signalling molecules such as cytokines from the immune system, hormones and growth factors, as well as chemicals such as nitric oxide, which has been linked to both induction and inhibition of apoptosis. If the control is not correct and a fault arises so that little apoptosis occurs and cell division is uncontrolled, tumours may form and develop. When there is too much apoptosis then the degeneration of tissues occurs.

Other external factors such as viruses and bacteria, harmful pollutants or ultraviolet light may upset the mitosis/apoptosis balance by damaging cells at a greater rate than they can be replaced or repaired. Too little cell death occurs. The protein cyclin D is produced by cells when they are affected by external growth factors. Cyclin D sets off a chain of events, effectively an enzyme cascade that activates the genes needed to produce two other cyclins, A and B. In order for the cell cycle to proceed all three cyclins must be stimulated.

Stress is now known to cause a large number of problems in the body and one of these is disrupting the cell signalling that is necessary to control the mitosis/apoptosis balance.

There are many different forms of stress and the type will determine how the cell responds. The cell's response to stress often includes the activation of pathways to increase the chance of survival, or to initiate apoptosis. To start with, the cell responds by defending itself and trying to recover from the stressful stimulus. This might involve the heat shock response to the unfolding of proteins. This response involves the production of special proteins that ensure the correct folding of newly-synthesised polypeptides. This response helps cells counteract any damage and increases the chance of cell survival. This response is seen in lower organisms as well and so is a conserved mechanism evolutionarily because it is essential to survival. If the stressful factors continue then cell death pathways are inevitably initiated.

Test yourself

10 Name the type of gene which when mutated gives rise to massive changes in the body plan.
11 Apoptosis was first described in 1842 but it was only fully investigated in 1965 when a distinction was made between necrosis and apoptosis. Explain the difference between the two processes.
12 How is apoptosis controlled?

Investigating a metabolic pathway

Red bread mould, Neurospora crassa, is a fungus that can be cultured on an agar medium containing inorganic salts, biotin (vitamin B7) and a suitable source of carbon, such as glucose. Mycelia of N. crassa are able to synthesise all the amino acids and other biological molecules that they require from the raw materials in this minimal medium. Following treatment with ultraviolet radiation some forms of this fungus fail to grow on minimal medium. Unlike the wild-type fungi, these strains of N. crassa are unable to synthesise certain amino acids and require them in the medium if they are to grow.

A metabolic pathway in N. crassa converts a precursor substance to three amino acids as follows.

precursor substance → ornithine → citrulline → arginine

Of these amino acids, only arginine is used in the production of proteins.

Spores were collected from wild-type fungi (W) and from three strains of N. crassa (X, Y and Z) derived from mycelia that had been irradiated. The spores were spread onto minimal media and media containing 0.005 mM of each of the amino acids in the pathway above. After 5 days of incubation any mycelia growing on the agar were removed, dried and weighed. The results are shown in the table.

Strain of N. crassa	Dry mass of N. crassa after 5 days growth on different media/mg			
	Minimal	Minimal + ornithine	Minimal + citrulline	Minimal + arginine
W	38.1	37.3	35.6	36.3
X	0.9	29.2	37.6	37.2
Y	0.0	0.0	0.0	20.4
Z	1.0	0.8	34.1	37.6

1 Explain what happens to the fungi that are irradiated so that they are unable to make certain amino acids.
2 a) Explain why dry mass was used to measure the growth of the different strains of N. crassa.
   b) Explain why strain W was included.
3 Explain how the results in the table provide evidence for the sequence of reactions in the metabolic pathway that produces arginine.
4 The product of a metabolic pathway often interacts with an enzyme that catalyses a reaction at the beginning of the pathway. State the term used for this form of control and explain the advantages for controlling enzyme activity in this way.

Exam practice questions

1 A sequence of bases in the middle of the coding strand of a gene is as follows:

TACAAAATGCTTGTCCC

Which of the following is a substitution mutation? *(1)*

A TACAAATGCTTGTCCC

B TACAAAACTGCTTGTC

C TACAAAATGCTTATCCC

D TACAAAATGCCTTGTC

2 A gene codes for a particular enzyme. A mutation occurs in the third triplet at the beginning of the first exon of the gene. The mutation involves the insertion of a pair of nucleotides. What is the most likely effect of this mutation on the enzyme?

A The active site of the enzyme remains unchanged.

B The only amino acid to change is in position 3 in the primary structure.

C The primary structure of the enzyme corresponding to the first exon is changed.

D There is no change to the primary structure of the enzyme. *(1)*

3 Bacteria synthesise the amino acid tryptophan from intermediates in the metabolism of glucose. Five different enzymes (E1–E5) catalyse the reactions in the metabolic pathway. The structural genes that code for these five enzymes are controlled by a repressor, which is inactive unless tryptophan is present.

Which would lead to the continuous production of tryptophan?

A A deletion in the structural gene for enzyme 1

B A frameshift mutation in the regulator gene

C The absence of glucose from the medium surrounding the bacteria

D The addition of tryptophan to the medium surrounding the bacteria. *(1)*

4 The following may occur during the life of an animal cell.

1 Cytoplasm divides into two new cells

2 DNA and protein in the nucleus is degraded

3 Nuclear envelope disintegrates and reforms

4 Cytoplasm bulges outwards to form cell fragments

Which occur during apoptosis?

A 1 and 3 only

B 2 and 4 only

C 2, 3 and 4

D 1, 2, 3 and 4. *(1)*

5 The diagram shows the mitotic cell cycle.

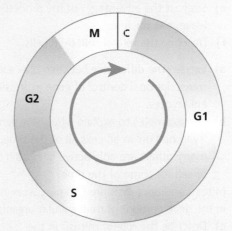

a) i) Identify the stages of the cell cycle indicated by G1, G2, S, M and C. *(5)*

   ii) What term refers to all the stages of the cell cycle except stages M and C? *(1)*

   iii) State what happens in a cell during stage G1 and stage C. *(4)*

Cells respond to external growth factors by producing the protein cyclin D. This cyclin initiates an enzyme cascade that activates genes for cyclins A and E. All three cyclins are needed to stimulate the movement from stage G1 to S and cyclin A stimulates the movement from G2 to M.

b) Outline what happens in a cell in response to an increase in the concentration of:

   i) cyclin E in G1

   ii) cyclin A in G2. *(6)*

c) Proto-oncogenes are genes that code for growth factors, growth factor receptors, cyclins and regulatory enzymes. Explain why they are categorised as proto-oncogenes. *(5)*

6 a) Outline the role of transcription factors in gene expression in eukaryotic cells. *(4)*

The diagram shows the bases on the template (non-coding) strand of a section of a eukaryotic gene.

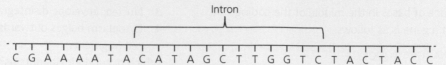

Intron

C G A A A A T A C A T A G C T T G G T C T A C T A C C

b) How is it possible to tell from the diagram that the strand is DNA and not RNA? *(1)*

c) Make diagrams to show
   i) the primary mRNA that is transcribed from the base sequence shown
   ii) the mature mRNA that leaves the nucleus. *(5)*

d) Use the diagrams you drew in answer to (c) to explain how primary mRNA is changed to mature mRNA. *(3)*

e) Suggest the advantages of the processing that occurs to mRNA following transcription and before it leaves the nucleus. *(3)*

f) Describe the role of mRNA in cells. *(3)*

7 a) Explain the difference between pre- and post-transcriptional control of gene expression in eukaryotes. *(5)*

b) Use examples to explain why some genes are 'switched on' in all cells all of the time, but many others are only active in certain types of cell for some of the time. *(5)*

Mitosis and apoptosis are two processes involved in the development of multicellular organisms.

c) Describe the role of mitosis in the development of the body forms of multicellular organisms. *(4)*

One of the most common regulator proteins that inhibits cell division is p53, which is sometimes called the 'guardian of the genome' since it stops the cell cycle until damage to DNA is repaired or, if this fails, stimulates apoptosis. The gene that codes for p53 is *TP53*. If this gene mutates it means that its protein does not inhibit cell division.

d) Outline the role of proto-oncogenes in regulating the cell cycle. *(3)*

e) Cells with damaged DNA produce large quantities of the protein p53. Explain the reasons for this. *(4)*

f) Outline the process of apoptosis in mammalian cells. *(5)*

Stretch and challenge

8 One of the homeobox genes in the fruit fly *Drosophila melanogaster* is *antennapedia* (*antp*) (see Figures 9.7 and 9.8).

The gene *antp* codes for a polypeptide that acts as a transcription factor. The polypeptide has a region that codes for the part that binds to DNA. Different regions of the tertiary structure of the polypeptide are known as domains.

The homeodomain of transcription factors is the region of the polypeptide that binds to DNA. The homeodomain of the antennapedia protein has three α-helices as shown in the figure. Homeodomains of other transcription factors have other shapes.

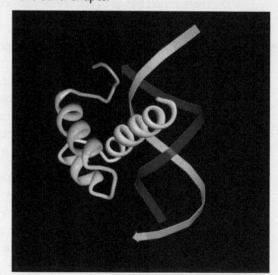

The primary structure of the transcription factor coded by *antp* is shown on the next page. Each letter represents an amino acid. The system (FASTA) for identifying the amino acids found in proteins uses single letters. The amino acid at the N-terminal of the protein is M, or methionine (Met in the three letter code).

```
N-terminal (at position 1)

1         10           20           30           40           50           60
MTMSTNNCES MTSYFTNSYM GADMHHGHYP GNGVTDLDAQ QMHHYSQNAN HQGNMPYPRF

          70           80           90          100          110          120
PPYDRMPYYN GQGMDQQQQH QVYSRPDSPS SQVGGVMPQA QTNGQLGVPQ QQQQQQQQPS

         130          140          150          160          170          180
QNQQQQQAQQ APQQLQQQLP QVTQQVTHPQ QQQQQPVVYA SCKLQAAVGG LGMVPEGGSP

         190          200          210          220          230          240
PLVDQMSGHH MNAQMTLPHH MGHPQAQLGY TDVGVPDVTE VHQNHHNMGM YQQQSGVPPV

         250          260          270          280          290          300
GAPPQGMMHQ GQGPPQMHQG HPGQHTPPSQ NPNSQSSGMP SPLYPWMRSQ FGKCQERKRG

         310          320          330          340          350          360
RQTYTRYQTL ELEKEFHFNR YLTRRRRIEI AHALCLTERQ IKIWFQNRRM KWKKENKTKG

         370
EPGSGGEGDE ITPPNSPQ C-terminal (at position 378)
```

a) i) How many amino acids are there in the protein?

ii) The transcription factor has a tertiary structure but no quaternary structure. Explain why.

iii) Why are the N-terminal and C-terminal of a polypeptide so-called?

The part of the gene that codes for the homeodomain is known as the homeobox.

The primary structure of the homeodomain of the antennapedia protein is shown in the first row of the table below. The table also shows the primary structures of the homeodomains of several homeotic genes from other animals.

Homeotic gene	Source	Sequence of amino acids in the homeodomain coded for by five homeotic genes
antp	Fruit fly	RKRGRQTYTR YQTLELEKEF HFNRYLTRRR RIEIAHALCL TERQIKIWFQ NRRMKWKKEN
Hox-B7	Mouse	RKRGRQTYTR YQTLELEKEF HYNRYLTRRR RIEIAHTLCL TERQIKIHFQ MRRMKHKKEN
bicoid	Fruit fly	PRRTRTTFTS SQIAELEOHF LQGFNRYAPR LADLSAKRAI QTAIVHALCK ERQRRHHIQS
goosecoid	Xenopus toad	KRRHRTIFTD EQLEALENLF QETKYPDVGT REQLARRVHL REEKVEVHFK MRRAKHRRQK
mab-5	Roundworm	SKRTRQTYSR SQTLELEKEF HYHKYLTRKR RQEIASETCH TERQVKIHFQ MRRMKHKKEA

Find the homeodomain of the antennapedia protein in the primary structure of the whole protein shown on the previous page and then answer the questions that follow.

b) i) State the amino acid positions where most of the homeodomain starts and where it ends.

ii) How many amino acids are there in the homeodomain?

iii) Explain why there are 180 base pairs in the homeobox coding for the homeodomain.

iv) Explain how it is possible to predict most of the base sequences of the homeobox region given the data in the table.

v) Suggest why the gene for the antennapedia protein has more base pairs than the 1134 required to code for the polypeptide.

c) i) Compare the homeodomain of the antennapedia protein with the homeodomains of the other proteins in the table.

ii) Suggest an explanation for the comparisons you made in (c)(i).

Chapter 10

Patterns of inheritance

Variation

You will be aware of the vast differences that exist in the living world, from the smallest to the largest organism. For example, notice the differences between the tiny shrew, the elephant and the whale, even though they are all mammals. The snowdrop and the giant redwood are both plants but they are vastly different in both size and appearance. Variation can best be described as the range of differences between organisms. This includes differences within a species, **intraspecific variation** (Figure 10.1) and across different species, **interspecific variation** (Figure 10.2). In this chapter we are mostly interested in intraspecific variation. Understanding how these differences occur requires a look at genetic and environmental factors as well as the roles played by meiotic cell division and sexual reproduction. These processes are significant because they help to create the genetic variation and differences within a species that make it possible for adaptations and changes to occur in response to a changing environment. Without these possibilities the opportunity to adapt is lost, evolution is limited or is prevented and species will become extinct.

> **Tip**
>
> You can revise cell division in Chapter 6 of *OCR A level Biology 1 Student's Book*.

> **Key term**
>
> Variation The range of differences in characteristics between organisms.

Figure 10.1 Intraspecific variation within the domestic dog species: (a) the Jack Russell Terrier, (b) the Border Collie and (c) the Bullmastiff.

Figure 10.2 Differences between species: interspecific variation among living organisms.

Phenotypic variation

All the features of an organism comprise the phenotype – this includes all of the features except its genome. Phenotypic variation is the variation of these features and the phenotype is the expression of an organism's genotype. The effects may be visual, for example colour of hair or eyes, or may be detected by chemical tests such as blood group or the presence of enzymes. Some features are only determined by the genotype, for example blood groups; some are determined by the interaction of genotype with the organism's environment and others may be the result of the environment alone, for example scars from past injuries. Environment may influence the final expression of the genotype to such an extent that two individuals with the same genotype raised in different enviroments may show phenotypes that are considerably different.

How the environment affects variation

Both the genotype and the environment contribute to the phenotype. The genotype is the genetic composition of an organism. The genotype determines the characteristics of the organism in terms of the alleles for a particular gene that it contains. In diploid cells, which carry two copies of every chromosome, the pair of alleles present at a given locus may be identical, which is called homozygous. When the alleles at a locus are different it is called heterozygous. Each allele in a pair of alleles may be dominant or recessive.

Sometimes the environment affects the phenotype in exactly the same way as the effects that would be seen by a genotypic change. This effect is

159

called phenocopy and can be observed in a number of different organisms. Some varieties of tomato plant have a purple stem which only develops in cold temperatures. In warmer temperatures these plants develop a normal green stem indistinguishable from varieties of tomato which have genetically determined green stems.

In the fruit fly *Drosophila melanogaster* normal body colour is grey with black edges on each segment, giving a striped appearance. A genetic mutant has yellow body colour, a genotypic character that is expressed regardless of the environment. However, if the larvae of normal flies are given a diet of silver salts they develop yellow bodies regardless of their genotype. So the genetically grey body flies can become yellow bodied if fed a different diet, probably as a result of disrupting the enzymes involved in the pigment production pathway.

In the Himalayan rabbit the normal coat colour in a moderate climate is a white coat with black points (tail, nose and ears), making them phenotypically different from genetically determined black rabbits. However, the markings change with both temperature and age. Cold temperatures increase the areas of dark markings, the depth of colour of the coat and the amount of 'smut' on the coat (smut refers to dark hairs growing through the middle of white areas of the coat). Baby Himalayans are especially sensitive to temperature and develop dark bands on their fur in cold temperatures. This effect is a result of the complete inability to produce pheomelanin, and the ability to produce eumelanin only when the temperature falls below a certain point. The rabbits resemble genetically determined black rabbits. Hence the environment has modified the phenotype of the Himalayan rabbit as a phenocopy of genetic black rabbits without any change in the genotype. This is likely to be a result of the action or inaction of an enzyme determining black fur.

Genetic factors

There are two types of variation in phenotype

- continuous variation
- discontinuous variation.

The difference between the two types of variation is a result of the number of genes involved in determining a characteristic and the expression of the alleles of those genes. The two types of variation are compared in Table 10.1.

Table 10.1 Comparing continuous and discontinuous variation.

Feature	Discontinuous variation	Continuous variation
Definition	Features cannot be measured across a complete range so they form distinct classes or categories: discrete or categorical data (qualitative data)	Features can be measured across a complete range from one extreme to the other; data collected are quantitative data
Gene locus	Usually only one but there may be a very small number	Many loci and may be on different chromosomes
Number of alleles	Often just one pair of alleles (monogenic) but there may be a very small number	Many genes contribute to the inheritance (polygenic); each has its own alleles
Effect on phenotype	The feature is either present or absent; the differences are discrete categories	There are many intermediates between the extremes, e.g. between tallest and shortest
Environmental influence	Environment has little influence	Environmental factors have a significant effect
Example	Ability to roll the tongue, human blood groups	Height in humans, milk yield in cattle

However, there are a number of different factors that affect expression of the genotype, which in turn will affect the phenotype. Although the genetic potential may be present, it may be modified or altered completely by environmental factors (Figure 10.3). As an example take the effect of diet on a child: a well-balanced diet should result in a child growing to his or her full potential, but a poor diet may limit growth due to a lack of proteins and vitamins. Plants that are kept in the dark or are deficient in magnesium become yellow because chlorophyll synthesis is halted or slowed down. This condition is known as **chlorosis**. Seedlings grown in the dark are chlorotic, but also have long stems with small, curled leaves yellow in colour, which is known as **etiolation**. They will be unable to produce chlorophyll without enough light. If given light, dark-grown plants will begin producing chlorophyll, become green and will show normal growth. The genetic potential is present in the plants but it may not be developed due to environmental conditions. Albino plants such as *Arabidopsis thaliana* mutant *ppi2* (thale cress) cannot produce chlorophyll because they do not have the genetic potential to do so. However, they may survive if there is an external supply of sucrose.

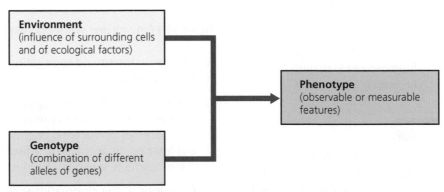

Figure 10.3 The phenotype is the result of an interaction between the environment and the genotype.

Genetic variation as a result of sexual reproduction

Genetic variation exists because there are differences in the genes due to the presence of different alleles. It is increased by sexual reproduction when DNA from one individual is combined with that from another organism during fertilisation. To confer any selective advantage, or disadvantage, the differences must be capable of being passed on and so the offspring must be fertile. For this to be so, the parents must be from the same species.

Meiosis during sexual reproduction generates variation by

- crossing over between maternal and paternal chromosomes in meiosis I

- independent assortment in meiosis I of maternal and paternal chromosomes in a homologous pair, which later separate

- segregation between sister chromatids in meiosis II, which is also random

- the whole process halving the chromosome number to produce haploid gametes that are then restored to a diploid cell by fertilisation with another gamete carrying different alleles

- the occurence of chromosome mutations

Tip

Revise the effect of random mating and fusion of gametes on variation in the section on meiosis in Chapter 6 of *OCR A level 1 Student's Book*.

Tip

You should revise nuclear division in Chapter 6 of *OCR A level Biology 1 Student's Book*.

Test yourself

1 Describe the best form of graphical representation for the two types of variation
   a) discontinuous variation
   b) continuous variation.

2 What are the main differences between cells produced by mitosis and those produced by meiosis that account for the different levels of variation in living things?

3 Why is variation so important in living things?

4 What is the significance of chiasmata in
   a) cell division
   b) variation?

During prophase I, crossing over occurs (Figure 10.4). This is an exchange of some alleles between the chromatids of the homologous chromosomes; essentially, they are redistributed. This happens because chromatids from opposing homologous chromosomes form attachments called chiasmata. These attachments act as breaking points when the chromosomes separate. The broken sections rejoin the non-sister chromatid from the other chromosome of the homologous pair, creating new combinations of alleles on the chromatids.

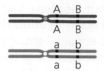 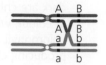 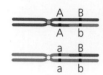

Pairing of bivalent in early prophase I

Chiasma forms between non-sister chromatids in prophase I

Breakage and exchange of parts of non-sister chromatids

Genotypes of gametes =   (AB) (Ab) (aB) (ab)

Figure 10.4 Crossing over during prophase I.

Random assortment and segregation occur at two different places in meiosis, as follows.

1 At metaphase I the homologous chromosome pairs align themselves alongside one another in random orientation, with no relationship to their paternal or maternal origins. This is called random assortment and later on, in anaphase I, it leads to random segregation when each chromosome is separated from its homologous pair.

2 At meiosis II, the chromatids line up independently at the equator and are also randomly assorted and segregated. The chromatids separate to opposite poles, creating four groups of chromosomes that will be contained in four haploid cells. If there is crossing over in meiosis I, then none of the chromatids separating is genetically identical to any other.

Both gene and chromosome mutations occur randomly and increase variation in offspring because mutations cause a permanent change in the genotype. During interphase the DNA replicates. During this process mutations may arise due to incorrect copying of DNA. It can occur in either type of cell division: mitosis or meiosis.

When a female gamete is fertilised by a male gamete the chromosome number is restored to the full diploid number. The maternal chromosomes from the egg cell form one half of each homologous chromosome pair, with the paternal chromosomes from the sperm forming the other half of each pair. Therefore, two different sources of genetic material are brought together, further increasing variation. The process of fertilisation is also random, with any one of millions of different sperm produced by the male of the species possibly being the one that fertilises the egg. This is another source of variation.

Monohybrid inheritance

Parental phenotypes	Long wing (wild type) × Vestigial wing	
Parental genotypes	WW	ww
Parental gametes	(W)	(w)

F₁ genotype	Ww
F₁ phenotype	All long wing (wild type)

F₁ phenotypes	Long wing × Long wing	
F₁ genotypes	Ww	Ww
F₁ gametes	(W) (w)	(W) (w)

		Male gametes	
		(W)	(w)
Female gametes	(W)	WW	Ww
	(w)	Ww	ww

F₂ genotypes	WW	Ww	Ww	ww
F₂ phenotypes	Long wing	Long wing	Long wing	Vestigial wing

F₂ phenotypic ratio	3 long wing : 1 vestigial wing

Figure 10.5 An example of a monohybrid cross showing inheritance of wing length in *D. melanogaster*.

Key term

Codominance When two alleles of the same gene influence the phenotype of a heterozygous organism because both are dominant over any other alleles of the gene but neither allele is dominant over the other.

Incomplete dominance When neither of two alleles of a gene dominates so there is blending of the two to form an intermediate.

A genetic diagram allows us to see the inheritance of genes and their alleles. In *Drosophila melanogaster* wing length is determined by two alleles at a gene locus: one for normal long wings (**W**) and one for vestigial or small, reduced wings (**W**). These two alleles separate during meiosis because they are on different chromosomes of a homologous pair. The allele for normal long wings is dominant and so in a genetic diagram of the cross we would use an upper-case **W** to represent it. For recessive alleles we use lower case; hence **W** for vestigial wings.

Monohybrid crosses

A monohybrid cross, or a monogenic cross, is the inheritance of a single pair of alleles of a single gene. In cases such as the example outlined above, where one allele is dominant and the other is recessive, the expected phenotypic ratio in the F_2 generation is 3:1 (see Figure 10.5).

A monohybrid cross may also involve incomplete dominance. In such cases neither allele is dominant, although they are both present at the same gene locus. The phenotype is a mixture of the effects of the two alleles. So, for example, if two alleles for petal colour in flowers were red and white, the heterozygote would be pink.

A monohybrid cross may also involve codominance. There are a number of examples of codominance, such as coat colour in cattle and human blood group. All of these are also examples of discontinuous variation. In such crosses the alleles are represented by a capital letter and different superscripts. Codominance changes the expected phenotypic ratio in the F_2 generation to 1:2:1 (see Figure 10.6).

Genetic diagrams

There are a number of basic rules to follow when drawing genetic diagrams.

- Always start with the parental phenotypes and follow with the parental genotypes where known.

- When there are two alleles at a locus, one dominant and one recessive, always aim to represent a gene by a single letter. In some cases a second may be added to avoid confusion. Use upper case for the dominant allele and lower case for the recessive allele. Choose a letter that is easily distinguished.

- If a gene has more than two alleles (multiple alleles) in the population use a single upper-case letter for the gene and superscript letters for the alleles. For example, the human ABO blood groups use I for the gene with the alleles given as I^o, I^A and I^B.

- Codominance may be shown with the gene in upper case and the codominant alleles as superscript letters (as for the A and B blood groups, I^A and I^B).

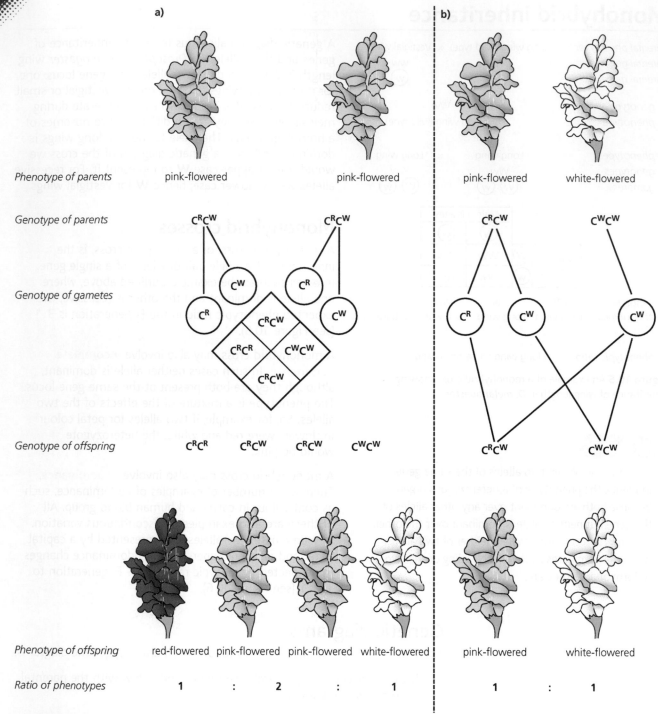

a)

Phenotype of parents	pink-flowered
Genotype of parents	C^RC^W
Genotype of gametes	
Genotype of offspring	C^RC^R C^RC^W C^RC^W C^WC^W
Phenotype of offspring	red-flowered pink-flowered pink-flowered white-flowered
Ratio of phenotypes	1 : 2 : 1

b)

Phenotype of parents	pink-flowered white-flowered
Genotype of parents	C^RC^W C^WC^W
Genotype of gametes	
Genotype of offspring	C^RC^W C^WC^W
Phenotype of offspring	pink-flowered white-flowered
Ratio of phenotypes	1 : 1

The letter C represents the gene; the alleles are represented by C with either the superscript ^{R} or ^{W}.
C^R = red
C^W = white
This indicates that the alleles are neither dominant nor recessive

Figure 10.6 Incomplete dominance in snapdragon plants. (a) Crossing two pink-flowered snapdragon plants and (b) crossing a pink-flowered snapdragon with a white-flowered snapdragon.

Dihybrid crosses

The inheritance of two different characteristics in an organism is called dihybrid inheritance. In many cases these characteristics are each controlled by a gene, consisting of a pair of alleles, inherited independently from each other because the two genes are located on different chromosomes. Such characteristics are said to be unlinked. At meiosis they separate independently from other characteristics. When the genes for two different characteristics are found on the same homologous pair of chromosomes, the genes are said to be linked, which means these will not separate independently at meiosis.

Genetic diagrams can be used to show inheritance of two different characteristics in an individual: this is called a dihybrid cross. We will first look at unlinked genes. These are genes that are on different non-homologous chromosomes. During meiosis I the alleles of one gene may segregate with either of the alleles of the other gene. In the example shown in Figure 10.7 body colour and wing type are the two characteristics being investigated, with **E** representing the normal, dominant grey or wild-type body colour and **e** representing the black or ebony body, which is recessive. Wings are represented by N for the wild-type normal wing, which is dominant, and n is used for the recessive, curly wing.

Wild type fruit flies have long wings and grey bodies. Some flies have curly wings and ebony (black) bodies. When pure bred flies of both types are crossed all the offspring show the wild type phenotype. When these F_1 flies are crossed all the combinations of features will appear in the offspring, in the following numbers (not always in these exact numbers):

Wild type (long wings and grey body)	**650**
Long wings and ebony body	**198**
Curly wings and grey body	**225**
Curly wings and ebony body	**68**

This approximates to a ratio of 9:3:3:1. But look carefully and you can see that the ratio of long wings to curly wings is about 3:1 and the ratio of grey body to ebony body is also about 3:1. This suggests that the two genes have segregated independently of one another and are therefore on different chromosomes. They are described as unlinked gene loci (linked gene loci are explained on next page).

The Punnett square shows nine different genotypes, but because of dominance there are only four different phenotypes in the ratio 9:3:3:1 which approximates to the numbers given above.

The explanation for the results in this dihybrid cross is **independent assortment**. The genes segregate in meiosis independently of one another because they are on different chromosomes.

Linkage is the term used to describe genes that are situated on the same autosomal chromosome (although at different gene loci). Linkage disturbs the usual phenotypic ratios in the offspring that are shown in the diagrams here. It also reduces variation because the linked genes do not separate at meiosis. Instead, they *behave* as if they are a single gene

Key terms

Wild type The phenotype of the standard form of a species as it exists in nature. Therefore, the wild type is considered to be the 'normal' allele at a locus, as against a non-standard, 'mutant' allele.

Key terms

Linkage When two or more genes are located on the same chromosomes.

Autosomal Referring to an autosome, any chromosome other than a sex chromosome.

Parental phenotypes Normal wing, grey body × Curly wing, ebony body
Parental genotypes   NNEE   nnee
Parental gametes   (NE)   (ne)

F_1 genotype   NnEe
F_1 phenotype   All normal wing, grey body

F_1 phenotypes   Normal wing, grey body   ×   Normal wing, grey body
F_1 genotypes   NnEe   NnEe
F_1 gametes   (NE)(nE)(Ne)(ne)   (NE)(nE)(Ne)(ne)

		Male gametes			
		(NE)	(Ne)	(nE)	(ne)
Female gametes	(NE)	NNEE	NNEe	NnEE	NnEe
	(Ne)	NNEe	NNee	NnEe	Nnee
	(nE)	NnEE	NnEe	nnEE	nnEe
	(ne)	NnEe	Nnee	nnEe	nnee

F_2 genotypes   N_E_   N_ee   nnE_   nnee
F_2 phenotypes Normal wing, Normal wing, Curly wing, Curly wing,
                grey body   ebony body   grey body   ebony body

F_2 phenotypic ratio   9   :   3   :   3   :   1

Figure 10.7 A dihybrid cross involving two unlinked genes on different chromosomes: body colour and wing type in *D. melanogaster*.

Tip

The dash indicates that the allele could be dominant or recessive

Tip

Independent assortment is shown in Figure 6.14 on page 108 of *OCR A level Biology 1 Student's Book*.

Key term

Sex linkage Referring to genes present on the X chromosome. It has nothing to do with sex determination.

Hemizygous The state of genes on the X chromosome in the sex with the XY chromosome pair. In humans and *Drosophila* this is the male. In these organisms, sex-linked genes, i.e. those on the X chromosome, in effect have only one allele for a characteristic, which will be expressed whether recessive or dominant.

and will not separate during meiosis I. This is true unless crossing over has occurred somewhere on the chromosome between the two gene loci. In this case the number of individuals showing the different pairing of characteristics will be much lower than the number showing the original linked characteristics.

A **test cross** is a general cross to find the genotype of an unknown fly by crossing it with a homozygous recessive fly (a known genotype), for example, the vestigial-winged ebony-bodied flies, which are homozygous recessive for both genes. In this case the two genes are not linked and so they are on different chromosomes. A cross between these flies with long-winged grey-bodied flies, heterozygous for both genes, such as those F_1 flies produced from a dihybrid cross, will give the expected ratio of 1 : 1 : 1 : 1.

However, a similar cross involving two linked genes will give a 1 : 1 ratio if there is no crossing over because the genes are on the same chromosome (see Table 10.2). If there is some crossing over, it means that there is some distance between the linked loci on the chromosome. We call this **partial linkage**.

Sex linkage occurs when genes are present on the X chromosome. (Note that *sex linkage* has nothing to do with *sex determination*.) The X chromosome contains hundreds of genes that code for many important characteristics. This is important because the Y chromosome carries few genes apart from those related to sex determination. Therefore, a male with a single recessive allele on the X chromosome will have that allele expressed as he is hemizygous. The female has two X chromosomes, and

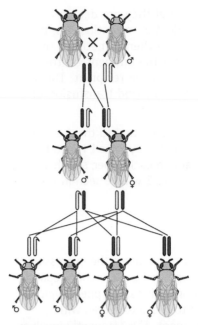

Figure 10.8 Diagram of sex-linked inheritance of the white-eyed mutant in *D. melanagaster*.

Table 10.2 Hypothetical results for test crosses involving female fruit flies heterozygous for two gene loci. In crosses 1, 2 and 3 the gene loci, A/a and B/b are linked on the same chromosome. In cross 4, the gene loci are on different chromosomes.

Cross	Genotypes and phenotypic ratios				Explanation
1	ABab 52		abab 48		Complete linkage: no crossing over
2	ABab 46	Abab 5	aBab 6	abab 43	Partial linkage: genes close together
3	ABab 34	Abab 12	aBab 16	abab 38	Partial linkage: genes are further apart than in cross 2
4	AaBb 24	Aabb 26	aaBb 29	aabb 21	Unlinked: independent assortment

so carries two alleles (one on each X chromosome) for each sex-linked characteristic. Thus there is a reduced chance that a recessive allele of a sex-linked gene will have an effect in females, because she would need to carry the recessive allele on both chromosomes for it to be expressed. The recessive allele will be masked if there is a dominant allele at the same locus on the other X chromosome. When females have a recessive allele masked by a dominant allele, they are referred to as **carriers**. The red-eyed female with the $X^R X^r$ genotype in Table 10.3 is a carrier.

Table 10.3 Genotypes and phenotypes for eye colour in *D. melanogaster*.

Males		Females	
Genotype	Phenotype	Genotype	Phenotype
$X^R Y$	Red eyes	$X^R X^R$	Red eyes
$X^r Y$	White eyes	$X^R X^r$	Red eyes
		$X^r X^r$	White eyes

Multiple alleles

So far we have looked at genes with two alleles. The pairs of alleles have either been dominant and recessive or may have been codominant. However many genes have more than two alleles, although a diploid individual will only inherit two of these many possible types.

Some crosses involve three or more alleles for a gene at the same locus. An individual will only have two of all the possible alleles. Human blood groups are an example of multiple allelism, as there are three different alleles: I^o, I^A and I^B (see Table 10.4).

There are three alleles at the human ABO blood group locus, which is on chromosome 9. The gene controls the production of antigens on red blood cells. They are called antigens because if red cells from one

Table 10.4 The alleles for human blood group.

Blood group	Possible genotypes
A	$I^A I^A$ or $I^A I^o$
B	$I^B I^B$ or $I^B I^o$
AB	$I^A I^B$
O	$I^o I^o$

person are injected into an experimental animal they are detected as non-self ('foreign') and stimulate the production of antibodies. If red blood cells are injected into another human then sometimes they stimulate the same response, but not always if the antigens on the red blood cells of the donor are compatible with the recipient. This is why it is important that blood is typed before a blood transfusion or a transplant operation.

The alleles I^A and I^B are both dominant to the allele I^o and so the A and the B characteristic will be the phenotypes of the heterozygotes $I^A I^o$ and $I^B I^o$ respectively. However, A and B are codominant to each other, so when they are present together characteristics of both A and B will be seen in the phenotype, forming the fourth blood group AB.

Epistasis

Epistasis is when two or more genes interact to influence the expression of the same phenotypic characteristic. Although these genes may separate independently during meiosis, the phenotypic ratios are different to those expected with independent assortment. This frequently involves an enzyme pathway. Each gene codes for a single polypeptide that becomes an enzyme. If an individual has two copies of a recessive allele that codes for a non-functioning enzyme the metabolic pathway stops and the characteristic is not expressed. Examples are given in Figure 10.9 and Table 10.5.

Table 10.5 The effect of epistasis on the phenotypic ratios in crosses between plants shown in examples 1 to 4 in Figure 10.9 when plants heterozygous for both loci (AaBb) are self-fertilised.

Example		Ratios of phenotypes in F_2			
		A_B_	A_bb	aaB_	aabb
Genes do not interact, i.e. not epistasis		9	3	3	1
1	Recessive epistasis	9 purple	3 red	4 white	
2	Recessive epistasis	9 purple	7 white		
3	Dominant epistasis	12 purple		3 red	1 white
4	Dominant epistasis	15 purple			1 white

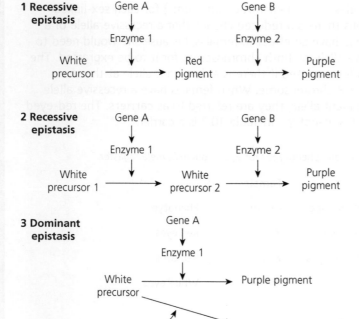

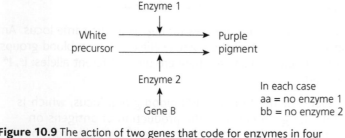

Figure 10.9 The action of two genes that code for enzymes in four different hypothetical pathways in production of flower pigments. In all cases the genes are not linked.

Table 10.5 shows the effect of different types of epistasis on the phenotypic ratios obtained in the F_2 generation and Figure 10.9 shows the enzyme pathways involved.

- Example 1: Both dominant alleles **A** and **B** are needed to produce the purple pigment, but in plants with the genotype **A_bb** a red pigment is produced as no functioning enzyme 2 is made. When there are only recessive alleles, **aa**, neither **B** (to give purple) nor **b** (to give red) is expressed.

- Example 2: Both dominant alleles **A** and **B** are needed to produce any pigment (purple). If one locus is homozygous recessive (**aa** or **bb**) then no pigment is produced.

- Example 3: Enzyme 1 competes more successfully than enzyme 2 for the precursor substrate. This has come about by gene duplication — there are two gene loci that code for variants of the same enzyme (there are many 'families' of enzymes like this).

- Example 4: The two duplicate enzymes coded by dominant alleles both act to produce the same pigment.

Using the chi-squared (χ²) test

The chi-squared (χ²) statistical test is used to test categorical data to see if there is any difference between expected data and observed data. It is a way of confirming experimental results against what is expected to happen, or of rejecting a null hypothesis if the results are too different from the expected based on that hypothesis. The expected data are those predicted in an experiment, whereas the observed data are the actual data that have been collected during the experiment. In order to do this a hypothesis is developed. The hypothesis (known as the null hypothesis) is a statement about the system we are studying which allows us to make predictions about the expected numbers in each category.

A student completed a test cross between two groups of fruit flies showing two different unrelated characteristics: wing length and body colour. The expected ratio among offspring for a dihybrid test cross is 1:1:1:1 assuming the loci are not linked and that independent assortment has occurred. A chi-squared test should be used to decide whether the observed ratio in the experiment is as expected based on the hypothesis that these genes are not linked.

The chi-squared test provides a probability that any differences that occur are due to chance. The first step is to calculate the chi-squared value using the formula

$$\chi^2 = \Sigma \frac{(O - E)^2}{E}$$

Where Σ means sum of...

O means the observed value

E is the expected value.

The calculated chi-squared value is then compared to a critical value provided in a table of probabilities. Usually biologists use the 5% (or 0.05) probability value so this is the column in the table to compare your calculated value against. The correct row is found using the number of degrees of freedom (df). These are found using $n - 1$ (where n is the number of categories).

In order to understand this better, follow the example of chi-squared calculations outlined in Table 10.6.

Table 10.6 Results of a test cross in *D. melanogaster* (O) and the subsequent chi-squared calculations to test whether the observed data match the expected results.

Category	O	E	O – E	(O – E)²	(O – E)²/E
Long wings/grey body	70	81	−11	121	1.49
Vestigial wings/grey body	91	81	10	100	1.23
Long wings/ebony body	86	81	5	25	0.31
Vestigial wings/ebony body	77	81	−4	16	0.20
Totals	324	324			$\chi^2 = 3.23$

In this example, the calculated χ^2 value is 3.23. There are four categories so the number of degrees of freedom ($n - 1$) is 3. You can look up the chi-squared probability in Table 15.4 on page 307. Tables like this give you a χ^2 value for each number of degrees of freedom and probability value.

At 3 degrees of freedom and a probability of 0.05 or 5% the critical value is 7.82. This is the critical value where there is a 5% probability that the results could have been obtained by chance. Our value is below this (by a good margin), so it is clear that the difference between the results of this experiment and the expected outcome is very likely to be due to chance or random influences. The difference between the observed and expected values is not statistically significant and the null hypothesis cannot be rejected.

If the calculated value had been greater than the critical value it would have been significant at the 5% probability level and the result considered significant. The null hypothesis would then be rejected and the decision made that the difference between the results and the expected outcome was due to a factor other than chance, such as autosomal linkage.

Tip

The steps to follow in the chi-squared test are described in full in Chapter 16.

Example

Genetics with fruit flies

Two mutant conditions in *D. melanogaster* are vestigial (small) wings and sepia (brown) eyes. Some students investigated the inheritance of the genes for wing length and eye colour. The letter **W** represents the allele for long wing and the letter **w** represents the allele for vestigial wing. The letter **E** represents the allele for red eyes and the letter **e** represents the allele for sepia eye colour.

The students crossed pure-breeding wild-type flies, which had long wings and red eyes, with flies homozygous for the mutant alleles. All the F_1 offspring had the wild-type phenotype.

Next the students crossed the F_1 flies among themselves. The F_2 offspring were scored and counted to give the results shown in Table 10.7.

Table 10.7 Phenotypes of the F_2 offspring.

Phenotype	Number of flies
Wild type	1577
Long wings, sepia eyes	568
Vestigial wings, red eyes	512
Vestigial wings, sepia eyes	158

1 Using the symbols given above, draw a genetic diagram to explain the results in the F_1 and F_2 generations. Construct a table to show the expected ratio in the F_2 generation if the genes are unlinked.

2 The students made the hypothesis that the two genes are on different chromosomes. Use a chi-squared test to see if these results conform to the results that would be expected if the genes are on different chromosomes.

Use the table of chi-squared probabilities in Table 16.4. Show all your working. Explain the results.

Next the students investigated the inheritance of the wing length gene and another eye-colour gene, cinnabar, **Cn/cn**. Pure-breeding fruit flies with long wings and cinnabar-coloured eyes were crossed with flies with vestigial wings and red eyes. All the F_1 offspring had the wild-type phenotype. The F_1 female flies were crossed with males homozygous for the mutant alleles.

3 Using the symbols given above, make a genetic diagram and draw a table of phenotypes to show the expected results.

The offspring were scored and counted to give the results shown in Table 10.8.

Table 10.8 Results of the crosses between F_1 female flies and males homozygous for the mutant alleles.

Phenotype	Number of flies
Wild type	136
Long wings, cinnabar eyes	15
Vestigial wings, red eyes	14
Vestigial wings, cinnabar eyes	119

4 A chi-squared test was performed on these data, giving a calculated value for χ^2 of 181.83. (If you have used a spreadsheet program or an online program to calculate χ^2 then you could use the figures to verify this.) Use this value of χ^2 (Chi squared) to estimate the probability that the results of the cross depart significantly by chance from the expected ratio. State the conclusion that can be drawn from this second cross. Explain your answer.

The students repeated the cross with male F_1 flies and female flies showing both mutant phenotypes. The results are shown in Table 10.9.

Table 10.9 Results of the crosses between F_1 male flies and females homozygous for the mutant alleles.

Phenotype	Number of flies
Wild type	144
Long wings, cinnabar eyes	0
Vestigial wings, red eyes	0
Vestigial wings, cinnabar eyes	141

5 Suggest why these results differ from the results with female F_1 flies.
6 Explain the advantage of using fruit flies with genotypes that are homozygous for the mutant alleles in these crosses.

Answers

1 The question asks for the $F_1 \times F_1$ cross, including the parental cross which gave rise to the F_1.
Table 10.10 shows the phenotypes for the genotypes shown in the Punnett square in Figure 10.10.

Parental phenotypes	Wild type $\times$	Vestigial wing, sepia eyes
Parental genotypes	WWEE $\times$	wwee
Gametes	(WE) +	(we)
F_1 genotype	WwEe	
F_1 phenotypes	All wild type (long wings and red eyes)	

F_1 genotypes   WwEe $\times$ WwEe
F_1 gametes   (WE) (We) (wE) (we)   (WE) (We) (wE) (we)

		Female gametes			
		(WE)	(We)	(wE)	(we)
Male gametes	(WE)	WWEE	WWEe	WwEE	WwEe
	(We)	WWEe	WWee	WwEe	Wwee
	(wE)	WwEE	WwEe	wwEE	wwEe
	(we)	WwEe	Wwee	wwEe	wwee

Figure 10.10 The $F_1 \times F_1$ cross.

Table 10.10 Phenotypes for the genotypes in the cross. An underscore (_) indicates that the allele can be **W** or **w** if in the first pair or **E** or **e** if in the second pair.

Genotype	Proportion out of 16	Phenotype
W_E_	9	Wild type (long wings, red eyes)
W_ee	3	Long wings, sepia eyes
wwE_	3	Vestigial wings, red eyes
wwee	1	Vestigial wings, sepia eyes

2 The null hypothesis is: if the gene loci are on different chromosomes, there will be no significant difference between the observed numbers of fruit flies of different phenotypes and the expected numbers derived from a 9:3:3:1 ratio.
$\chi^2 = 5.39$ (Table 10.11, 3 significant figures). There are three degrees of freedom ($n - 1$). The critical value for $p = 0.05$ and three degrees of freedom is 7.82; 5.39 is less than 7.82 which indicates that the probability of

getting this result by chance is between 0.5 and 0.1. Therefore the observed results are not significantly different from the expected results and we cannot reject the null hypothesis.

The two gene loci are on different chromosomes/ are not linked on the same chromosome. These two genes show **independent assortment**. There are two different arrangements of the homologous chromosomes in metaphase I of meiosis. The probability that a cell dividing by meiosis will show one of these arrangements of homologous chromosomes is 0.5 (50%, ½). This gives rise to four different types of gamete (as shown in the Punnett square in Figure 10.10). The probability of each type of gamete is 0.25 (25%, ¼). Fertilisation of gametes occurs at random. The probability of each type of fusion is shown by the cells in the Punnett square = 0.25 × 0.25 = 0.0625 (6.25%) or 1/16. There is dominance at each locus so there are only four different phenotypes (as shown in Table 10.10).

Table 10.11 Chi-squared calculations to test whether the observed data match the expected results.

Category	O	E	$O - E$	$(O - E)^2$	$(O - E)^2/E$
Wild type	1577	1583.438	−6.4375	41.44141	0.026172
Long wings, sepia eyes	568	527.8125	40.1875	1615.035	3.059865
Vestigial wings, red eyes	512	527.8125	−15.8125	250.0352	0.47372
Vestigial wings, sepia eyes	158	175.9375	−17.9375	321.7539	1.828797
Totals	2815	2815			$\chi^2 = 5.39$

3 F_1 genotypes WwCncn × wwcncn

F_1 gametes (WCn) (Wcn) (wCn) (wcn) + (wcn)

	Female gametes			
	(WCn)	(Wcn)	(wCn)	(wcn)
Male gametes (wcn)	WwCncn	Wwcncn	wwCncn	Wwcncn

Figure 10.11 The $F_1 \times F_1$ cross.

The cross in Figure 10.11 shows that the expected ratio is 1:1:1:1. Table 10.12 shows the phenotypes for the genotypes shown in the Punnett square in Figure 10.11.

Table 10.12 Phenotypes for the genotypes in the cross.

Genotype	Proportion out of 4	Phenotype
WwCncn	1	Wild type (long wings, red eyes)
Wwcncn	1	Long wings, cinnabar eyes
wwCncn	1	Vestigial wings, red eyes
wwcncn	1	Vestigial wings, cinnabar eyes

4 $\chi^2 = 181.83$ (3 sf). There are three degrees of freedom ($n - 1$). The critical value for p = 0.05 and three degrees of freedom is 7.82; 181.83 is greater than 7.82 which indicates that the probability of getting this result by chance is less than 0.05. Therefore the observed results are significantly different from the expected results and we can reject the null hypothesis.

The two gene loci **W/w** and **Cn/cn** do not show independent assortment. This suggests that they are not on different chromosomes and must be linked on the same chromosome (they are in fact both on chromosome 2). This is an example of autosomal linkage. The recombinant groups that are present in smaller numbers (15 and 14) are the result of crossing over. Crossing over occurs in prophase I of meiosis.

5 These results are different because they all fall into two classes: wild type (long wings and red eyes), and vestigial wings and cinnabar eyes. These are the same as the parental generation. There has been no crossing over because crossing over does not happen in male *Drosophila*.

6 A cross between an organism with an unknown genotype and an organism showing a recessive trait or traits is called a test cross. The advantage of performing a test cross is that the alleles from the unknown genotype are revealed in the offspring of the test cross. For example, a fruit fly could have the phenotype long wing and red eyes. Its genotype can be represented as **W_Cn_** as we do not know whether it has any recessive alleles. When crossed with a fly that is **wwcncn,** if some of the offspring have vestigial wings and some have cinnabar eyes it means that the fly with the unknown genotype is definitely heterozygous for both genes, **WwCncn**. As you have seen, test crosses with male fruit flies will allow the genotype to be determined but not allow linkage to be worked out because crossing over does not occur in male *Drosophila*. No one knows why this is the case: crossing over occurs in males of other species.

Test yourself

5 One form of haemophilia is sex linked. What is the probability that a boy will inherit this condition from his father? Explain your answer.

6 Explain what epistasis is and what impact it has on variation.

7 In snapdragon flowers the alleles for red and white flower colour show incomplete dominance. What phenotypic ratio would you would expect in the F_2 generation for a cross between a red-flowered plant and a white-flowered plant?

8 Explain why the human ABO blood group system is described as multiply allelic.

9 Outline how the chi-squared test is used to analyse the results of a genetic cross.

Factors affecting the evolution of a species

Key terms

Selection pressure A factor that gives a greater chance of surviving to some members of the population than others.

Environmental resistance The combination of both biotic and abiotic factors that limits increases in population size.

Carrying capacity The maximum population size of a species that a particular habitat can support over time.

Tip

Refer to Chapter 15 in *OCR A level Biology 1 Student's Book* for more detail on natural selection.

Key term

Stabilising selection Type of selection that operates against the extremes of the range of phenotypes so the population remains the same over time.

Directional selection Type of selection that favours one extreme of the range of phenotypes so the population changes over time.

Organisms in a population will reproduce and population size will increase unless there are factors that limit the increase. Predation, competition and disease are some of the biotic factors that act as selection pressures preventing a population from increasing by limiting the number of individuals that survive each generation. Environmental resistance to population increase is caused by environmental factors, which may be either biotic, such as predation, availability of food or incidence of disease, or abiotic, such as space to live, minerals and water supply. These factors play a significant role in limiting any increase in a population because the environment cannot support the numbers.

When a population reaches the maximum that a particular environment can support the environment is said to have reached its carrying capacity. The inevitable result is that some of the young die before they can reproduce. Clearly some organisms in a population are more able to survive than others because they are better adapted and can compete more successfully. Their relative rate of survival is greater. This is because there is variation in the population which allows selection of those best suited for the environment.

Stabilising selection

A population remains stable for a particular character or trait if selection acts on variation and removes the extremes. A study of human birth weight in 1973 showed that there is higher mortality (death) in very small and very large babies. In other words they are less likely to survive. This means that the birth weight remains fairly stable with only a narrow range around the mean birth mass. This is an example of stabilising selection. If a new, extreme phenotype (i.e. one that lies far from the mean) arises it will have no selective advantage and so will not be selected. The norm is maintained. See Figures 10.12 and 10.13.

Directional selection

In a changing environment individuals with features more suited to the new environment will be selected and be more likely to survive long enough to breed and have offspring. This gives them a competitive advantage and the features (and the alleles producing them) have a greater chance of passing to future generations. This is directional selection, where an increasing proportion of individuals shows the advantageous feature and fewer show any of the other features (Figure 10.14). As a result, the population changes.

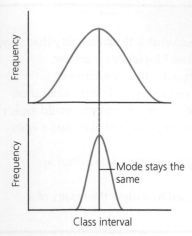

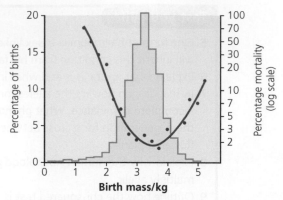

Figure 10.12 Stabilising selection in birth mass in human babies. The red line represents mortality rate.

Figure 10.13 Stabilising selection reduces the variation of one phenotype in a population. The upper graph shows the range of phenotypes before natural selection, the lower graph shows the range of phenotypes after stabilising selection has occurred.

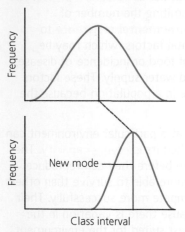

Figure 10.14 The graphs show variation in one characteristic in a population. The upper graph shows the range of phenotypes before natural selection, the lower graph shows the range of phenotypes after directional selection has occurred.

The ground finches on an island of the Galápagos archipelago in the Pacific Ocean lived on a range of seeds until a drought in the 1970s destroyed many of the plants. Only the hardier and larger-seeded plants were left. A lot of the finches were at a disadvantage because they were unable to eat the larger seeds. However, those with a genetic mutation causing slightly larger beak size were able to eat and survive on the larger seeds. The consequence was a shift in the population towards slightly larger beak size (see Figure 10.15). This is one of the evolutionary aspects of natural selection.

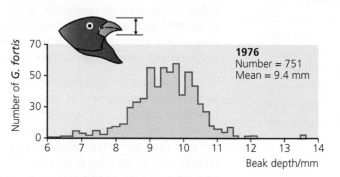

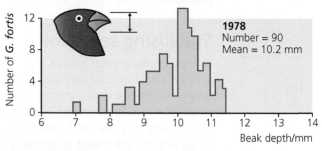

Figure 10.15 Directional selection in the ground finch, *Geospiza fortis*, on Daphne Major island in the Galápagos between 1976 and 1978

Genetic drift

Changes in the frequency of an allele in a population can occur randomly, and it may be quite by chance whether they are passed on to the next generation or not. This effect is greatest in small populations. Those alleles that do get passed on may result in a large change if the population is indeed small. This random change in allele frequency that occurs without natural selection is called genetic drift. In a large population it only has a small impact. However, in some instances genetic drift may result in the

Key term

Natural selection The mechanism for evolution. It is the survival to reproductive age of those organisms with characteristics best suited to their environment. This increases the chances of their alleles being passed on to subsequent generations.

Genetic drift A change in allele frequency in the absence of natural selection.

Genetic bottleneck A genetic or population bottleneck is a sharp reduction in population size due to environmental events such as fire, earthquake, flood or drought or due to disease or human activity. Any subsequent population increase is then based on the limited gene pool available from the few surviving individuals.

Founder effect When a small population colonises a new area any population descending from this small colonising 'ancestor group' will have a much reduced genetic diversity and therefore a different allele frequency from other similar populations.

complete elimination of an allele from a population in just one or two generations. This will reduce genetic variation and could lead to extinction or indeed to the development of a new species.

Genetic bottleneck

A natural disaster such as a volcanic eruption or excessive hunting by humans may result in a population being sharply and severely reduced to a very small number. All subsequent generations will be derived from the few survivors. The variety of alleles at each gene locus in this new, reduced, gene pool will be very small indeed compared to other populations. This has been seen in many animal populations, such as the Hawaiian goose (né né, or *Branta sandvicensis*), which evolved after the Canada goose arrived on Hawaii 500 000 years ago. Because they were restricted to the islands by isolation this became a genetic bottleneck. Analysis of mitochondrial DNA of modern né né shows that they are closely related to the giant Canada goose and the dusky Canada goose and DNA analysis has shown they have a very limited genetic diversity.

The Northern elephant seal, *Mirounga angustirostris*, is a classic example of a genetic bottleneck caused by overhunting by humans. The population numbers fell to just 30 in 1890. Now the population has increased to hundreds of thousands but it still has very limited genetic variation. The giant panda is another example, with its genome showing a severe genetic bottleneck that occurred about 43 000 years ago. Genetic drift is very likely to follow genetic bottlenecks leading to loss of more genetic diversity.

Founder effect

A founder effect occurs when a small population colonises a novel area, such as a newly formed volcanic island. Clearly, until mutations occur in the new population, only the alleles present in the colonisers can occur in the developing population. Allele frequencies may be quite different from the original population and there will be reduced genetic diversity. Small populations of the brown anolis lizard, *Anolis sagrei*, were released on islands in the Bahamas to test the hypothesis that the founder effect could be responsible for long-term genetic differences between populations. Male and female lizards were chosen at random from one large population and transferred in small groups to seven small islands. Over time, all the populations evolved the same adaptations for life on these small islands, but retained the differences that they had inherited from the founder populations.

Huntington's disease is caused by a dominant allele for the protein huntingtin. As the allele is dominant a person with only one copy is at high risk of developing the disease. The allele causes a longer huntingtin protein to be made, which accumulates in the brain and causes brain cell death. The disease occurs twice as frequently in Tasmania than in the rest of Australia due to a single individual with the disease settling in Tasmania and passing on the mutant allele. Another condition known as tibial muscular dystrophy is caused by the mutant allele of the gene *TTN* that codes for the skeletal muscle protein titin. It is most common in Finland where it affects 10 in every 100 000 people due to a mutation in the founder population that colonised Finland and have given rise to the Finnish population. Another example of the founder effect is seen on a small island in the western Pacific, where there is a disproportionate number of individuals suffering from an inherited eye defect that was present as a recessive mutant allele in just one of the original colonisers.

Using the Hardy–Weinberg principle

Gene pool All the alleles for a particular gene in the whole population.

Two scientists, G.H. Hardy and Wilhelm Weinberg, used allele frequencies in populations to predict how the frequencies would change over subsequent generations. For the next section you will need to understand the concept of the gene pool.

Cystic fibrosis is an inherited condition in human populations causing thickened mucus production. It makes breathing and digestion difficult. It is the result of a recessive allele affecting 1 in every 2500 individuals. Given that there are two alleles at this gene locus, one dominant and one recessive, the frequency of the two alleles can be calculated using the Hardy–Weinberg principle. According to this principle, with two alleles at a locus it must be that

$$p + q = 1$$

Here, p is the frequency of the dominant allele, **F**
q is the frequency of the recessive allele, **f**
1 represents the whole population.

So, as 1 in every 2500 people are affected, the proportion of people who are homozygous recessive, **ff**, is 0.0004 this is represented by q^2, making $q = \sqrt{0.0004} = 0.02$ and therefore the dominant allele $p = 0.98$. So the frequency of **f** = 0.02 and that of **F** = 0.98 in the human population.

It follows, if p represents the frequency of the dominant allele, **F**, that the genotype of the homozygous individuals in the population, **FF**, must have a frequency of $p \times p$ or p^2. The frequency of the recessive allele, **f**, is q so the frequency of the genotype for the homozygous recessive individuals, **ff**, must be q^2. The heterozygous genotype, **Ff**, has a frequency of $2pq$.

In a single population the genotype frequency can therefore be represented by another of the Hardy–Weinberg equations

$$p^2 + 2pq + q^2 = 1$$

Here, p^2 represents the frequency of the homozygous dominant genotype, **FF**; q^2 represents the frequency of the homozygous recessive genotype, **ff**; the frequency of the heterozygous genotype, **Ff**, is represented by $2pq$.

So, to calculate the frequency of the heterozygous individuals

$$2pq = 2 \times 0.02 \times 0.98 = 0.0392$$

So, approximately 4% of the population will be carriers of the recessive allele, **f**.

The Hardy–Weinberg principle predicts that allele frequencies will remain the same from generation to generation. However, it is based on the following assumptions

- there will be no migration, either emigration or immigration
- there is no gene mutation
- the population is large and all individuals can breed freely and randomly within the population with equal success
- there is no selection
- there is no genetic drift.

You will always be given the two equations in any exam questions involving the Hardy-Weinberg principle. Check to see whether you are given information about allele frequencies (equation 1) or genotype frequencies (equation 2).

If a population is found to meet these criteria then it is likely that the population will also be in **Hardy–Weinberg equilibrium** for that locus. This may be tested by observing allele frequencies in populations.

Using the Hardy–Weinberg principle

To understand the Hardy–Weinberg principle better it often helps to carry out a practical activity using coloured beads to model the different alleles. For this you need a large number of identically sized and shaped beads in two different colours.

The human blood group system MN has two codominant alleles for the gene L: L^M and L^N. This is a different system from the ABO blood group system. In a population, 36% of people are homozygous $L^M L^M$ with the phenotype MM, 48% are heterozygous $L^M L^N$ (MN) and 16% are homozygous $L^N L^N$ (NN).

So in a population of 100 individuals we know there will be 200 alleles (since each individual will have two alleles for this gene). Let's say you have red beads to represent the M alleles and yellow beads to represent the N alleles. You will need 120 red beads because the 36/100 MM individuals each carry two alleles, so there are 72 M alleles for this group, and the 48/100 MN individuals will carry another 48 M alleles.

The frequency of M in the population is 120/200 = 0.6.

Given that the total number of alleles is 200 then the number of N alleles must be 200 − 120 = 80. So you will use 80 yellow beads.

Using the equation $p + q = 1$, if p = frequency of M allele and q = frequency of N allele, $p = 0.6$ so $q = 1 - 0.6 = 0.4$.

Place the 200 coloured beads into an opaque bag: you can use anything that cannot be seen through. This is your gene pool; the pool of all the possible alleles in the population.

Now use one hand to remove two beads at a time without looking at what you are selecting: it must be completely random. Record the pairs of colours you have removed using a tally in a table and then replace the beads in the bag. Continue until you have 100 pairs. This is one generation.

Your table of results will look something like Table 10.13.

Table 10.13 Tally table of results.

	Red/red	Red/yellow	Yellow/yellow
Tally	IIII	II	IIII
...			
Example results (partial)	4	2	4

To check your results for this activity you should be able to match the frequency of each *phenotype* given above. You can now introduce another aspect by removing 10% of the NN phenotypes from the gene pool at each generation and repeating the sampling for five generations.

1 Explain what we are modelling by removing 10% of the NN phenotypes for each generation.
2 What impact does this have on the allele frequency after five generations when compared to the original sampling where no phenotypes or alleles were removed? (Assume that the Hardy–Weinberg principle is upheld for the initial sampling and that no changes in allele frequency occur.)
3 Iron overload disorder is a monogenic disorder controlled by a gene on chromosome 6. People with the disorder are homozygous recessive. It is estimated that 1 in 200 people in the UK have the disorder in which iron levels increase with age. What proportion of the population is heterozygous?

Isolating mechanisms in forming new species

If a large population is split into smaller groups, for whatever reasons, this will tend to create sub-groups or sub-populations that become isolated from each other. Such reasons are called isolating mechanisms. The resulting effect will be different changes in the allele frequencies for each group, depending on the frequencies present in the founding populations and on the selection pressures that are experienced subsequently by each population. Some alleles may be eliminated completely and others may increase in frequency. In the long term the two sub-populations may become sufficiently different that they will not be able to interbreed and so they will become different species.

The mechanisms can be sub-divided into two sorts, as outlined below.

- Allopatric speciation (from the Greek *allos*, meaning 'other') is any geographical mechanism that physically separates the population into distinct and separate sub-groups. Examples are rivers or mountain ranges forming a physical barrier so that the habitat is divided and the result is genotypic or phenotypic divergence. This is caused by different selection pressures, genetic drift or differences in mutations in the two groups. Or it may be the result of climate changes creating seasonal differences, such as is seen in the terrestrial and marine iguanas of the Galápagos Islands which differ because they have been separated by geographical and climatic differences. The Galápagos tortoises are another example; the differences between them are so significant that they can be identified as belonging to a particular island. If these populations come back into contact they have changed to such an extent that they are reproductively isolated and are no longer able to exchange genes in reproduction.

- Sympatric speciation does not involve geographical separation, but changes occur so that some individuals are not able to interbreed with the rest of the population. This is a result of mechanisms in which reproductive differences cause individuals to become isolated within a population. Two or more species may form from a single ancestral species all within the same geographical location. Sympatric speciation can be seen in the cichlids of East Africa, which live in the lakes of the Rift Valley; for example, Lake Victoria, Lake Malawi and Lake Tanganyika. Well over 800 different species have descended from the ancestral fish *Oryzias latipes* millions of years ago and are now all reproductively isolated. The hawthorn fly, *Rhagoletis pomonella*, seems to be in the process of sympatric speciation, with different populations feeding on different fruit. One population in North America appeared soon after apples were introduced. Apples are not a native species and once this fruit had been introduced one group of the population began feeding on apple instead of the preferred hawthorn fruit. The remaining hawthorn fly population still feeds on hawthorn and does not feed on apples. The two groups do not interbreed; the hawthorn-feeding flies mature later in the season and take longer to do so. Recent DNA studies have shown that six out of 13 gene loci for enzymes are different in the two groups of flies. Other examples of reproductive isolating mechanisms include shifted breeding cycles in animal groups; different courtship behaviours, such as in swans and geese, which although related have been isolated for so long that any offspring are infertile; and altered physical features, such as in species where the reproductive organs are no longer compatible or when polyploidy occurs.

Hybridisation sometimes occurs between different species of plant. Hybrids are often sterile and are unable to breed with either parent. However, if the chromosome number doubles because of a failure during meiosis they become fertile. This is an example of polyploidy allowing speciation and can be seen in cord grass, a plant that grows in saltmarshes.

A foreign species of cord grass, *Spartina alterniflora*, from the East and Gulf coasts of the USA invaded salt marshes in Southampton Water in the UK and interbred with the native species of cord grass, *S. maritima*, to give rise to a sterile hybrid known as *S. townsendii*. This hybrid, although sterile, was able to reproduce asexually. Then a chromosome mutation occurred to double its chromosome number making it fertile to form English cord grass, *Spartina anglica*. This is a distinct species because it does not interbreed with any of the others, including *S. maritima* and *S. alterniflora* – it is reproductively isolated. This type of speciation happens in plants, but rarely in animals.

Artificial selection

Artificial selection is where humans direct selection by choosing the features of plants or animals and selecting the individuals that are allowed to reproduce. This is a different mechanism of evolution from that seen in natural selection, where selective agents in the environment favour those best suited to survive. Artificial selection has been carried out by humans for thousands of years. Selective breeding of wild plants and animals is thought to have begun about 10 000 years ago when humans began to settle in larger groups and started farming. A good example of artificial selection is modern bread wheat, *Triticum aestivum*, which arose from the genus *Triticum* (wild wheat) and the genus *Aegilops* (wild grass) about 10 000 years ago.

- Introducing new genes: modern bread wheat, *T. aestivum,* evolved from the characteristics of the two original genera by humans selecting the features they required over a period of time, i.e. selecting those crosses that provided the features they needed by artificial pollination.

- Hybridisation: this created sterile hybrids. Although both wild grass and wild wheat have diploid chromosome numbers of 14 (in seven pairs), the chromosomes were not compatible and so would not pair up in the hybrid, which means meiosis is not possible.

- Polyploidy: a chance chromosome mutation in the sterile hybrid caused the chromosome number to double. This results in the nuclei containing more than the diploid chromosome number after fertilisation and so the plants could now form a new fertile species, although it cannot cross with the other species including the parents.

- Further chance mutations resulted in plants increasing the haploid chromosome number to four making tetraploid individuals that produce diploid gametes instead of haploid gametes.

- Later hybridisation: crosses of the tetraploid with a wild grass (with a normal diploid number of 14 chromosomes) resulted in a sterile hybrid with 21 chromosomes (triploid).

- Further chromosome mutations doubled the number again making another fertile hybrid with 42 chromosomes (hexaploid) called spelt wheat, *Tritcum aestivum spelta.*

- Further selection by man resulted in modern wheat, *Tritcum aestivum aestivum.*

Artificial selection in animals is illustrated well by modern dogs, where intensive selective breeding has given rise to a large variety of different dog breeds (see Figure 10.1). Modern domestic dogs most certainly evolved from the ancestral grey wolf at around the time of settling and farming, or just

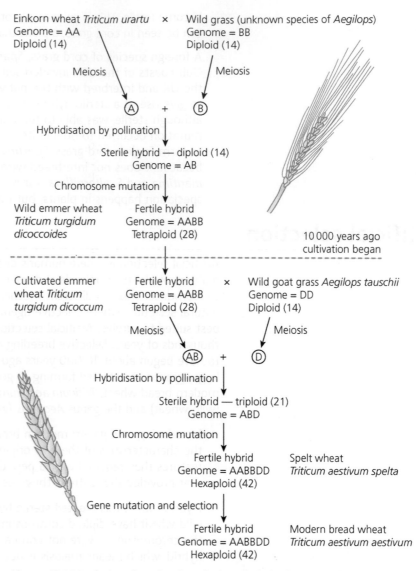

Figure 10.16 Evolution of modern bread wheat by hybridisation and polyploidy. Each genome (A, B and D) has seven different chromosomes.

Tip

Question 11 on page 297 of *OCR A level Biology 1 Student's Book* is about assessing the genetic and environmental contributions of variation in milk yield.

before. Originally the wolf was selected for its ability to be tamed and used in tasks such as hunting. Those that could not be tamed were not selected. However, at the same time selection for the ability to be tamed also caused physical changes that were later ideal for specific tasks. The tamed wolf dogs were used to hunt prey and run them down until the humans were able to catch up with them and use tools to perform the kill. Some breeds, such as terriers, were selected for running into burrows after animals such as rabbits, and so were selected for small size and speed. Gun dogs were selected for the ability to learn, point and run after game, so size, speed and intelligence were important. At present all dog breeds are the same species but in time the differences may result in speciation between breeds.

Selective breeding has been an important factor in cattle where improving milk yield is economically important. However, this is a difficult task because milk yield is determined by many genes and as such is an example of continuous variation. It is also significantly affected by environmental conditions such as type of food, temperature, type of grass and soil, and living conditions. Recognising this, farmers may also use techniques such as music and human contact to improve milk yield.

The need to maintain a resource of genetic material

To improve milk yield in dairy cattle it has been important to maintain a resource of genetic material including types that are close to the original wild types. In this way the gene pool doesn't become too small and so weaken the population by reducing variation. This was seen in Holstein cows where inbreeding reduced the gene pool with a consequent drop in milk yield.

In some instances a lack of variation could lead to loss of a whole population in the event of disease or an environmental change. There have been many other instances where a gene pool has been severely reduced. After a bottleneck, genetic variation is reduced, leading to a reduced ability to withstand environmental change.

Ethical considerations

Unfortunately some intensive selection processes have resulted in conditions and diseases that are extremely damaging or deleterious for the animals, which must be considered from an ethical position. These are a result of selecting desired features and, as an unintended consequence, also selecting damaging genes. For example, breathing problems in Bulldogs and Pekinese dogs are common because of their extremely shortened snouts. Giant dogs such as Saint Bernards and Great Danes show a strong correlation between their large size and frequency of hip dysplasia. They also tend to overheat as it is difficult to cool a large body, and are prone to malignant bone tumours in their legs, probably as a result of their huge weight. Inbreeding in order to maintain a pedigree often results in inheritance of two recessive alleles for a genetic disease. As a result, autoimmune conditions are more common in some breeds, such as the Basset Hound. Addison's disease, an autoimmune disease affecting the adrenal glands, occurs frequently in Bearded Collies and Standard Poodles. Diabetes mellitus, another autoimmune disease affecting the islets of Langerhans in the pancreas and the ability to control blood glucose concentrations, occurs more in Samoyeds and Australian Terriers. Other inherited conditions include many cancers, such as malignant blood vessel tumours which are common in dogs like Golden Retrievers. All of these inherited diseases raise ethical issues about the extreme selective processes that take place to maintain these breeds.

In each case the limited number of alleles in the gene pool of the population makes the survival of these breeds very tenuous. Retaining a resource of genetic material will assist this as well as being important as part of the selection programme.

> **Tip**
>
> You should research similar problems in other domesticated animals, such as cattle. Also research the consequences of inbreeding in crop plants.

> **Test yourself**
>
> 10 The Hardy–Weinberg principle is represented by the following equations:
>
> $p^2 + 2pq + q^2 = 1$
>
> $p + q = 1$
>
> Albino mice have white coats because they cannot produce melanin. Melanin production is controlled by a single gene **C**. The dominant allele **C** gives a brown coat but the recessive allele **c** gives an albino coat. Out of 60 mice in a laboratory, 45 were brown. Use the Hardy–Weinberg principle to estimate the frequency of the dominant allele for these mice. Show your working.
>
> 11 There are small populations of tigers spread across parts of India, Nepal, China and South-East Asia including the island of Sumatra in Indonesia. These populations are classified as sub-species based on differences in colour, markings and body size.
>
> a) Explain how these different sub-species may have arisen.
>
> b) Suggest why these different populations are described as sub-species rather than as different species.
>
> 12 Suggest why artificial selection in animals is more difficult than in plants.

Exam practice questions

1 Albinism in rodents is controlled by a recessive allele. Mice of the species *Apodemus flavicollis* have a diploid number of 48. A rare mutation in the species results in albinism. How many copies of this allele will be found at one of the poles of a cell at telophase I of meiosis in an albino mouse?

   A 48

   B 24

   C 2

   D 1   *(1)*

2 In fruit flies, *D. melanogaster*, ebony body colour and vestigial wing are recessive to the wild type and are not linked. Crossing fruit flies that were heterozygous for both characteristics gave 320 offspring. How many offspring would be expected to show both recessive features?

   A 16

   B 20

   C 60

   D 180   *(1)*

3 Chickens of a particular breed are white because they have a dominant allele, **I**, of a gene that acts to inhibit a gene, **F/f**, that controls pigment production in feathers. The ratio of chickens in the offspring of the cross between two white birds that are heterozygous, **IiFf**, is:

   A 15 white feathers : 1 coloured feathers

   B 13 white feathers : 3 coloured feathers

   C 12 white feathers : 4 coloured feathers

   D 8 white feathers : 8 coloured feathers   *(1)*

4 a) i) Explain the term *multiple alleles*. *(2)*

     ii) Explain the effect that multiple alleles have on variation. *(3)*

  The locus of the gene that determines the ABO blood group is on chromosome 9.

  b) i) A couple have four children. Each of their children has a different blood group. Use a genetic diagram to show how this is possible. *(5)*

     ii) The woman is expecting the couple's fifth child. What is the probability that the child will be a boy with blood group O? *(1)*

  c) Explain why the human ABO blood group system is an example of discontinuous variation. *(3)*

  d) Explain the genetic basis of continuous variation. *(5)*

5 The Northern elephant seal, *Mirounga angustirostris*, was hunted during the nineteenth century. The population in 1900 was unlikely to have been more than about 100 individuals. Since hunting ended, the population has increased to 170 000. A genetic analysis of samples of the population found that all the gene loci studied were homozygous and none were polymorphic.

  a) Explain why there is no polymorphism in the gene loci that have been studied in the population of *M. angustirostris*. *(5)*

  b) Explain why organisms with little or no polymorphism are at risk of extinction even though they may have a large population. *(5)*

The house mouse, *Mus musculus*, was introduced accidently to the island of Skokholm off the coast of south Wales in the late nineteenth century. The mice on Skokholm have an interfrontal bone which is a very rare phenotypic trait in populations of house mice in Britain. House mice in Dale, one of the places on the mainland close to Skokholm, also have a high frequency of this condition.

  c) Explain the significance of these observations on the phenotypes of mice on Skokholm and in Dale. *(3)*

The population of *M. musculus* on Skokholm was studied intensively in the 1960s. The graph shows the changes in the population over the study period.

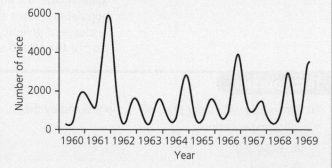

  d) Outline a way in which the size of a population of *M. musculus* on a small island could be estimated. *(6)*

  e) Discuss the consequences of the population changes shown in the graph for survival of the mice on Skokholm. *(5)*

6 The Arkhar-Merino breed of sheep was produced between 1934 and 1950 in what is now Kazakhstan. This involved crossing domestic sheep with the wild Arkhar sheep, *Ovis orientalis* which lives at high altitude in the mountains of western and central Asia. The aim was to develop a new breed that would combine the superior qualities of the Merino breed of domestic sheep, *Ovis aries*, with the hardiness of the Arkhar, including its ability to live at high altitudes. The breeding process began with the insemination of Merino ewes with semen from dead Arkhar rams. Forty three hybrid lambs were obtained (hybrid 1). Ewes of two domestic breeds, Precocé and Rambouillet, were inseminated with sperm from hybrid 1 rams resulting in about 400 offspring (hybrids 2 and 3). The sheep from these crosses were in turn crossed with domestic breeds for several generations (see diagram below). The vast majority of the offspring combined the desired features of both species. They were well adapted to high altitude (2500–3000 m), showed good increase in body weight and produced a good quantity and quality of fleece.

Step 1   Arkhar ram   ×   Merino ewe

Step 2   Precocé ewe   ×   Hybrid 1   ×   Rambouillet ewe

Step 3   Precocé ewe   ×   Hybrid 2   Hybrid 3   ×   Rambouillet ewe

Hybrid 4   Hybrid 5

a) i) State the proportion of the genes derived from the Arkhar rams that are present in hybrid 5 sheep. *(1)*

ii) Explain why so many steps were required in the process of breeding the Arkhar-Merino sheep. *(3)*

Tests found that the blood profile of the hybrids was superior to that of the three breeds of domestic sheep with respect to suitability for breeding at high altitudes. Subsequent selection for desirable characteristics produced the stable new breed.

b) i) State two problems that face animals living at high altitude and suggest the features of the Arkhar sheep that helped the new breed to survive at high altitude. *(5)*

ii) Use the information provided to explain how artificial selection differs from natural selection. *(5)*

iii) Explain why wild relatives of domesticated livestock, such as Arkhar sheep, should be conserved. *(4)*

iv) Suggest why many domesticated species cannot be improved in the ways described for the Arkhar-Merino breed. *(3)*

7 Coat colour in cats provides examples of different patterns of inheritance. One of the genes that controls coat colour has two alleles: black (C^b) and ginger (C^g). There are three types of female cat – ginger, black and tortoiseshell (a mixture of ginger and black fur) – but only two types of male cat – ginger and black. Two crosses were carried out:

Cross 1   black female × ginger male
Cross 2   ginger female × black male

Kittens from cross 1 were either tortoiseshell females or black males. Kittens from cross 2 were either tortoiseshell females or ginger males.

a) Copy and complete the table to show the genotypes of the kittens from the two crosses. *(4)*

Cross	Offspring	
	Phenotypes	Genotypes
1 Black female × ginger male		
2 Ginger female × black male		

b) Use a genetic diagram to predict the phenotypes of the kittens in a cross between tortoiseshell females and ginger males and the ratio between them. *(7)*

Siamese cats have a pale coat with dark ears, faces, feet and tails. The dark colour is caused by the black pigment, melanin. The enzyme tyrosinase is involved in the metabolic pathway to produce melanin. Siamese kittens are all white at birth with the pigment developing only in the cooler extremities some days later.

c) Explain the appearance of Siamese cats. *(3)*

d) Suggest the effect on the phenotype of keeping Siamese kittens in different temperatures. *(2)*

e) Suggest how the phenotype of cats can be influenced by environmental factors alone. *(3)*

Stretch and challenge

8 In a population of ground beetles there are two alleles of a gene for body colour. The dominant trait is black. A survey discovered that out of 4000 individuals 3000 were brown and the rest were black.

a) Calculate the frequency of the two alleles of the gene for body colour.

Another trait in these ground beetles is the length of the hind wing. Some beetles have long wings that extend beyond the abdomen (long wing) and others have short wings that are totally enclosed by the front wings (short wing). A population of ground beetles was sampled and there were 3000 long-winged beetles and 1000 short-winged beetles.

b) Which trait is dominant? Explain your answer.

The population genetics of sickle cell anaemia has been studied all over the world. Using electrophoresis it is possible to determine the genotypes of individuals as **Hb^{N}Hb^N**, **Hb^{N}Hb^S** and **Hb^{S}Hb^S**. Analysis of a small population on an island found that the frequencies of the three genotypes were:

Hb^{N}Hb^N	75
Hb^{N}Hb^S	24
Hb^{S}Hb^S	1

c) Test these data using a chi-squared test to see if this population is in Hardy–Weinberg equilibrium.

Three alleles of the *alpha2* gene influence blood platelet function. Nucleotide substitutions account for these alleles known as 1, 2 and 3. The frequencies of these three alleles in a population are

allele 1	0.5
allele 2	0.3
allele 3	0.2

d) What are the likely frequencies of the different genotypes in the population?

e) What assumptions have you made in calculating the genotype frequencies?

9 Some students investigated the inheritance of the gene known as bar eye in fruit flies. The diagram shows the phenotypes.

A third phenotype was observed: wide bar. Flies with this phenotype have eyes with a width larger than bar eye, but smaller than round eye.

Wild-type round eye phenotype

Mutant narrow bar eye phenotype

The students set up two crosses

cross 1:
female wild-type flies crossed with bar-eyed males

cross 2:
female bar-eyed flies crossed with wild-type males.

The results of the crosses are shown in the table.

Eye shape	Sex	Cross 1	Cross 2
Round eyes	Male	147	
	Female		
Narrow bar eyes	Male		127
	Female		
Wide bar eyes	Male		
	Female	156	132

a) Draw genetic diagrams to explain the results shown in the table for cross 1 and for cross 2.

b) Predict the results that are likely to be obtained in the following crosses involving the F$_1$ fruit flies.

Cross 3:
wide-bar females × round-eyed males

Cross 4:
wide-bar females × narrow-bar males

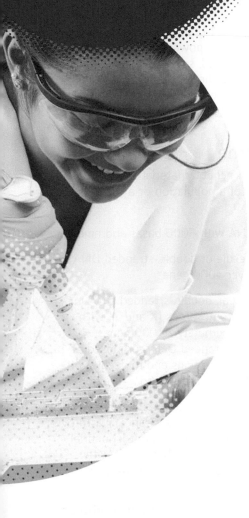

Chapter 11

Manipulating genomes

Prior knowledge

- DNA is composed of two antiparallel polynucleotide strands that are held together by hydrogen bonds between pairs of nitrogenous bases.
- Each base pair in DNA is composed of a purine base and a pyrimidine base: adenine pairs with thymine and guanine pairs with cytosine.
- DNA is copied by DNA polymerase in the process of semi-conservative replication.
- The genetic code is a triplet code in which groups of three nitrogenous bases in DNA and RNA code for the 20 amino acids used to make polypeptides (see Figure 11.25).
- One polynucleotide strand in DNA is the coding strand, in that it has the same sequence of triplets as mRNA; the other strand of complementary bases is the template strand on which mRNA is produced by RNA polymerase during transcription.
- The genetic code is described as degenerate as most of the amino acids are coded by more than one triplet.
- The genetic code is universal in that all organisms and viruses use the same code.
- During protein synthesis, DNA is transcribed to form mRNA which is then translated on ribosomes to form polypeptides.
- Gene expression in eukaryotes is controlled at three levels: at transcription, post-transcription and post-translation.
- There are different forms of gene mutation: some mutations have observable effects on the phenotype; most mutations are recessive, some are dominant; most are harmful, some are lethal.
- Genetic polymorphism is the existence of two or more alleles of a gene in a population, each with a frequency of at least 1% in the gene pool.

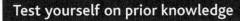

Test yourself on prior knowledge

1 Outline how a gene codes for a polypeptide.
2 State the three DNA triplets from the coding strand that code for STOP. Explain why these codons have this effect.
3 Describe the effect of different types of gene mutation on a polypeptide.
4 Distinguish between regulatory genes and structural genes.
5 Explain why some genes are always expressed in cells, but others are not.

Figure 11.1 A forensic scientist takes a blood sample for DNA analysis.

Uniqueness of DNA

The data held on a DNA database is a small snapshot of people's total DNA. The data held by the police is enough to be unique to one person, except for those who are identical twins. Much of the genome is shared with other people. Much our DNA is similar and even identical to that of other species. After all, we use the same enzymes in processes such as respiration; we have also seen how the homeobox genes involved in early development are common to many groups of eukaryotes (see Chapter 9). However, there are some aspects of our DNA that we do not share with anyone.

Various techniques of DNA analysis (Figure 11.1) and sophisticated systems of data storage and retrieval, coupled with the power of the internet, allow databases and researchers to be interconnected, so information about the natural world has grown exponentially since the 1980s. In *OCR A level Biology 1 Student's Book* Chapter 14 you were introduced to some of these techniques and the way they are used in classification.

The structure of the genome

The genome is the minimum quantity of genetic material that contains one copy of all the genes of an individual or of a population or a species. Some examples of genomes are

- HIV: a single-stranded molecule of RNA with 5000 bases and nine genes

- *Escherichia coli*: a single, circular molecule of double-stranded DNA with nearly 5 million base pairs and just over 4000 genes

- mouse, *Mus musculus*: molecules of linear, double-stranded DNA in 19 different autosomes and two sex chromosomes (X and Y) with a total of 2.8×10^9 base pairs and approximately 23 000 genes; mitochondrial genome: circular, double-stranded DNA composed of 16.3×10^3 base pairs with genes coding for 13 proteins and molecules of tRNA and rRNA, which are all essential for mitochondrial functions

- human, *Homo sapiens*: molecules of linear, double-stranded DNA in 22 different autosomes and two sex chromosomes (X and Y) with over 3.0×10^9 base pairs and approximately 21 000 genes; mitochondrial genome: circular, double-stranded DNA 16.5×10^3 base pairs with genes coding for proteins and molecules of tRNA and rRNA as in the mouse genome.

The science of genomics began with very simple genomes that proved easy to investigate and to sequence. The principles of DNA sequencing are explained later in this chapter. Genomes of many species have now been sequenced, which means that the whole of the base sequence is known. All of them have double-stranded DNA so their size is measured in base pairs (bp), kilobase pairs (kbp) or megabase pairs (Mbp).

Viral genomes are the simplest and prokaryotic genomes are smaller than eukaryotic genomes. We have already seen in Chapter 9 that the control of gene expression in prokaryotes is much simpler than the control of eukaryotic genes. The genomes of eukaryotes have large quantities of DNA that are not transcribed and translated into proteins, but act to control gene expression. Also, eukaryotic genes have introns that are transcribed, but not translated. Figure 11.2 shows an outline of the structure of the genome of thale cress, *Arabidopsis thaliana* (Figure 11.3), which has a diploid number of 10 as well as DNA in its chloroplasts and mitochondria. The diploid number refers to the number of chromosomes in the nucleus.

Key terms

Genome The minimum quantity of genetic material that contains one set of all the individual genes; in humans this is the DNA in 22 autosomes, the X chromosome and the Y chromosome plus mitochondrial DNA. Plant genomes also include the DNA in chloroplasts.

Genomics The application of the techniques of genetics and molecular biology to the mapping of genes on chromosomes and the sequencing of genes or complete genomes of organisms and viruses.

Tip

This is a good place to remind yourself of the differences between prokaryotes and eukaryotes. Use the index of *OCR A level Biology 1 Student's Book* to find differences in their cell structure.

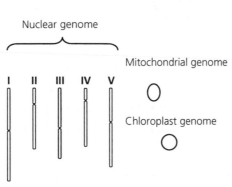

Genome		Size / Mbp	Number of genes
Nuclear	Chromosome I	29.1	6543
	Chromosome II	19.6	4036
	Chromosome III	23.2	5220
	Chromosome IV	17.6	3825
	Chromosome V	25.9	5874
Total for nuclear genome		115.4	25 498
Mitochondrion		0.37	58
Chloroplast		0.15	79

Figure 11.2 The structure of the genome of thale cress, *Arabidopsis thaliana*. This is sometimes called the haploid genome as only one chromosome of each type is shown. The chromosomes are not drawn to scale.

Figure 11.3 Thale cress, *Arabidopsis thaliana*.

Nuclear DNA in eukaryotic organisms is divided into regions that have different functions. There are structural genes that code for the assembly of amino acids to make polypeptides. In a gene there are coding sequences or exons separated by non-coding introns. The gene for the β polypeptide of human haemoglobin has three exons and two introns and is 1605 bp in length. The exons have a total length of 438 bp and so code for a polypeptide of 146 amino acids (triplet code). Structural genes are transcribed as mRNA which is edited and then translated. Other non-structural genes code for tRNA and rRNA. Some regulatory genes code for proteins that act as transcription factors and many code for forms of RNA that also control transcription and hence gene expression. Promoters are control sequences found upstream of genes and are the site of binding of RNA polymerase at the start of transcription. There are long lengths of DNA that separate structural and regulatory genes. These used to be considered as 'junk' DNA with no function, although now they are considered to have functions in gene control.

> **Tip**
>
> Do not confuse the *karyotype* of an organism with the genome; the karyotype is the depiction of all the chromosomes in a cell of an individual.

Also included in the genomes of all eukaryotes is the DNA in mitochondria (mtDNA). The genomes of plants and some protoctists include the DNA in chloroplasts as well (ctDNA). These two organelles originated from prokaryotes and still possess what are essentially prokaryotic genomes with little non-coding DNA.

Techniques in DNA analysis

> **Key term**
>
> **Polymerase chain reaction (PCR)** An automated process that amplifies selected regions of DNA using alternate stages of polynucleotide separation (denaturation of DNA) and polymerisation catalysed by DNA polymerase.

Two of the most important techniques employed by scientists in analysing DNA are those of the polymerase chain reaction and gel electrophoresis.

Polymerase chain reaction

The small quantity of blood on the knife in Figure 11.1 may not have yielded enough DNA to test, in which case the quantity would need to be increased so it can be analysed. The polymerase chain reaction (PCR) is used to increase the quantity of DNA using its ability to be replicated by DNA polymerase. The process of PCR also involves the analysis of DNA as well as increasing its quantity. Figure 11.4 shows the stages of PCR.

187

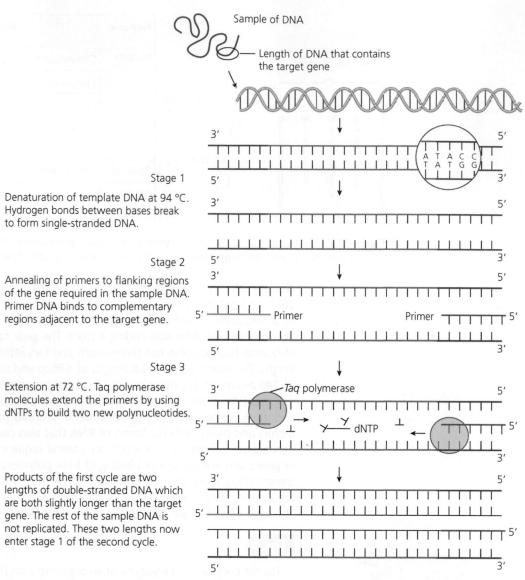

Sample of DNA

Length of DNA that contains the target gene

Stage 1

Denaturation of template DNA at 94 °C. Hydrogen bonds between bases break to form single-stranded DNA.

Stage 2

Annealing of primers to flanking regions of the gene required in the sample DNA. Primer DNA binds to complementary regions adjacent to the target gene.

Primer

Primer

Stage 3

Extension at 72 °C. Taq polymerase molecules extend the primers by using dNTPs to build two new polynucleotides.

Taq polymerase

dNTP

Products of the first cycle are two lengths of double-stranded DNA which are both slightly longer than the target gene. The rest of the sample DNA is not replicated. These two lengths now enter stage 1 of the second cycle.

Figure 11.4 Amplification of a sample of DNA using the polymerase chain reaction (PCR). The diagram shows the three stages of the first cycle in which two copies of the template DNA from the original sample are copied.

Tip

This is a good time to revise the process of semi-conservative replication of DNA. See Chapter 3 of *OCR A level Biology 1 Student's Book*; also watch an animation of replication. You can find one at the DNA Learning Center (www.dnalc.org). As you watch, remind yourself that in most organisms the stages of replication occur at low temperatures (at or below about 40 °C).

PCR involves continuous cycles of replication to produce multiple copies only of the DNA of interest. Each cycle lasts about 5 minutes and has three stages

1 denaturation

2 annealing

3 extension (also known as elongation).

Denaturation

PCR starts with a denaturation process that breaks the hydrogen bonds holding the two polynucleotide strands together so that primers can gain access to the sequence of bases that are now exposed. This occurs at 94 °C for about 3 minutes and forms the two template strands that are to be replicated.

During the denaturation stage of all the remaining cycles the samples are heated to 94–95 °C for only about 45 seconds.

Annealing

The temperature in the cycler now decreases so that annealing can take place. The temperature is kept at 50–65 °C to allow hydrogen bonds to form between the two types of oligonucleotide primer and the complementary DNA sequence of bases.

In PCR, two primers are designed to bind to the region of DNA that is to be copied. Through complementary base pairing, one primer attaches towards the 5′ end of one strand and a different primer attaches towards the 5′ end of the other strand. Because the strands are antiparallel the primers attach at the 'left' of one strand and at the 'right' of the other (see Figure 11.4). The sequence of bases in each of the two primers is chosen carefully so that they bind to sequences just outside the region of DNA that is to be amplified. Primers that are 20 or so nucleotides long will target just one place in the entire genome, which means that the original DNA sample may be quite large and contain much DNA that is not going to be copied.

Primers are necessary to identify sites where synthesis will take place and because DNA polymerase functions by adding nucleotides to an existing piece of double-stranded DNA.

The primers attach to the start of the DNA strand, and then DNA polymerase will add nucleotides all along the rest of the DNA template strand as happens in semi-conservative replication during the cell cycle. Since the primers are added at a high concentration, they anneal to the exposed single strands before these can come back together. Primers may be tagged with fluorescent molecules so that the DNA can be visualised and the progress of PCR monitored. This also allows the DNA to be analysed.

The microbiologist Tom Brock discovered *Thermus aquaticus* in a hot spring in Yellowstone Park in 1966. This bacterium is the source of the heat-stable DNA polymerase *Taq* polymerase. *Taq* polymerase was the first thermostable DNA polymerase to be used in PCR.

Extension

The temperature is next increased to 72 °C for 90 seconds so that extension or elongation occurs with DNA polymerase building up newly synthesised polynucleotides complementary to the template strands. The DNA polymerase uses deoxynucleotide triphosphates (dNTPs) to make the new polynucleotide, adding nucleotides to the 3′ end of the primer and synthesising the polynucleotide strand in the 5′ → 3′ direction. The hydrolysis of a bond between the first and second phosphate groups of each dNTP provides the energy for the formation of a phosphodiester bond between the newly added nucleotide and the growing strand. The other two phosphates leave the active site of the polymerase as pyrophosphate (P-P).

Once the DNA has been copied, the mixture is heated again, which once more separates the two strands in each DNA molecule, leaving them available for copying. Once more, primers fix themselves to the start of each strand of unpaired nucleotides, and DNA polymerase makes a complementary copy.

A single DNA molecule can be used to produce billions of copies of itself in just a few hours. Figure 11.4 shows the first cycle in the amplification

Figure 11.5 A student loading a thermal cycler for PCR. The DNA sample is placed into a plastic tube together with the primers, free deoxynucleotide triphosphate molecules (dNTPs), a buffer solution and DNA polymerase. The thermal cycler automatically changes the temperature of the mixtures in each tube.

of one region of DNA using two primers. The multiplex PCR involves the simultaneous amplification of numerous DNA sequences in a single reaction mixture by using more than one pair of primers.

Some compounds inhibit PCR reactions; examples are substances associated with the stages of extracting and purifying the DNA. These include ionic detergents and gel loading dyes and the enzyme proteinase K, used in extracting DNA from cellular material which, if left in the mixture, will break down polymerases. Similarly, certain substances present in blood can inhibit PCR, such as haemoglobin and the anti-clotting agent heparin.

Example

Figure 11.6 shows the changes in temperature during the first and second cycles in PCR of a sample of DNA.

1 Name the enzyme that carries out the synthesis of DNA during PCR.
2 Explain why
   a) a high temperature is required for stage 1
   b) enzymes do not need to be added for each cycle.
3 a) Describe what happens in stages 2 and 3.
   b) Suggest why stage 1 of the first cycle is longer than stage 1 of the second and all subsequent cycles.
4 Explain why one pair of primers is not suitable to use for amplifying all samples of DNA when using PCR.
5 Explain how PCR differs from DNA replication during the cell cycle.

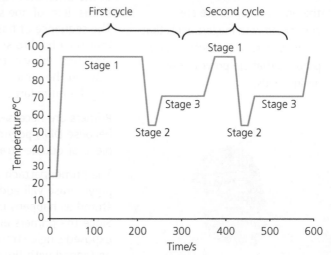

Figure 11.6 The changes in temperature that occur in the first two cycles of PCR.

Answers

1 DNA polymerase; for example, *Taq* polymerase.
2 a) At high temperatures the double-stranded DNA is denatured. The hydrogen bonds break so that the two strands separate exposing the bases.
   b) The polymerase is thermostable. It is not denatured at high temperature.
3 a) Stage 2 is annealing. The primers, short sequences of DNA, bind with the target sequence on DNA where there are complementary bases. Stage 3 is the elongation or extension stage when DNA polymerase adds nucleotides to the primer to form double-stranded DNA using the original strand as a template.
   b) The DNA that is separated into two strands in the first cycle is the DNA from the whole sample and is much longer than the strands that separate during all subsequent cycles. More hydrogen bonds have to break during the first cycle.
4 Different target DNA has different base sequences, so different primers are required to anneal with it.
5 PCR occurs at temperatures between 50 and 90°C, not at temperatures below 40°C. The primer is DNA not RNA. Replication copies the entire DNA in a cell; PCR only copies small stretches of DNA. The DNA polymerase used in PCR is thermostable; in most organisms it is not. The exceptions are the prokaryotes that live in very hot environments. PCR occurs in small plastic tubes.

Gel electrophoresis

Electrophoresis is a technique for separating and identifying substances in a similar way to paper and thin-layer chromatography. Electrophoresis is used to analyse proteins and DNA and is carried out on paper or more often with gels. Gel electrophoresis involves placing a mixture of molecules into wells cut into a gel, adding a suitable buffer solution and applying an electric field. The movement of charged molecules within the gel in response to the electric field depends on a number of factors.

Tip

It is important to remember the difference between the anode and the cathode. The anode is the positive pole that attracts negatively charged molecules; the cathode is the negative pole that attracts positively charged molecules. DNA is always negatively charged in gel electrophoresis. Protein can be negatively or positively charged.

- Composition of the gel: common gels are polyacrylamide gel (polyacrylamide gel electrophoresis, PAGE) for proteins and agarose (a pure form of agar) for DNA.

- Net (overall) charge of the molecules: negatively charged molecules move towards the anode (+) and positively charged molecules move towards the cathode (−); molecules with a higher charge move faster than those with less overall charge.

- Size: smaller protein molecules and fragments of DNA move through the 'holes' in the gel faster than larger ones.

Electrophoresis of proteins

The charge on proteins is dependent on the ionisation of the R groups on the amino acid residues. Some amino acids have R groups that are positively charged ($-NH_3^+$), some have R groups that are negatively charged ($-COO^-$) and some have ones that are not charged. Whether these R groups are charged or not depends on the pH of the solution. When proteins are separated by electrophoresis the procedure is carried out at a constant pH by using a buffer solution. Usually the proteins are denatured in a reducing agent (mercaptoethanol) that breaks disulfide bonds. The proteins are often added to sodium dodecyl sulfate (SDS) which converts the proteins to negatively charged rod shapes so that they move through the gel according to their size. The smaller proteins move faster through the gel and therefore travel further than larger proteins towards the anode (see Figures 11.7 and 11.8).

Gel electrophoresis is used to separate the polypeptides produced by different genes; it is also used to separate the variant forms of enzymes produced by different alleles of the same gene.

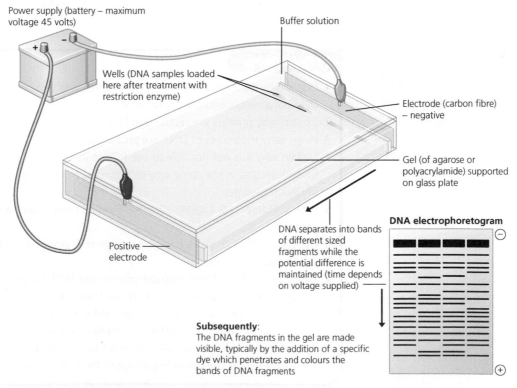

Figure 11.7 An electrophoresis tank containing a gel ready to be switched on. The gel is like a molecular sieve that allows small molecules to travel further than larger molecules.

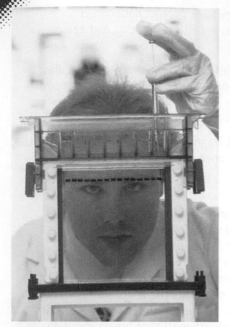

Figure 11.8 Gel electrophoresis of proteins as part of research into a vaccine for HIV/AIDS. The gel was placed in the tank containing a suitable buffer solution. Protein samples with a red tracking dye have been added to wells along the top of the gel. They will migrate downwards towards the anode. Protein gels are usually run vertically, but gravity has nothing to do with the separation.

The β globin in haemoglobin molecules normally has the amino acid valine at position 6. This amino acid has a non-polar R group so is uncharged. In sickle cell anaemia (SCA), glutamic acid is replaced by valine, which has an R group that is charged. These two variants of β globin (normal and sickle cell) can be separated by electrophoresis because they have different net charges. This means that haemoglobin molecules in people who have sickle cell anaemia have a slightly lower negative charge than normal haemoglobin and so they do not move as far through the gel as normal haemoglobin. People that have a single allele for this variant form of β globin do not have sickle cell anaemia. The test to find out whether someone carries the sickle cell allele makes use of the difference in charge on the two molecules to reveal both forms of the globin on the electrophoretogram.

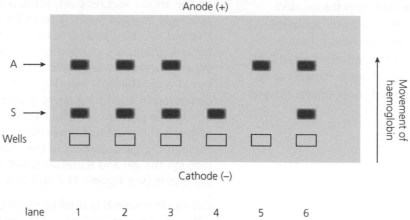

Figure 11.9 Separation of haemoglobin by gel electrophoresis. This analysis was carried out on a family in which one child has sickle cell anaemia. Lane 1 contains haemoglobin standards (A = normal haemoglobin, S = sickle cell haemoglobin); lanes 2 and 3 are the haemoglobin samples from the parents; lanes 4, 5 and 6 are haemoglobin samples from their children.

Test yourself

1 Explain the difference between each of these pairs: exons and introns, DNA polymerase and RNA polymerase.
2 Explain why primers are required in PCR.
3 How many molecules of DNA are produced after eight cycles of PCR?
4 Explain why it is not possible to use PCR to increase the number of RNA molecules in the same way as it is used to increase the number of DNA molecules.
5 Look at Figure 11.9. Explain why there are no more than two bands per individual.
6 Use Figure 11.25 to show how the change to β globin is the result of a substitution mutation.
7 The enzyme alcohol (ethanol) dehydrogenase (ADH) is an enzyme composed of one polypeptide. There are three alleles of the gene that codes for ADH, known as *ADH1*, *ADH2* and *ADH3*. The three polypeptides have the same length, but different overall charges and can be separated by electrophoresis. Explain how electrophoresis can show the different phenotypes of all the possible genotypes for this enzyme.

Gel electrophoresis of DNA

All DNA fragments carry a small charge thanks to the negatively charged phosphate groups in the sugar-phosphate backbone of each polynucleotide. In DNA electrophoresis, these fragments move through the gel towards the anode. The equipment used is shown in Figure 11.10. The tracking dye moves through the gel slightly in front of the smallest DNA fragments so that the progress of electrophoresis can be followed and stopped before reaching the end of the gel. The distance travelled by a length of DNA is inversely proportional to its length: the shorter it is the further it travels through the gel. Molecules of the same size migrate the same distance and form a band as you can see in Figure 11.11. Electrophoresis separates fragments of DNA from samples. The distances that they travel through the gel are determined by visualising the DNA with stains such as ethidium bromide and Azure A.

Figure 11.10 Apparatus for carrying out gel electrophoresis. Here you can see a power pack with black and red leads to the electrophoresis tank, several tubes containing DNA samples and a stand with micropipettes.

Melt agarose gel in buffer solution

Insert a toothed comb at one end of the tank to make the wells to take DNA samples

Pour in molten agarose gel

Leave gel to set. Place electrodes at either end of the tank

When gel is set pour in buffer solution and remove the comb

Add blue dye to each DNA sample

Add DNA and dye mixture to the wells

Connect electrodes to the power supply

When the blue dye is within 10 mm of the end of the gel, disconnect the power supply

Pour away the buffer and add DNA stain (Azure A) for 4 minutes

Rinse with water and analyse the fragments of the DNA, which will appear blue

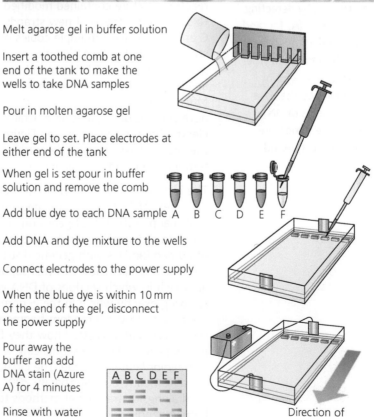

Figure 11.11 A flow chart diagram showing the stages in electrophoresis of DNA.

Tip

Ethidium bromide stains DNA but is only visible with UV light. Ethidium bromide is also a mutagen, so it is unlikely you will use it at A level unless you do some electrophoresis at a university or a research centre.

The separation and detection of DNA is automated by combining the DNA with fluorescent markers. In capillary flow electrophoresis, the fragments of DNA are detected with a laser and a sensor that detects the fluorescence as each band passes towards the anode (see Figure 11.12). It is this method that was used in the first generation of DNA sequencing machines and was responsible for increasing the speed of sequencing (see below).

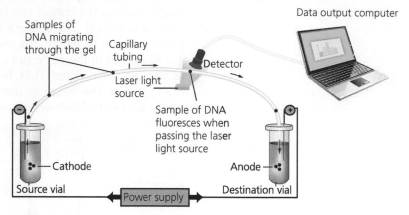

Figure 11.12 Capillary flow electrophoresis. As the separated fragments of DNA pass the laser beam they fluoresce. The fluorescence is detected by a sensor and the information integrated.

DNA sequencing

Key terms

DNA sequencing The sequence of nucleotide bases in samples of DNA is determined by detecting the nucleotide bases (A, T, C and G) as they are added to a template strand by a polymerase enzyme.

High-throughput sequencing Any method in which the base sequences in samples of DNA, including whole genomes, are determined in a short period of time.

The principle of DNA sequencing is to find the order of the nucleotide bases along regions of DNA. The chain-termination method of sequencing developed in the 1970s involved replicating short fragments of single-stranded DNA much in the same way as described for PCR. However, the reaction mixture contained modified dNTPs known as dideoxynucleotides to stop the synthesis of new strands of DNA at each of the bases, A, T, C and G. This gave multiple copies of DNA that had a complete template strand but with varying lengths of newly synthesised strands. These partially replicated DNA molecules could be separated by electrophoresis as they were all of different mass. To begin with this was a slow process. Much faster, high-throughput sequencing became possible with the development of sequencing machines containing 96 sets of capillary flow electrophoresis apparatus (see Figure 11.12). These machines increased the speed at which the different lengths of DNA synthesised by chain termination could be sequenced.

The increase in speed opened up the opportunity to sequence whole genomes. At first the process began with traditional methods of genetics that mapped the loci of genes on chromosomes. These methods use the data provided by genetic crosses and from studying the pedigrees of human families with genetic diseases, such as Huntington's disease and haemophilia. Later this was supplemented by the so-called 'shotgun' approach in which sections of DNA were sequenced at random without knowing exactly where in the genome they came from. Once sequenced, the data from these random fragments was compared to find common sequences that suggested how they fitted together like a very long, linear jigsaw puzzle. Shotgun sequencing was used first to sequence the genome of *Haemophilus influenzae*. The circular chromosome of this bacterium contains nearly 2 million bp. By 2001, shotgun sequencing had been used alongside traditional methods to produce a draft sequence of the human genome.

Next-generation sequencing Any method of sequencing that has replaced the chain-termination (Sanger) method of sequencing.

Tip

Luciferase is the enzyme used by fireflies to emit flashes of light. See page 60 in *OCR A level Biology 1 Student's Book* for more about this enzyme. Pyrosequencing is another application of luciferase.

The chain-termination method, also known as Sanger sequencing, has now been replaced by new methods of sequencing known as next-generation sequencing (NGS) (see Figure 11.13). These are processes that enable the sequencing of thousands to millions of molecules of DNA at the same time without the need for electrophoresis. Other methods are used to identify the bases that are added to the extending chain during the sequencing process; for example, pyrosequencing uses the enzyme luciferase to emit flashes of light as pyrophosphate (P-P) is released when individual nucleotides are added to DNA templates during the synthesis of polynucleotides.

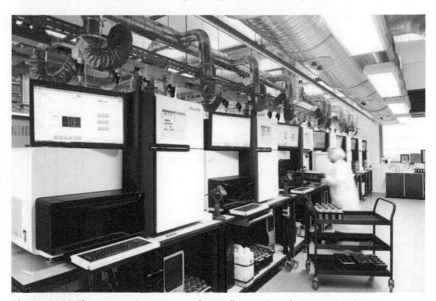

Figure 11.13 The sequencing room at the Wellcome Trust's Sanger Institute near Cambridge. These are high-throughput next-generation sequencing machines.

Table 11.1 shows a timeline of the history of DNA sequencing, with some key events between 1975 and 2015.

Table 11.1 Timeline of DNA sequencing.

Date	Event in the history of DNA sequencing
1975	Chain-termination method developed by Fred Sanger (Figure 11.14) and Andrew Coulson in the UK and the chemical degradation method by Alan Maxam and Walter Gilbert in the USA.
1977	The first genome, that of the 5386 nucleotide, single-stranded bacteriophage φX174, was completely sequenced using the chain-termination method.
	The shotgun approach was used to sequence the genome of *Haemophilus influenzae*.
1990s	The development of capillary flow electrophoresis and appropriate detection systems for DNA.
1996	Genome of the yeast *Saccharomyces cerevisiae* was sequenced: this was the first eukaryote to be sequenced.
1998	Development of sequencing machines with 96 capillary sequencers, a technique that was termed high-throughput sequencing.
2001	First draft of the human genome published (complete genome published in 2004).
2005	The first next-generation sequencing machine became available.
2012	The publication of the genome of the tomato, *Solanum lycopersicum*, sequenced using the chain-termination technique and next-generation methods. Probably the last genome to be sequenced using chain-termination (Sanger) sequencing.
2015	Genome of an extinct mammal, the woolly mammoth, sequenced.

Figure 11.14 Sir Fred Sanger, winner of two Nobel Prizes – in 1958 for developing a method to sequence amino acids in proteins and in 1980 for the chain termination method for sequencing DNA. The Sanger Institute near Cambridge, where the genomes of many organisms have been sequenced, is named after him.

These next-generation methods can be a hundred to a thousand times faster than previous methods, so reducing the cost of sequencing 1 million nucleotides (1.0 megabase pairs) to as little as 0.1% of the cost of using chain termination. A complete genome can be sequenced for less than US$1000. The development of the nanopore method of sequencing promises to revolutionise the process so it can be carried out rapidly almost anywhere, not just in expensive laboratory facilities (see Figure 11.15).

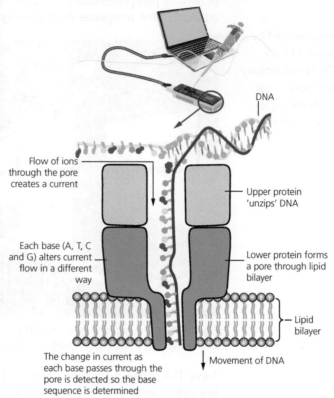

Figure 11.15 Nanopore sequencing. The diagram shows a vertical section through a nanopore. This method is still in development and being trialled, but when widely available the technique should make sequence data available cheaply and easily for all sorts of applications in biology.

Whole-genome sequencing

The reduced costs of next-generation sequencing allow the genomes of different species to be sequenced. Comparisons of whole genomes allow scientists to find the location of the same or similar genes in different species and also provide data to analyse evolutionary relationships between species. Next-generation sequencing also allows the sequencing of genomes from many individuals of the same species. This gives information on the genetic variation within a species.

As well as the human genome, the genomes of some other mammals have been sequenced. Particularly important are the genomes of our domesticated animals and those at risk of extinction, such as giant pandas. The genomes of horse, cattle, pig and yak have been sequenced. Crop plants have also been sequenced; examples are rice, papaya, cabbage, potato, wheat and tomato.

By understanding the genome, researchers have been able to 'knock out' the genes in a simple organism known as *Mycoplasma* to find out the minimum number required to keep the organism alive. Similar studies have been done with other prokaryotes and with yeast. Scientists in the USA have also assembled a genome to make a synthetic organism (see page 207 in *OCR A level Biology 1 Student's Book*). Gene sequencing has given scientists opportunities to design and make new molecules by writing completely new base sequences and inserting them into bacterial or eukaryotic cell using the techniques of genetic engineering. This is possible because they can use the genetic code to predict the amino acid sequence encoded by any length of DNA. It is also possible to predict how these new proteins will fold to give specific tertiary structures. This allows for the production of new molecules, such as enzymes involved in medicinal drug production. Also possible is changing the sequence of bases in existing genes to improve the way in which certain proteins work as in the various forms of human insulin that are now available for the treatment of diabetes. It is also possible to redesign cellular systems so that organisms, such as those that are genetically modified for the production of proteins, can work more efficiently. Synthetic biology is the term coined for all the various applications of molecular and cellular biology, much of which relies on a knowledge of the structure and function of nucleic acids.

Epidemiology is the study of the spread of diseases in populations. Whole-genome sequencing of pathogens has improved the ability of epidemiologists to identify the particular strains that infect people, livestock and crops and determine the most appropriate control methods to use. For example, genome sequencing of the Ebola virus contributed to understanding the transmission of the disease during the epidemic in West Africa in 2015.

Key term

Synthetic biology is a new science that involves the design of new biological systems and the redesign of existing systems using a knowledge of nucleic acids and cell biology.

DNA profiling

Sequencing has revealed the existence of many ways in which the DNA of individuals in a species differ from one another. This makes it possible to identify individuals with a very high degree of accuracy since the chances of any two individuals within a species having exactly identical DNA is very small, unless they are clones. One of the strange features of DNA is the existence of nucleotide sequences that are repeated, often many times. The regions of DNA that have these repeats are considered to be 'genes' with the variable number of repeats being treated as if they are alleles. As these regions are highly variable these 'genes' are another example of genetic polymorphism.

Before the era of high-throughput and next-generation sequencing, establishing the identify of someone depended on looking for differences in phenotypes; for example, the identification of carrier status in sickle cell anaemia by using protein electrophoresis (see Figure 11.9) and the use of antibodies to detect people's blood groups. The discovery of sequences that are repeated in DNA in structural genes and in regions between them led to methods of genetic profiling (often called genetic fingerprinting). Among the first markers to be used were **minisatellites**, which are sequences repeated at various points along chromosomes. These are 10–100 bp in length and are inherited from both parents. They can be cut by restriction enzymes and the lengths of the fragments determined by gel electrophoresis.

The use of minisatellites has now been largely replaced by that of **short tandem repeats** (STRs), also known as **microsatellites**. STRs are repeated sequences of nucleotides which are much shorter than minisatellites. STRs are regions of the genome composed of 2–5 nucleotides repeated 10–30 times. For example, the sequence CACACA may be repeated between five and 20 times. The number of repeats in any one STR is variable. The variability in STRs is caused by the inaccuracy of DNA polymerase in copying these regions during replication. The error rate in replication that leads to mutation is very low, but one effect is an increase in the number of these repeated base pairs.

Tip

The role of restriction enzymes is described below, in the section on genetic engineering. Minisatellites are also known as variable number tandem repeats (VNTRs) and these were used in the first form of genetic fingerprinting to be used to give the ladder-like bands on gels, such as the one in Figure 11.11.

Tip

There is more about the gene for the protein huntingtin in Chapter 10.

One person may have the triplet ACA repeated 65 times at one position in their genome and 118 times at another position. STR markers can be simple (identical-length repeats), compound (two or more adjacent repeats) or complex (several different length repeats). They are found on the 22 autosomal chromosomes as well as on both X and Y sex chromosomes. The CAG repeat sequence in the *huntingtin* gene is an example of an STR. In DNA profiling, the different versions of selected numbers of STRs are determined. Many of the most useful STRs are in regions of non-coding DNA between structural genes.

DNA profiling in forensics

DNA profiling is used by forensic scientists to identify individuals such as the victims of crime, suspects in criminal cases, victims of disasters and missing people.

Only a select number of STR markers are used in forensic DNA profiling: the European DNA 17 profile consists of 16 loci and markers that identify gender. These STRs show a high degree of polymorphism, making them of particular use to the forensic scientist, and the STR regions are in non-coding DNA so there is no selective pressure against any of the alleles, with the result that there is much variation between different people.

STRs used in forensic science tend to be four- or five-nucleotide repeats, as they are robust, suffer only low levels of environmental degradation and provide a high degree of error-free data. STR loci are ideal for use in forensic science for a number of reasons. They represent discrete alleles that are distinguishable from one another, they show a great power of discrimination, only a small amount of sample is required due to the short length of STRs, PCR amplification is robust and multiple PCR can be used, and there are low levels of artefact formation during amplification.

A typical print out shows the sex of the individual and two peaks for each of the STR genes. Most people are heterozygous because they have two alleles of each gene with different number of repeats in each (see Figure 11.16).

STR profiles are displayed in the green, blue and yellow channels of a four-colour fluorescent system, with the red channel being used for a size marker.

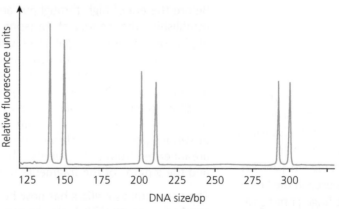

Figure 11.16 Part of an STR profile showing three of the loci that are shown in the green channel. The peaks represent the number of repeat sequences and these are indicated in boxes beneath the peaks for each gene as shown in Figure 11.17.

PCR is used to make many copies of the DNA with these STRs. After amplification by PCR, the sequences are separated by gel electrophoresis. The original DNA sample could be as little as 100 pg from about 16 cells. The STRs are labelled with fluorescent dyes and detected by a laser scanner. More modern analysis dispenses with the separate electrophoresis stage

Tip

Remember that these 'genes' and their 'alleles' are not like those discussed in Chapter 10. They do not code for polypeptides, but they are inherited in the same way.

and detects the fluorescence during PCR with four colours, such as red, green, yellow and red. A print out from part of such an analysis is shown in Figure 11.17. A full analysis using 10 or 13 STR loci will include results like those for the green loci (as shown in Figure 11.17) and the results for the other loci and will therefore have many more peaks.

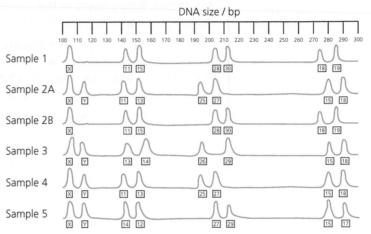

Figure 11.17 The results of an STR analysis. The results for the four green loci are shown for six samples in a sexual assault case. Samples 2A and 2B were taken from a vaginal swab of the woman; samples 1, 3, 4 and 5 were prepared from buccal swabs taken from the woman and three possible suspects in the case.

Tip

Test yourself question 10 refers to the results for this STR analysis.

DNA profiling and disease risk

DNA testing or genetic testing can be used to detect the risk of a person having a genetic disease or disorder. The sample of DNA may come from a sample of blood or a buccal swab, which involves wiping a cotton bud on the inside of the cheek to remove some epithelial cells. Almost all of these molecular genetic tests use PCR. As PCR is highly selective it allows copies to be made of just the section of DNA needed for the test. However, it can only amplify a very small length of DNA (typically up to 2000 bases) and, as most genes are larger than this, multiplex PCR reactions are needed to copy a whole gene. There are genetic tests for many monogenic diseases, such as cystic fibrosis, Huntington's disease, sickle cell anaemia, β-thalassaemia, haemophilia A and B, severe combined immunodeficiency syndrome and some forms of breast cancer.

All these diseases are the result of mutations that are inherited and so run in families. If someone is confirmed as having the mutation, genetic tests are often offered to other members of the family. Some people know that a particular disease runs in their family and opt for genetic testing. However, in the case of the two types of haemophilia up to 40% of people who develop these diseases come from families in which the disease is previously unknown. Also people who inherit a dominant allele for Huntington's disease, for example, do not always develop symptoms of the disease. Similarly people who inherit the *BRCA1* or *BRCA2* genes, which are autosomal dominant, inherit an increased risk of developing breast cancer but do not necessarily develop the disease itself.

Single nucleotide substitutions are the cause of most genetic diseases. Single nucleotide polymorphisms (SNPs; pronounced 'snips') are the simplest and most common type of genetic variation, comprising around 90% of genetic variation in humans. They occur during meiosis when DNA is replicated. For example, a SNP may replace a cytosine with a thymine in a stretch of DNA. This will not change the length of the DNA so, if the SNP

Test yourself

8 What is the difference between a single nucleotide polymorphism (SNP) and a short tandem repeat (STR)?

9 Explain why SNPs and STRs are examples of genetic polymorphism.

10 Explain why there can be four alleles of an SNP, but no more than four.

11 Use the data in Figure 11.17 to identify the victim and the perpetrator of the crime. Explain your answers.

falls in the exon of a gene, there is no effect on the gene's reading frame during protein synthesis. There are about 10 million SNPs in the human genome, with one found every 100–300 bp. They act as useful biological markers of disease. Each SNP can consist of up to four alleles, because there are four possible bases at each nucleotide position. In the analysis of SNPs the specific base present in the SNP must be established. This is achieved using DNA sequencing techniques.

The different human genome sequencing projects have revealed that individuals have between 3 and 4 million SNPs between one another and the reference sequence. Thus far, a total of over 30 million SNPs have been discovered from human genome sequencing projects.

The availability of cheap next-generation sequencing techniques, such as nanopore sequencing (see Figure 11.14), will allow doctors to find out the sequence of specific regions of the genome and prescribe drugs that will be effective on people with specific genotypes. This is personalised medicine, as described in Chapter 11 in *OCR A level Biology 1 Student's Book*.

Activity

Molecular caretakers of DNA

The proteins BRCA1 and BRCA2 are tumour-suppressor proteins that play a central role in DNA repair. They move to the site of DNA damage and recruit other proteins to initiate DNA repair complexes. BRCA1 also activates p53, a key protein involved in activating apoptosis in response to irreparable DNA damage.

The genes that code for these proteins are *BRCA1* and *BRCA2*. Mutations of these genes are involved in the development of breast cancer. When either *BRCA1* or *BRCA2* are absent as the result of a mutation, these DNA repair complexes do not form after DNA damage.

1 Outline the roles of the proteins BRCA1 and BRCA2 in protecting the genome.

2 Suggest how mutations of *BRCA1* and *BRCA2* increase the risk of breast cancer.

3 Suggest how BRCA1 and BRCA2 activate p53.

BRCA1 is 81 kb in length and comprises 24 exons; the introns vary in length between 403 bp and 9.2 kb. More than 1000 mutations of this gene are known. These include an 11 bp deletion, a 1 bp insertion, a stop codon and a missense substitution. Particular mutations are much more common in particular groups of people; for example, Ashkenazi Jews are much more likely to have a deletion of two nucleotides in exon 2 which results in a premature stop signal at codon 39 in the BRCA1 protein. This deletion is known as 185delAG.

This shows the deletion:

DNA sequence ... AAA ATC TTA GTG TCC C ...

185delAG DNA sequence ... AAA ATC TTA GTG TCC C ...

4 Mutations to *BRCA1* are inherited as autosomal dominant disorders. Explain how the inheritance of this disorder differs from the inheritance pattern for an autosomal recessive disorder, such as cystic fibrosis.

5 a) Explain what is meant by a missense substitution.

   b) State the effect of a mutation that results in a stop codon.

   c) Explain why an 11 bp deletion may result in a frameshift mutation.

Women with an abnormal *BRCA1* or *BRCA2* gene have an extremely high risk of developing breast cancer and ovarian cancer.

6 a) Explain how particular mutations of *BRCA1* are more likely to occur in particular groups of people.

   b) Suggest an advantage of this for screening the human population for breast cancer.

Mole rats live in burrows and have many adaptations to a subterranean way of life. Professor Eviatar Nevo from the University of Haifa in Israel has studied blind mole rats, *Spalax galili*, for over 50 years and has never observed any spontaneous tumours, even though these animals may live for 20 years or more. Naked mole rats, *Heterocephalus glaber*, also never develop tumours. When these animals are treated with carcinogens, their immune system attacks any tumour cells that develop. Mole rats have a variant of p53, thought to be an adaptation to the low-oxygen environment in which they live. The genomes of *S. galili* and *H. glaber* have been sequenced.

7 Suggest how the immune system of the rats detects and destroys cancer cells.

8 Explain how studies of the genomes of mole rats may prove useful in research on cancer.

Tip

Look back to Chapter 10 to remind yourself about how the events of the cell cycle are controlled.

Bioinformatics

Bioinformatics is a research tool involving a combination of biological data with computer technology and statistics. It builds up databases and allows links to be made between them. The databases hold gene sequences, sequences of complete genomes, amino acid sequences of proteins and details of protein structures. Computer technology facilitates the collection and analysis of this mass of information in databases and allows access to them via the internet (see Figure 11.18).

In 2014, over a thousand databases held over 6×10^{11} bp of sequencing data. Databases that hold the coordinates required to show three-dimensional models held details of over 100 000 different proteins and nucleic acids. The amount of data is vast and growing at an exponential rate. The information needs to be in a form that can be searched so software developers play an important role in developing suitable systems that allow access to all the data.

Databases also hold information on macromolecules: their structures and functions, gene expression patterns, metabolic pathways and control cascades. A particular problem for the future is collecting and organising data on the variety of proteins that are synthesised by eukaryotic cells. Human cells splice together exons of structural genes in different combinations to give polypeptides with different primary and hence different secondary and tertiary structures. Also, polypeptides can be assembled in different ways to form proteins with different quaternary structures as in the case with lactate dehydrogenase. Polypeptides can be glycosylated differently too.

There are many databases that hold different types of information, as in the following examples.

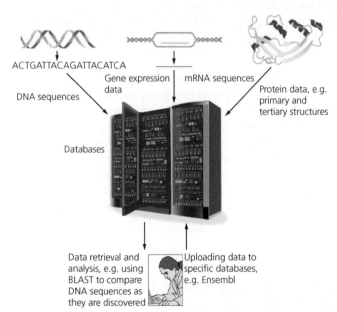

Figure 11.18 Bioinformatics: the storage, retrieval and analysis of data from studies of DNA, RNA, proteins, genotypes and phenotypes.

- The Genomes OnLine Database (GOLD) is a comprehensive online resource to catalogue and monitor genetic studies worldwide. GOLD provides up-to-date status on complete and ongoing sequencing projects along with a broad array of curated metadata.

- Nucleotide sequences are held by the Nucleotide Sequence Collaboration between GenBank (USA), the European Nucleotide Archive and the Center for Information Biology and DNA Data Bank in Japan. These databases compare sequences between the different organisations on a daily basis.

- The database Ensembl holds data on the genomes of eukaryotic organisms. Among others it holds the human genome and the genomes of zebrafish and mice that are model organisms used a great deal in research.

Tip

Bioinformatics is the web. You can find out how information stored in databases is used in research by investigating a particular protein.

Without good search and retrieval tools all the information stored would be of little value. The search tool BLAST (which stands for Basic Local Alignment Search Tool) is an algorithm for comparing primary biological sequence information, such as the primary sequences of different proteins or the nucleotide sequences of genes. Researchers use BLAST to find similarities between sequences that they are studying and those already saved in databases.

When a genome has been sequenced, for example the human genome, comparisons can be made with other known genomes, such as that of the fruit fly, *Drosophila melanogaster*. Sequences can be matched and degrees of similarity calculated. Similarly, comparisons can be made between amino acid sequences of proteins or structures of proteins. Very close similarities indicate recent common ancestry.

When human genes, for example those concerned with development, are also found in other organisms such as *Drosophila*, these organisms can provide useful models for investigating the way in which the genes have their effect. Being able to read gene sequences is providing valuable information in the development of vaccines for malaria. All the information about the genome of the malarial parasite, *Plasmodium*, is available in databases for researchers to access. This information is being used to find new methods to control this parasite. Sequencing malarial parasites from different areas of the world shows that Cambodia is the area where mutations giving rise to drug resistance tend to arise.

Bioinformatics allows the comparison of genomes in different organisms to investigate evolutionary relationships. At one level, sequence data confirms the division of life into three domains. At another level it shows the similarities between genes of organisms with very different phenotypes and ways of life. Many genes in model organisms such as yeast, fruit fly and zebrafish are the same as in us. These genes code for proteins that fulfil the same roles, such as the enzymes of respiration. At the smallest level – the single base pair – the differences can be used to investigate the evolutionary relationships between closely related populations and species.

Tip

Chapter 14 in *OCR A level Biology 1 Student's Book* has some examples of the way in which sequence data is used to investigate relationships between species.

Genetic engineering

Key term

Vector Any structure that is used to deliver a gene into a host organism in genetic engineering. Examples are viruses, plasmids and liposomes. In Chapter 11 in *OCR A level Biology 1 Student's Book* we used the term vector for an organism that transmits a pathogen from one host to another. In genetic engineering a vector transfers a piece of DNA into an organism.

Genetic engineering is the term usually applied to the process of genetic modification by means that are not possible using selective breeding (artificial selection). It involves the removal of a gene or genes from one organism and placing them into another. At one extreme this involves transferring genes between species. But it can also involve taking a gene from an individual of a species and transferring it into other individuals of the same species. The DNA corresponding to a gene is obtained in one of a number of ways and inserted into a vector, such as a virus, plasmid, bacterial artificial chromosome (BAC) or liposome, which is used to transfer the gene into host cells. Alternatively the DNA can be inserted directly into cells without using a vector.

The genetic code is universal so it is possible to make transfers between widely different species; for example, from a human into a bacterium, a jellyfish into a bacterium or a bacterium into a plant. All cells give this genetic code the same meaning so the protein originally encoded by the gene will be the same protein that is produced in any cell to which it is transferred.

These are some of the reasons for wanting to transfer genes between genomes

- modifying bacteria and eukaryotic cells to make special chemicals that are only produced in small quantities by other methods

- making crop plants resistant to diseases, pests and herbicides

- making livestock resistant to diseases and pests

- improving the yields from crop plants and livestock

- improving the nutritional qualities of crop plants

- modifying animals to make human proteins for medicines that are difficult to obtain by other methods

- modifying bacteria so they absorb and metabolise toxic pollutants.

Scientists researching molecular and cell biology have made many discoveries that have found applications in the genetic modification of organisms. These are the 'tools' that are available in the genetic engineer's 'toolkit' for the transfer of genes between genomes: some are listed in Table 11.2.

Table 11.2 'Tools' for the genetic engineer.

'Tools'	Source	Role in genetic engineering
Restriction enzymes	Bacteria	Cut DNA at specific sequences
Ligase enzymes	Viruses (e.g. T4 bacteriophage) and bacteria	Seal cut ends of DNA by forming phosphodiester bonds
Reverse transcriptase enzyme	Retroviruses (viruses with an RNA genome not a DNA genome), e.g. HIV	Uses RNA template to make cDNA
Plasmids	Small circular pieces of DNA found in bacteria and yeasts	Vectors to transfer DNA into bacteria and yeasts
Liposomes	Artificially produced from phospholipids	Vectors to transfer DNA into cells by fusing with cell membrane
Bacterial artificial chromosomes (BAC)	Artificially produced	Vectors to transfer DNA into bacteria
Viruses	Retroviruses with RNA as genetic material	Vectors to transfer genes into human cells
	Bacteriophages	Vectors to transfer genes into bacteria
CRISPR-Cas9	Bacteria	Cuts DNA in specific places for genes to be inserted

Tip

Obtaining mRNA from protein-synthesising cells is much easier than trying to find and isolate the length of DNA on one chromosome that codes for the protein. Reverse transcriptase was used to produce the DNA used to modify bacteria for the production of insulin (see Figure 11.22).

Reverse transcriptase is an enzyme that catalyses the reverse of transcription. The enzyme uses RNA as a template to synthesise a polynucleotide using deoxynucleotide triphosphate molecules (dNTPs). The complementary DNA strand that is produced has a base sequence that is complementary to the sequence of bases on RNA. When a molecule of mRNA isolated from the cytoplasm of cells is used as the template, the cDNA has the same base sequence as the coding strand of the nuclear DNA from which it was formed in transcription. The great advantage of this is that the cDNA has no introns, only exons.

Restriction enzymes are used to cut genes from lengths of DNA. They cut across both strands of DNA at specific sites known as restriction sites (as they are restricted to cut only at these sites). These sites are typically about 6 bp in length. The restriction enzymes come from bacteria that are too difficult or dangerous to culture so the genes that code for them have been removed and inserted into safer strains of bacteria that are modified to produce high quantities of them.

Each restriction site has a specific nucleotide sequence which is palindromic, so it reads the same in the 5′ to 3′ direction as in the 3′ to 5′ direction. Restriction enzymes are named after the bacteria from which they were first isolated. They are present in bacteria to defend against attack by viruses; this is called restricting an infection by viruses. Table 11.3 shows the restriction sites for five restriction enzymes and Figure 11.20 shows the action of two of these enzymes in cutting DNA.

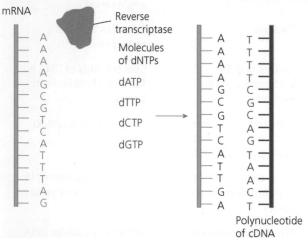

Figure 11.19 Reverse transcriptase assembles nucleotides into a sequence that is complementary to RNA. The cDNA produced is used as a template strand in making a complete double-stranded molecule of DNA by DNA polymerase.

Table 11.3 The recognition sites for five restriction enzymes. Vertical lines indicate the cutting site.

Restriction enzyme (source bacterium and strain)	Restriction site
*Bam*HI *Bacillus amyloliquefaciens* H	5′ … G|GATC C … 3′ 3′ … C CTAG|G … 5′
*Hpa*I *Haemophilus parainfluenzae*	5′ … GTT|AAC … 3′ 3′ … CAA|TTG … 5′
*Eco*R1 *Escherichia coli* RY13	5′ … G|AATT C … 3′ 3′ … C TTAA|G … 5′
*Hind*III (pronounced Hindy 3) *Haemophilus influenzae* Rd	5′ … A|AGCT T … 3′ 3′ … T TCGA|A … 5′
*Hae*III *Haemophilus aegyptius*	5′ … GG|CC … 3′ 3′ … CC|GG … 5′

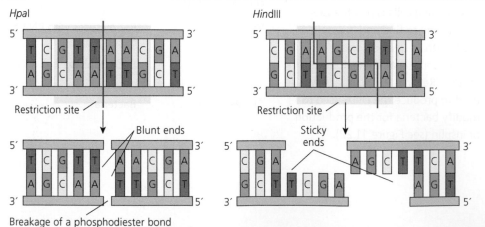

Figure 11.20 The effects of two restriction enzymes, *Hpa*I and *Hind*III, in cutting DNA at specific restriction sites.

Some restriction enzymes cut DNA to give it free, unpaired 'ends'. You can see that *Bam*HI, *Eco*RI and *Hind*III do this. These ends are known as 'sticky ends' because they will form base pairs with complementary sequences of bases. This is how a gene cut by a restriction enzyme can be inserted into a plasmid or into viral DNA. Others, such as *Hpa*I and *Hae*III, cut straight across DNA to give blunt ends.

Once a sequence of DNA is removed it is often necessary to produce large quantities of it. This is done by inserting the gene into bacteria (Figure 11.21) and using the bacteria to replicate the gene. This may be done by inserting the gene into a vector, such as a plasmid or a virus, which then takes it into the bacterium. The plasmid or viral DNA must have the same restriction site as the gene so that the same restriction enzyme can cut it to give a short length of complementary bases. DNA cut by a restriction enzyme such as *Hpa*I can have a length of nucleotides added to give it sticky ends. Copies of the gene are mixed with the plasmid. Some plasmids will take up the gene by complementary base pairing of sticky ends; other plasmids will simply reform without taking up the gene. Hydrogen bonding between the sticky ends attaches the two and the enzyme ligase is added to form the covalent phosphodiester bonds of the sugar-phosphate backbone of DNA.

Key term

Ligase An enzyme that can be used to catalyse the formation of phosphodiester bonds between the terminal nucleotides of DNA fragments.

Tip

Look at diagrams of DNA to see the hydrogen bonds between the base pairs and the position of phosphodiester bonds in each polynucleotide. Note the three hydrogen bonds between C and G and the two between A and T.

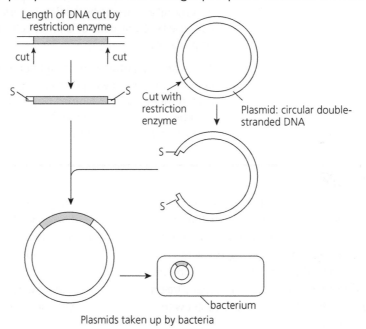

Figure 11.21 Plasmids are used as vectors in genetic engineering to insert genes into bacteria. The regions labelled S are sticky ends.

Recombinant DNA (rDNA) is produced once the gene is inserted into a vector. The host bacteria are treated with calcium ions, then cooled and given an electric shock to increase the chances of plasmids passing through the cell surface membrane. This is the process of electroporation. The bacteria are now described as **transformed** because they contain foreign DNA.

Some plasmids do not take up the foreign gene and some bacteria do not take up plasmids. There are a variety of ways to identify the transformed bacteria from those that do not have the recombinant plasmids. Similar techniques are used to identify eukaryotic organisms that have been genetically modified. Vectors may therefore contain marker genes, such as

- antibiotic-resistance genes: the foreign genes are inserted into these resistance genes so that they cannot be expressed; transformed cells are therefore sensitive to the antibiotic and untransformed cells are not

Key terms

Recombinant DNA (rDNA) Formed by combining DNA from two different sources, such as human and plasmid.

Electroporation Giving an electric shock to stimulate cells to take up pieces of DNA such as plasmids and bacterial artificial chromosomes.

- the gene for green fluorescent protein (GFP), which is a small protein that emits bright green fluorescence in blue or ultraviolet light
- the gene for β glucuronidase (GUS), an enzyme from *E. coli* that converts colourless substrates into coloured products; it is used in plants to indicate that they have been genetically engineered.

DNA polymerase in bacteria copies the plasmids; the bacteria then divide by binary fission so that each daughter cell has several copies of the plasmid. The bacteria transcribe and may translate the foreign gene. If the bacteria produce a foreign protein they are described as transgenic. For transcription to occur in the host organism a promoter must be inserted with the foreign gene. The length of DNA that is constructed from promoter, foreign gene and marker gene, and maybe other genes as well, is known as a **gene construct** or **construct** for short.

Bacteria can make many copies of genes very quickly. This gives many copies that can be used in research. Alternatively the bacteria are grown in large quantities to make specific proteins, such as enzymes for the food industry. However, this is not always possible as bacteria do not carry out the complex post-translation processes of eukaryotic cells to cut, fold and modify proteins by adding sugars. As a result genetic engineers may use yeasts, plant cells or animal cells as the host cells for the production of proteins.

There are advantages in using bacteria, yeasts and cultures of mammalian cells to produce these proteins. These cells have simple nutritional requirements, large volumes of product are produced, the production facilities do not require much space and the processes can be carried out almost anywhere in the world.

Test yourself

12 Explain why predicting amino acid sequences from nucleotide sequences is more accurate than predicting nucleotide sequences from amino acid sequences.

13 The bacteriophage lambda has a genome of 48 502 base pairs. There are seven restriction sites for *Hind*III and five for *Bam*HI. State how many fragments are produced when the lambda DNA is cut with

a) *Hind*III alone

b) *Bam*HI alone

c) both restriction enzymes.

14 Suggest why restriction enzymes do not cut bacterial DNA.

15 In an investigation of the chloroplast genome of *Arabidopsis thaliana* the DNA was cut with restriction enzymes and the fragments produced analysed by electrophoresis. Outline the practical steps involved in this investigation.

Activity

Producing insulin

Insulin was the first recombinant product approved for use in treating humans. It has been produced by biotechnology companies for many years in *Escherichia coli* and *Saccharomyces cerevisiae*. Figure 11.22 shows two images of the human insulin molecule.

1 Describe the structure of an insulin molecule as shown in Figure 11.22.

Insulin is a hormone involved in cell signalling. The two polypeptides have a specific three-dimensional shape to fit into a cell membrane receptor. Figure 11.23 shows how bacterial cells were genetically modified to produce insulin.

2 How do plasmids and bacterial chromosomes resemble each other and how are they different?

3 Why is it a better idea to use two different restriction enzymes to cut plasmids before the insertion of foreign DNA rather than using just one type?

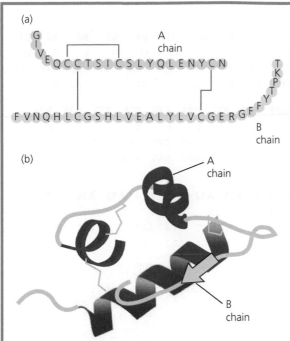

Figure 11.22 (a) The primary structure of human insulin and (b) a computer-generated image of the three-dimensional structure of insulin. In (a) the lines between chains indicate disulfide bridges between cysteine residues.

4 Make a diagram to show what happens at the annealing stage.

5 Bacterial cells are treated with calcium ions. What else may happen to them at this stage?

6 Explain why the gene *INS* taken directly from human β cells cannot be used in this process.

In 2010, researchers in India and Germany published the results of a study into improving the yield of insulin produced by GM microorganisms by using the yeast *Pichia pastoris*. The gene for insulin precursor (IP) was incorporated into a vector used for transforming yeast cells. Figure 11.24 shows the nucleotide sequence of the construct consisting of the *IP* gene including a nine-nucleotide joining section and two flanking sections at the 5′ and 3′ terminals. The diagram also shows the amino acid sequence of the protein that is translated by *P. pastoris*.

7 Explain why the tripeptide sequence is included in the gene.

8 With reference to Figure 11.24, explain how the tripeptide has been designed to be cut by trypsin. State the other place in the protein where trypsin acts.

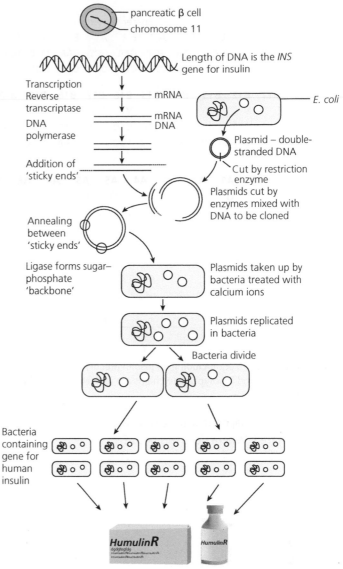

Figure 11.23 The steps involved in genetically modifying cells of *E. coli* to produce insulin on a commercial scale.

9 Suggest why the nucleotide sequences at X and Y have been included.

10 Explain the advantages of using genetically modified microorganisms for the production of insulin.

The major advantage of *P. pastoris* over *E. coli* is that *Pichia* is capable of producing disulfide bonds and glycosylation in proteins.

X ⌐

				spacer sequence									**B1**	**B2**	**B3**	**B4**	**B5**	**B6**	**B7**	**B8**
L	E	K	R	E	E	A	E	A	E	A	E	P	K	F	V	N	Q	H	L	C	G
CTC	GAG	AAG	AGA	GAA	GAA	GCT	GAA	GCT	GAA	GCT	GAA	CCA	AAG	TTT	GTT	AAC	CAA	CAT	TTG	TGT	GGT
GAG	CTC	TTC	TCT	CTT	CTT	CGA	CTT	CGA	CTT	CGA	CTT	GGT	TTC	AAA	CAA	TTG	GTT	GTA	AAC	ACA	CCA

B9	**B10**	**B11**	**B12**	**B13**	**B14**	**B15**	**B16**	**B17**	**B18**	**B19**	**B20**	**B21**	**B22**	**B23**	**B24**	**B25**	**B26**	**B27**	**B28**	**B29**
S	H	L	V	E	A	L	Y	L	V	C	G	E	R	G	F	F	Y	T	P	K
TCT	CAT	TTG	GTT	GAA	GCT	TTG	TAC	TTG	GTT	TGT	GGT	GAA	AGA	GGT	TTC	TTC	TAC	ACT	CCA	AAG
AGA	GTA	AAC	CAA	CTT	CGA	AAC	ATG	AAC	CAA	ACA	CCA	CTT	TCT	CCA	AAG	AAG	ATG	TGA	GGT	TTC

tripeptide sequence			**A1**	**A2**	**A3**	**A4**	**A5**	**A6**	**A7**	**A8**	**A9**	**A10**	**A11**	**A12**	**A13**	**A14**	**A15**	**A16**	**A17**
A	A	K		G	I	V	E	Q	C	C	T	S	I	C	S	L	T	Q	L	E
GCT	GCT	AAG		GGT	ATT	GTT	GAA	CAA	TGT	TGT	ACT	TCT	ATT	TGT	TCT	TTG	TAC	CAA	TTC	GAA
CGA	CGA	TTC		CCA	TAA	CAA	CTT	GTT	ACA	ACA	TGA	AGA	TAA	ACA	AGA	AAC	ATG	GTT	AAC	CTT

A18	**A19**	**A20**	**A21**			A	A	A
D	Y	C	D					
AAC	TAC	TGT	AAC		TAA	GCG	GCC	GCA
TTG	ATG	ACA	TTG		ATT	CGC	CGG	CGA

Y ⌐

Figure 11.24 A construct including an insulin precursor (IP)-encoding gene, designed for expression in *P. pastoris*. The amino acid positions in the A and B polypeptide chains are indicated. Note that the one-letter code is used for amino acids here, to save space (see Figure 11.25).

11 Explain what happens when a protein is glycosylated. Where does this occur?

12 Suggest how stem cells could be used to treat people with insulin deficiencies.

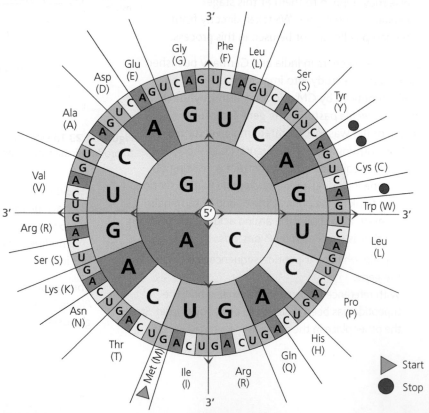

Figure 11.25 The genetic code. The amino acids are identified by their three-letter and one-letter codes. If you are using this to find DNA codons, then change U to T.

GM crop plants

The first GM crop plants had genes to improve cultivation by incorporating pest and herbicide resistance. Pest resistance reduces losses to insect pests like the cotton boll weevil and herbicide resistance allows farmers to spray herbicides during the growth of the crop to kill weeds that compete with the crop. A more recent development involves crops that are designed to improve human health. The nearest to reach the market is GM rice known as Golden rice™. Another is a GM banana which will act as a vaccine for hepatitis B. See Table 11.4.

Table 11.4 The features that have been inserted into genetically modified crop plants.

Feature	Example of crop	Foreign gene inserted into genome of host plant
Disease resistance	Papaya, squash	Resistance to ring spot virus
Pest resistance	Cotton, soya	Toxin coded for by a gene from *Bacillus thuringiensis* to kill pests such as boll weevil
Herbicide resistance	Soya, corn (maize), cotton	Gene that gives resistance to effects of herbicide glyphosate so allowing weed control by spraying during growth of crop
Drought resistance	Corn (maize), sugar cane	Genes to control loss of water vapour
Improved nutritional qualities	Rice	Several genes to produce precursors of vitamin A in endosperm
	potato	Increased starch content

Soya bean plants are highly susceptible to grazing by a variety of insect pests, such as cotton bollworm, *Helicoverpa armigera*, fall army worm, *Spodoptera frugiperda*, soybean looper, *Pseudoplusia includens*, velvetbean caterpillar, *Anticarsia gemmatalis*, soybean podworm, *Helicoverpa gelotopoeo* and the tobacco budworm, *Heliothis virescens*. Losses to these pests are estimated to run into billions of dollars each year. In response, a biotechnology company inserted a gene for pest resistance into its herbicide-resistant variety known as Roundup Ready™ (RR1). The plants are modified to express a gene from the bacterium *Bacillus thuringiensis* which codes for a toxin (Bt toxin) that is poisonous to insect pests. The toxin binds to receptors located on the microvilli of epithelial cells in the gut of the larvae. It inserts into the membrane, causing formation of pores or ion channels and a water potential imbalance that eventually kills the insects. This new variety was introduced on Brazilian farms in 2013 after 11 years of development and field trials. The new variety, INTACTA RR2 PRO™, is resistant to insect pests, tolerates spraying with glyphosate herbicides and has improved yield by having more beans per pod.

Tip

This list of pests of soya shows how vulnerable crops are to pest atttack. Different species evolve resistance to pesticides, which makes control very difficult.

Patenting and technology transfer

Many biotechnology companies have patented the genetic modifications that they have developed. To gain an economic return for the years of investment in research and development, they charge farmers a higher price for GM seed than for non-GM seed. Farmers who grow GM crops such as Roundup Ready must buy the herbicide Roundup from the same company.

Almost all the soya grown in the three biggest producers, Brazil, Argentina and the USA, is herbicide resistant. Increasingly new GM varieties will become available that combine that feature with others, such as pest resistance, high yield and efficient use of water. The development of GM cereal crops may be too expensive for subsistence and small-scale farmers in countries such as India.

There are many objections to the development, use and release of GM organisms. Some of these are

- antibiotic-resistance genes used to identify GM organisms could 'escape' and be transferred to pathogenic organisms, making it impossible to treat infections using such antibiotics

- herbicide-resistance genes could be transferred in pollen to weed species and lead to the development of 'superweeds' that are resistant to herbicides

- an increase in infection by plant pathogens has been correlated with an increase in the use of glyphosate herbicides on herbicide-resistant crops; this has been particularly the case with alfalfa in the USA

- foreign genes could be transferred to wild relatives of crop plants, or to non-GM and organic crops, so changing their genomes; this may 'pollute' those species

- foreign genes could mutate once they have escaped and have unforeseen consequences; their insertion into the genome could activate or silence genes

- making crops resistant to herbicides enables farmers to use more herbicides than on non-GM crops; this increases costs to farmers

- farmers cannot keep seed for sowing for the following crop as GM crops do not 'breed true'; this favours large-scale commercial farmers and not many farmers in developing countries

- genetic material from viruses used in genetic engineering could become incorporated with natural viruses so that an animal virus can infect humans or other animals or become more harmful; entirely new pathogenic viruses may evolve by this means

- we are dependent on seven types of grain-producing crop plant, each of which is becoming genetically uniform and losing its genetic diversity; this means that the major food sources of the planet are increasingly dependent on our manipulation of their genomes to meet future environmental challenges.

The growth of GM crops and the use of imported GM products, such as animal feed, have to be approved by authorities in the European Union. About half the countries, including France and Germany, have banned the growth of GM crops. There is only one GM crop that has been approved for growth in the EU that is grown commercially. This is a type of maize that is resistant to the European corn borer, *Ostrinia nubilalis*. However, countries in the EU, even those that ban the growth of GM crops, import products made from GM crops, such as animal feed made from GM soya and maize.

There are many campaign groups who object, often violently, to developments in genetic modification. Developments will continue and supporters say that it represents one way in which we will be able to feed the world's growing population and produce crops adapted to the consequences of climate change because the pace of change is too fast for traditional methods of plant breeding.

GM livestock

GM animals can be divided into two groups. The first group has been genetically modified to enhance overall performance and the whole animal will become available for the food market. As of 2015 no GM animal has been approved by regulatory bodies for human food. GM salmon may be the first to be given approval. The second group of GM animals has been transformed to produce specific substances in milk, eggs or blood – so that such substances can be used as therapeutic agents to treat disease, for example – or to serve as medical research models (see Table 11.5).

Table 11.5 Features that are being improved by genetic engineering.

Feature	Example of animal	Feature that has been added
Increase in growth rate	Atlantic salmon	Gene from Pacific Chinook salmon and a promoter the fish species Ocean Pout give salmon the ability to grow all year round
Decrease in pollution Reduction in dietary supplements	Pigs ('enviropigs')	Enzyme phytase to digest phytins so decreasing need for phosphorus supplements and reduce phosphate pollution
Prevention of disease transmission	Chickens	Catch but do not transmit bird 'flu
Resistance to bacterial disease	Dairy cattle	Human gene for lysozyme that is expressed in mammary glands to give protection against mastitis

Some proteins are produced by transgenic animals. Sheep and goats have been genetically modified to produce human proteins in their milk

- human antithrombin is produced by goats: this protein is used to stop blood clotting

- human α-antitrypsin is produced by sheep: this is used to treat people with emphysema.

The use of livestock to produce pharmaceuticals like these is known as 'pharming' and the animals as 'biopharm' animals. Some transgenic goats produce spider silk protein in their milk for research into new fibres. There would seem to be no limit to the types of protein that could be produced in this way.

GM microorganisms

Microorganisms were the first organisms to be genetically engineered and there are now many producing a very wide range of chemicals (see Table 11.6).

Table 11.6 Features that have been developed in microorganisms by genetic engineering.

Purpose of transformation	Example of microorganism	Feature that has been added
Production of animal protein	Bacteria	Chymosin (rennin) for cheese making Bovine somatotrophin (BST) for injecting into cattle to improve milk production
Wine production	Yeast	Improvement in the taste and colour stability of wine as well as reduction in the production of undesirable compounds, e.g. histamines
Production of human proteins for medical use	Yeast	Insulin, human growth hormone, vaccines
Dietary supplements	Bacteria	Tryptophan (an amino acid)

About 25% of commercial pharmaceuticals are biopharmaceuticals. GM organisms are used to produce human and animal medicines that are difficult to produce in other ways. This is a list of some of the many products for treatment

- insulin production for the treatment of diabetes (see Activity, above)

- human growth hormone to encourage growth in children with a deficiency of this hormone

- thyroid-stimulating hormone for treatment of thyroid cancer

- factors 8 and 9 for treating people who have a deficiency of one or other of these blood-clotting proteins

- vaccines, e.g. for influenza

- monoclonal antibodies, e.g. for diagnosis and treatment of cancers.

Most of the proteins used in genetic engineering, such as restriction enzymes and ligases, are themselves produced by GM microorganisms.

The advantage of using organisms is the large-scale production and therefore cheaper prices of the substances concerned. For example, in a year each GM goat can produce as much antithrombin as can be collected from 90 000 blood donations. Microorganisms and eukaryotic cells can be cultured on a large scale and production does not rely on other factors, such as the availability of insulin from dead animals.

GM pathogens

The techniques of genetic engineering have been used to study different aspects of plant and animal pathogens. *Mycobacterium tuberculosis* has been modified to investigate its metabolism, drug resistance and the ways in which it causes disease. It has been modified so that it will grow more rapidly, making it easier to study *in vitro*. Such research also gives clues to ways in which vaccines can be developed and drugs produced for treatment.

Viruses, such as adenovirus, have been genetically modified to deliver genes in gene therapy (see below). Adenovirus is a very suitable vector as it will infect human and other mammalian cells; it is not species- or cell-type-specific. These viruses have had two genes removed so that they do not replicate once they infect host cells. This removal gives space in the viral genome to insert a gene or genes that are up to 7 kbp in length. The recombinant human adenovirus type 5 expresses enhanced green fluorescent protein (GFP) under the control of a promoter. GFP is a protein originally isolated from *Aequorea victoria*, a bioluminescent jellyfish that fluoresces green when exposed to blue or ultraviolet light. Enhanced GFP (eGFP) is a GFP mutant with improved fluorescence and stability which is used as a marker. A green glow in ultraviolet light indicates that the virus has successfully delivered its genes.

Tobacco mosaic virus (TMV) has been modified to deliver the gene for a decapeptide hormone known as TMOF into the cells of crop plants. The GM viruses are sprayed on crop plants and invade cells. The host cells transcribe and translate the gene to produce TMOF that inhibits the production of the enzyme trypsin by insect pests without having serious effects on the host plant. The DNA sequence for TMOF was combined with that for TMV coat protein so that the decapeptide and coat protein are translated as one molecule. Trypsin cuts the peptide bond that joins to the two together so releasing active TMOF which then inhibits further production of trypsin by insects. After harvest, the leaves of the GM plants can be processed into a powder to be used as a spray to protect against insects such as mosquitoes.

There are few practical and ethical problems in these examples as proteins do not have to be extracted from animal sources or by collecting blood from many donors. The disadvantage of using bacteria to produce human proteins is that bacteria do not modify their proteins in the same way. It is much better, therefore, to use eukaryotic cells to produce human proteins. One example of this is the use of genetically modified hamster cells to produce factor 8. This protein is essential for blood clotting and is taken by those with haemophilia A.

Tip

TMV infects tobacco plants and many other broad-leaved crops, such as tomatoes, peppers and potatoes. TMOF acts in many insect pests and also in many insect vectors of disease, such as mosquitoes. This gives GM TMV a wide range of uses.

There are potential hazards from the possible release of genetically modified organisms into the environment from the laboratories and factories where they are used for commercial production. Some of the precautions taken to avoid this include

- transgenic microorganisms do not compete well in the natural environment as they are engineered to produce substances that give them no advantage and they require much energy that is not readily available to produce these substances

- containment facilities, such as filters on air conditioning and air locks on doors are fitted to prevent the escape of organisms

- lethal genes are added to the microorganisms so that they die if removed from the conditions of the culture.

In 2012, researchers discovered that the CRISPR-Cas9 system in prokaryotes, which cuts DNA in specific places, could be used to edit genomic DNA. One application of this system is to alter specific genes without changing other parts of an animal's genome. Now, instead of inserting genes from distantly related species, researchers can improve crops and livestock by replicating small genetic variations found naturally in different varieties of the same species.

Gene therapy

Gene therapy is an application of the principle of genetic engineering. It involves several different methods to correct a genetic 'error':

- replacing a mutated gene that causes disease with a functioning version of the gene

- introducing a new gene into the body to help fight a disease

- repairing mutated genes by using an enzyme to edit the DNA and inserting a functioning gene

- inactivating, or 'knocking out', a mutated gene that is not functioning properly.

There are two types of cell which can be treated by gene therapy: somatic cells which do *not* give rise to gametes and germ-line cells which produce gametes so transmitting genes to the next generation.

- Somatic cell gene therapy: somatic cells are body cells; all the cells in the body except gamete-forming cells and gametes. These cells all die when the individual dies so any genetic changes are not passed on to the next generation. However, gene therapy can be a treatment and maybe even a cure, for genetic conditions. Long-term treatments are possible by inserting a functioning allele into stem cells, such as those in bone marrow. It is most successful in cells that have a long life, such as those in the retina. It is less successful in cells that are replaced every few days, such as those in the airways and in the epithelium lining the gut.

- Germ-line gene therapy: the germ line is the term given to cells that differentiate to form gametes. In mammals, the germ line forms fairly early in development and this group of cells populates the gonads: the ovaries and testes. These cells increase in number by mitosis and then some will divide by meiosis to form gametes. If a correctly functioning allele is inserted into a fertilised egg, it means that all the cells formed by mitosis from that cell, once fertilised, will be genetically altered. The

Test yourself

16 State the roles of the following enzymes in genetic engineering: restriction enzymes, DNA ligase and reverse transcriptase.
17 Name three different types of vector as used in genetic engineering.
18 Outline how the bacterium *Bacillus subtilis* can be genetically engineered to produce bovine somatotrophin (BST), which is composed of 217 amino acids.

Key terms

Gene therapy The insertion of genes into cells to correct a genetic fault. This usually involves inserting a dominant allele into an individual who has two copies of a mutant recessive allele.

Somatic cell gene therapy The insertion of genes into body cells to correct a genetic fault in the individual concerned. The gene that has been inserted cannot be transmitted to the individual's children as it is not in the cells (germ-line cells) that give rise to gametes.

Germ-line gene therapy The insertion of genes into cells that can give rise to gametes so that the gene inserted is transmitted to future generations. Germ-line cells are the zygote, early embryonic cells and all the cells that can develop into gametes in the ovaries and testes.

change is permanent and will be inherited by future generations. This is not legal in the UK and many other countries, but there are people who consider it a development that will happen in the future.

Examples of gene therapy

Gene therapy was first used successfully to treat children with severe combined immunodeficiency syndrome (SCID). In ADA-SCID the enzyme adenosine deaminase (ADA) does not function. This happens in children who are homozygous recessive for faulty alleles of this gene. ADA is involved with the breakdown of adenosine, which is toxic to white blood cells. Children with the condition were treated by removal of the white blood cells known as T cells. These cells were given the dominant allele for this enzyme and then returned into the blood. This was first done successfully in 1990 and has been repeated since, but using modified stem cells from bone marrow rather than differentiated white blood cells. Further treatments led to complications but using lentiviruses as vectors for delivery of the dominant allele to the stem cells has led to improved results. Lentiviruses, such as modified human immunodeficiency virus (HIV), can deliver genes to dividing and non-dividing cells.

There are two forms of haemophilia. Both are sex-linked disorders of blood clotting. Factor 8 and factor 9 are two of the proteins involved in the blood clotting cascade and are coded by genes on the X chromosome. Both conditions have been treated by adeno-associated-virus-based gene therapies in which the vector is injected directly in the body and targets liver cells where the genes are expressed.

A person with β-thalassaemia in which the β globin of haemoglobin is not produced has been treated by incorporating the dominant allele into bone marrow stem cells. This treatment has proved successful. The patient was treated with chemotherapy to kill some bone marrow cells before inserting the treated stem cells. However, this success was thought to be very lucky and unlikely to be repeated on others with this condition. Trials of gene therapy of this disease, common in populations of Mediterranean origin in the UK, are in progress as of 2015.

Some of the most advanced work in this area is an approach to cancer treatment in which immune cells called T cells are removed from a patient's bloodstream, given a gene encoding a protein called a chimeric antigen receptor that targets tumour cells, before being reinfused back into the patient's bloodstream.

Benefits and hazards of gene therapy

Clinical trials of gene therapies have been successful in that children with SCID no longer need to live in a sterile environment and they no longer need to be treated with injections of adenosine deaminase and antibodies. As far as anyone can predict, the children who have been treated successfully can expect to lead a normal life and have a normal life span.

The autosomal recessive sight disorder Leber's congenital amaurosis, or LCA, was treated successfully with gene therapy in 2008. It is a rare disorder, apparent at birth, that affects 1 in 80 000 people. People given gene therapy for LCA can have their sight restored.

However, there are hazardous side effects.

- Virus vectors have caused (viral) diseases.

- The immune system may be stimulated to respond to the vector, e.g. a virus, and/or against a protein that is coded for by the gene inserted; each disease poses a unique set of problems with delivery of the gene treatment.

- Some boys with X-linked SCID have developed leukaemia; this happened because there was no control over where the retrovirus vector inserted the new gene into the genome. The insertion of the new gene caused the activation of proto-oncogenes that control cell division (see Chapter 9).

There are also practical problems with gene therapy.

- Gene therapies may be temporary because the cells that are genetically modified have a sort life span, such as those lining the airways that are targeted in treating cystic fibrosis.

- It has proved difficult to direct vectors to the specific cells where the allele is to be expressed. Even if the allele is inserted into the genome of the intended cells it is not always expressed.

- Only recessive conditions, such as haemophilia and LCA, can be treated. It is not yet possible to switch off a dominant allele such as the one that is the cause of Huntington's disease.

Problems about directing genes into appropriate places in the genome may soon be solved by the use of CRISPR-Cas9 technology, in which an enzyme can be directed to a specific DNA sequence to insert the gene. It is possible that once gene therapies for monogenic conditions become accepted there will be similar therapies for the more intractable multifactorial diseases, such as heart disease, dementias and cancers.

The types of gene therapy described so far are examples of somatic cell gene therapy. The solution to the problem of delivering genes to all the cells that require them could be overcome by using germ-line therapy which involves placing the gene or genes concerned into an egg or into a zygote. It would mean that the gene would be present in every cell of the body and may be passed on to the next generation. At present germ-line gene therapy is not legal in any country.

Tip

You should read more about the issues surrounding gene therapy. Try starting with two web sites that deal with these issues: BioEthics Education Project and Learn Genetics.

Tip

It is worth remembering that the first forms of gene therapy that will become widely available will be for the treatment of single-gene defects by adding a gene for a functional protein. This will involve targeting cells that are easily accessible; and they only need to affect a relatively small number of cells in order to be successful. SCID, haemophilia and some forms of blindness are the most likely examples.

Test yourself

19 Explain why LCA, SCID, SCA and CF are all suitable for treatment by gene therapy, but Huntington's disease is not.
20 Explain why almost all of the children who develop sex-linked SCID are boys.
21 Adenovirus causes conjunctivitis, gastroenteritis and infections of the airways. Suggest two ways in which the genome of the adenovirus used as the vector for gene therapy for LCA would differ from that of normal adenoviruses.

Exam practice questions

1 Which enzyme catalyses the breakage of phosphodiester bonds?
   A DNA ligase
   B DNA polymerase
   C restriction enzyme
   D reverse transcriptase (1)

2 DNA taken from the nucleus of a human hepatocyte could not be used for which of the following?
   A Sequencing the nucleotides in genes expressed in the eye
   B A probe for locating a gene in hepatocytes of other mammals
   C Inserted into a bacterium for the production of a human protein
   D The identification of loci that are polymorphic (1)

3 Woolly mammoths, *Mammuthus primigenius*, became extinct between 10 000 and 4 000 years ago. They were adapted to life in the cold tundra of northern latitudes.
   A frozen, 43 000-year-old woolly mammoth was found in Siberia. Its DNA was extracted and sequenced. The sequences of the genes coding for the α and β polypeptides of haemoglobin were compared with those of Asian elephants, *Elephas maximus*. The results showed that there was only one amino acid that differed between the α globins of the two animals and three differences between the amino acid sequences of the β globin.
   a) Describe the structure of a molecule of haemoglobin. (4)
   b) Suggest why gene sequencing was more likely to identify differences between the haemoglobin of the two species than electrophoresis of the proteins. (3)
   c) Suggest the likely effects of these differences on a molecule of mammoth haemoglobin as compared with that of Asian elephants. (3)
   d) Outline how genome-wide comparisons are made between two species, such as *M. primigenius* and *E. maximus*. (3)
   e) Discuss the benefits of sequencing the genomes of extinct and endangered animal. (6)

4 *Taq* polymerase is a thermostable DNA polymerase. The graph shows the effect of temperature on the rate of the reaction catalysed by this enzyme.

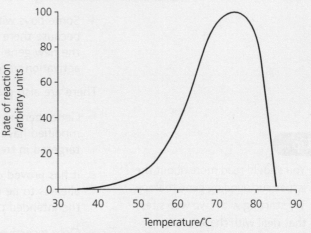

   a) With reference to the graph, explain the role of *Taq* polymerase in PCR. (4)
   A blood sample is taken for use in forensic analysis.
   b) i) State two sources of DNA in a blood sample. (2)
      ii) Suggest how DNA is extracted from a blood sample. (3)
   c) Explain why during the process of PCR
      i) primers do not anneal together
      ii) template strands do not anneal together. (4)
   d) Discuss the advantages of high-throughput sequencing. (5)

5 The gene *F9* codes for blood clotting factor 9. This protein has a trypsin-like region that hydrolyses a bond between arginine and isoleucine in factor 10 as part of the cascade involved in blood clotting.
   a) i) What is meant in this context by a *cascade*? (2)
      ii) Name the bond hydrolysed by factor 9. Suggest what is meant by the phrase 'trypsin-like region'? (2)
   Mutations in *F9* can result in the blood-clotting disorder haemophilia B. This is an X-linked recessive disorder with an incidence of about 1 per 30 000 live male births. As many as one third of these males have no affected family members.
   b) Explain why haemophilia B affects males, yet all the male offspring of an affected male will be normal. You may use a genetic diagram in your answer. (4)

c) i) How is the presence of mutant alleles for *F9* detected in females who may be carriers of the condition? *(2)*

ii) Suggest what advice may be given to females who are carriers. *(3)*

d) In mid-2015 there were 1095 unique variants of the *F9* gene listed in the database for this gene maintained by University College London (UCL). Discuss the benefits of maintaining databases like the one at UCL. *(3)*

e) Trials of gene therapy for haemophilia B involve adding a dominant allele to liver cells. Explain why it is easier to perform gene therapy when the mutant allele is recessive rather than when it is dominant. *(3)*

f) Use examples to discuss the benefits and hazards associated with gene therapy. *(5)*

Stretch and challenge

6 Ethanol is metabolised in the liver by two enzyme-catalysed reactions.

$$\text{ethanol} + \text{NAD} \xrightarrow{\text{Ethanol dehydrogenase}} \text{ethanal} + \text{reduced NAD}$$

$$\text{ethanal} + \text{NAD} \xrightarrow{\text{Ethanal dehydrogenase}} \text{acetate} + \text{reduced NAD}$$

a) Describe the role of NAD in cell metabolism.

b) Suggest the likely effect of excessive alcohol consumption on energy metabolism in liver cells.

The gene *ADH* codes for ethanol dehydrogenase and *ALDH* codes for ethanal dehydrogenase. Both genes have single nucleotide polymorphisms (SNPs) that influence the activity of the enzyme. These are summarised in the table.

Enzyme and position in primary sequence	Amino acid residue	Enzyme Activity	Allele	Abbreviated symbol for allele
Ethanol dehydrogenase, position 47	Histidine	Fast	*ADH2h*	A^H
	Arginine	Slow	*ADH2a*	A^A
Ethanal dehydrogenase, position 487	Glutamic acid	Fast	*ALDH2g*	B^G
	Lysine	None	*ALDH2l*	B^L

People with the genotype B^{L}B^L are unable to drink alcohol without feeling ill.

c) Explain why people with this genotype are unable to drink alcohol without feeling ill.

Researchers in Japan investigated a random sample of 454 people aged between 35 and 85 to find the frequencies of the two alleles of each gene. They extracted DNA from their volunteers and used PCR and gel electrophoresis to identify their genotypes. They did this by using primers of different length for each allele. The shortest primer was: GGC TCC GAG CCA CCA.

d) i) Using the sequence above, state the role of primers in PCR.

ii) Explain why the primers were of different lengths.

The electrophoretogram of nine of the volunteers is shown below. The numbers in brackets are the lengths of the DNA fragments produced in PCR in base pairs (bp).

e) i) Use the abbreviated symbols for the four alleles to state the genotype of each person, 1–9.

ii) State the phenotype of each person.

f) State those people who are most at risk of alcoholism. Explain your answer.

The table shows the numbers of people in the sample with the different genotypes.

ADH	*n*	ALDH	*n*
ADH2a ADH2a	19	*ALDH2g ALDH2g*	221
ADH2a ADH2h	149	*ALDH2g ALDH2l*	187
ADH2h ADH2h	286	*ALDH2l ALDH2l*	46
Total	454	Total	454

g) Is this population in Hardy–Weinberg equilibrium for these two genes? Use a suitable statistical test to assess the probability that these results are not due to chance.

M 1 2 3 4 5 6 7 8 9

500 bp →

← His (280)

200 bp →

← Arg (219)

← Glu (119)

100 bp →

← Lys (98)

Chapter 12

Cloning and biotechnology

Prior knowledge

- The two types of nuclear division are mitosis and meiosis.
- When cells divide by mitosis the resulting daughter cells are genetically identical to each other and to the parent cell.
- A clone is a group of individuals that have formed from one parent organism and are all genetically identical.
- Stem cells have the ability to divide and to give rise to cells that differentiate into specialised cells.
- In mammals, the zygote and the cells of the early embryo are totipotent, embryonic stem cells are pluripotent and tissue (adult) stem cells are multipotent.
- In plants, cells in the meristems are the equivalent of stem cells.
- Apical meristems are at shoot tips, root tips and in buds. These produce new cells to give growth in length. Lateral meristems produce cells that give growth in width of stems and roots.
- When cells stop dividing by mitosis they differentiate into specialised cells.
- Somatic cells are body cells. Germ-line cells are cells that give rise to gametes.
- Prokaryotes have a simple cellular structure without nuclei or membrane-bound organelles.
- Enzymes are biological catalysts. Each enzyme catalyses one specific type of chemical reaction in the organism.

Test yourself on prior knowledge

1 Name the stages of the mitotic cell cycle.
2 Distinguish between sexual and asexual reproduction.
3 List three features of mitosis that distinguish it from meiosis.
4 Define the term *stem cell*.
5 State three places in a mammal where multipotent (tissue) stem cells are found.
6 Distinguish between totipotent and pluripotent stem cells in animals.
7 Explain how germ-line cells differ from somatic cells.
8 State three structural features of eukaryotic cells that are not found in prokaryotic cells.
9 Explain why a eukaryotic cell may have up to 2000 different enzymes.

Figure 12.1 Orchids growing in a commercial greenhouse. In the 1970s, orchids used to sell for thousands of pounds. Now the same species can be bought for as little as £10.

Of all flowering plants, orchids must be among the most specialised (Figure 12.1). Many species have complex growth habits and life cycles involving both sexual and asexual reproduction. Orchids also have very specialised flowers for sexual reproduction. They are popular house plants because they have spectacular flowers, often highly specialised for pollination.

Orchid seeds are very small and are poor at germinating. To germinate in the wild they need to form associations with soil fungi, known as mycorrhizae, and rely on them to provide nutrients. This made growing them from seed very difficult until the 1920s when seeds were germinated in a medium that did not require mycorrhizae. The development of tissue culture later in the twentieth century gave horticulturalists the opportunity to propagate orchids more easily and to produce them for the mass market.

Some orchids produce swollen roots that grow into new plants. These separate from each other to form a clump of individual plants. This is asexual reproduction that does not involve meiosis or the production and fusion of gametes. The new plants develop solely by mitosis.

Cloning in plants

Tip

There is more about stem cells on pages 116–117 of *OCR A level Biology 1 Student's Book*.

Key terms

Meristem Plant tissue where growth occurs. Meristematic cells are stem cells that divide by mitosis to produce daughter cells some of which differentiate into specialised cells.

Vegetative propagation A form of asexual reproduction in which a plant grows new parts which eventually become separated from the parent. Many flowering plants reproduce vegetatively and many commercially important species are propagated artificially.

Adventitious roots Roots that grow from any part of a plant *other* than the main root. The roots that grow from a garlic bulb are an example.

Natural cloning

Vegetative reproduction is a form of asexual reproduction in which new plants develop from the meristematic regions, which are areas of undifferentiated cells in vegetative organs of a plant. These are the stems, roots and leaves. Specifically

- apical buds: in the tips of shoots and roots

- axillary buds: in the angle between the leaves and the stem

- vascular cambium: a tissue that forms xylem and phloem (and lies between them in vascular bundles)

- cork cambium: a cylindrical tissue that produces bark in shrubs and trees.

Meristematic cells are stem cells that divide to form more stem cells and other cells that differentiate into different types of specialised cell. They are small cells with a large ratio of nucleus to cytoplasm. They have thin cell walls, small vacuoles and no large central vacuole.

Vegetative reproduction is the natural way in which flowering plants reproduce asexually. Asexual reproduction produces miniature plants or plantlets which are clones. They remain attached to the parent plant, gaining a supply of energy through the phloem. There follows a brief survey of some forms of natural vegetative propagation in flowering plants.

Horizontal stems or runners creep or 'run' over the surface of the soil (see Figure 12.2). Roots form at the nodes, which grow leaves, buds and possibly stems. Roots that form on any structure other than the main root are called adventitious roots. The plantlets become separated from the parent as the horizontal stem dies. These form a clump of plants with small distances between them so they are not in direct competition for light, water and mineral nutrients.

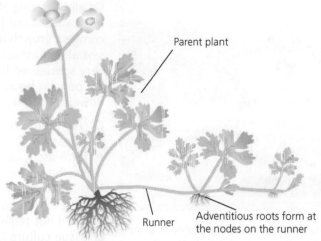

Figure 12.2 Runners of creeping buttercup, *Ranunculus repens*. The runners eventually wither and die to leave a clone of plants all derived from the same parent. As with many plants, *R. repens* also produces flowers and reproduces sexually as well as vegetatively.

Lateral stems or suckers form below ground or at the base of upright stems. These eventually become independent plants (see below, Figure 12.5).

Stolons are underground stems that grow horizontally. In potatoes these also grow downwards and swell with energy stores to form stem tubers (see Figure 12.3). Rhizomes are thick horizontal stems that send out shoots at intervals and branch. Ginger is the rhizome from *Zingiber officinale*.

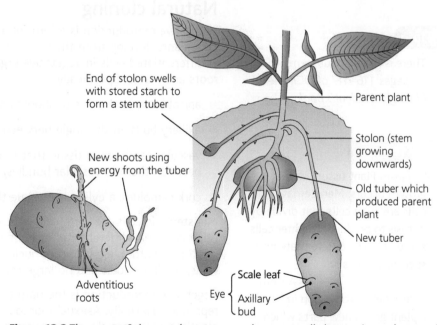

Figure 12.3 The potato, *Solanum tuberosum*, reproduces asexually by growing underground stems, or stolons, that swell up with stored starch to form stem tubers. If left in the ground these grow stems and adventitious roots as shown in the drawing on the left.

Short, swollen upright stems grow buds which develop into separate plants. Bulbs of onions, daffodils, tulips and garlic and corms of crocus are examples. Garlic, *Allium sativum*, reproduces almost entirely by asexual means. Each garlic bulb contains several cloves, each of which will grow into a new plant.

Dahlias have swollen roots or root tubers which if left in the ground grow new shoots and adventitious roots when conditions for growth are favourable. The leaves of some plants develop small plantlets around the margin; for example, *Kalanchoe daigremontianum*. Bluebells, *Hyacinthoides non-scripta*, reproduce sexually to produce seeds and asexually by producing daughter bulbs. Bluebells compete well with other plants that grow in woodlands before the trees come into leaf.

Artificial cloning

Gardeners and horticulturalists propagate plants by assisting the natural processes of vegetative reproduction. There are a variety of ways to do this including taking cuttings, layering, division, grafting or budding. Many cultivated varieties, known as cultivars, were initially propagated from a single seedling or from a single mutated plant or organ that was found, or from a mutant artificially produced by plant breeders or geneticists. Since all subsequent plants were propagated from this single seedling or mutated plant or organ, the entire cultivar is a clone. Some clones originated centuries ago: the Bartlett pear originated in 1770 and the Bramley apple from 1809.

Cuttings are taken from plants by removing stems, roots or leaves from a parent plant with good features that are worth propagating. This can be done by cutting a leaf at the point where it meets the stem, as in African violet, *Saintpaulia ionantha*, or across a side shoot just below the point where a leaf joins the stem, as in Joseph's coat, *Coleus*.

Stem cuttings must form adventitious roots as they already have shoots; root cuttings must form adventitious shoots as they already have roots; and leaf cuttings must form both adventitious roots and adventitious shoots as they have neither.

Stem and leaf cuttings (Figure 12.4) must be given sufficient water as they are now removed from the supply in the xylem from the parent and have no roots to absorb water. Most methods to prevent water loss of cuttings involve decreasing light and temperature and increasing the humidity in the immediate environment. This can be done by keeping cuttings in pots within polythene bags or on a commercial scale by spraying a fine mist and keeping the plants in the shade.

> **Tip**
>
> Contrast this with domestic dogs that were domesticated from wolves in different parts of the world and so have much more genetic diversity (see *OCR A level Biology 1 Student's Book* page 214), although this has been reduced in some breeds (see Chapter 10 in this volume).

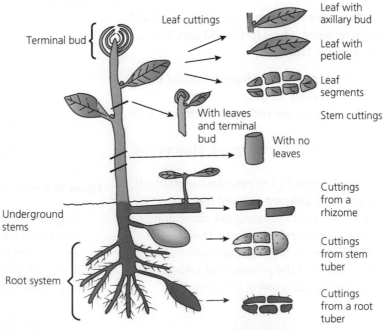

Figure 12.4 Different ways to propagate plants by taking cuttings.

Figure 12.5 A banana plantation. The plant in the foreground has a sucker growing beside it. This can be dug out and replanted or left in place when the adult plant has produced fruit.

Stem cuttings may be dipped into the synthetic auxins indolyl butyric acid (IBA) or 1-naphthaleneacetic acid (NAA) and then placed into a suitable soil or compost. After a while adventitious roots grow from the base of the cutting and the new plant can support itself. After a period of 'hardening off' in which the new plants are exposed to the unprotected environment, they can be planted out.

Division is a method of propagation in which clumps of plants are dug up and separated into smaller plants and replanted further apart. Division most commonly is used for the propagation of plants that naturally grow as clumps.

Plants that naturally produce structures for vegetative propagation can be divided. Banana trees (Figure 12.5) have suckers at their base which are removed for propagation. This is the only way cultivated bananas, which are triploid and therefore sterile, can be propagated.

Activity

Propagating *Pelargonium*

Pelargonium graveolens, also known as rose geranium, is a source of oil used in the perfume industry. Researchers in India followed the instructions in Figure 12.6 to take cuttings and clone plants of rose geranium. The cut stems were treated with lanolin (a cream-like substance) containing different concentrations of the synthetic plant hormone IBA. The cuttings were taken in August and in November.

To assess the success of the procedure, the cuttings were assessed after 60 days for the following parameters

- percentage of the cuttings with adventitious roots
- mean number of roots
- length of the longest root.

Rooted cuttings were transplanted into larger pots and their survival recorded after a further 30 days. The results are shown in Table 12.1.

1 Cuttings are taken with an oblique cut across the stem. Suggest reasons for this.
2 Why are cuttings treated with plant hormones, such as IBA and NAA?
3 Explain why some of the cuttings were treated only with lanolin.
4 Use the data in Table 12.1 to describe the effect of taking cuttings at different times of the year.
5 Describe the effect of increasing IBA concentration on the parameters chosen by the researchers.
6 Suggest two other parameters that could be used to assess the success of the procedure of taking cuttings for use in commercial growing of *P. graveolens*.

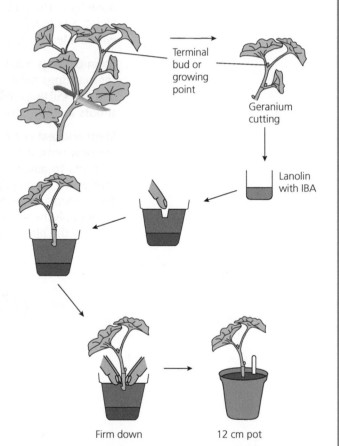

Figure 12.6 Instructions for taking cuttings of pelargoniums.

7 Suggest other features of *P. graveolens* that the researchers could investigate that might prove useful for an Indian perfume industry.
8 *P. graveolens* is a hybrid plant. Explain how it would be possible to find out its ancestry.

Table 12.1 Results of the experiment to assess the success of taking cuttings in *P. graveolens*. ppm = parts per million.

Concentration of IBA/ppm	Percentage of cuttings that rooted		Mean number of roots per cutting		Length of longest root/mm		Percentage survival of cuttings 30 days after transplanting	
	Aug	Nov	Aug	Nov	Aug	Nov	Aug	Nov
0 (Lanolin only)	16	38	8	20	53	183	45	58
500	39	80	17	40	101	250	60	88
1000	42	85	18	40	130	269	71	90
1500	46	90	19	56	131	304	73	94
2000	53	98	21	59	178	335	73	98
2500	52	89	20	49	160	320	72	92

Tissue culture

Many commercially important plants are propagated using methods of tissue culture, also known as **micropropagation**. This involves removing small pieces of tissue from a plant and culturing in a liquid medium or on a solid medium. The media contain sucrose as a carbon source and as a source of energy, mineral nutrients so the plant cells can make all the biological molecules that they need, and plant hormones to regulate the type of growth that occurs. At the beginning hormones are added to stimulate cell division, and later hormones stimulate cells to differentiate to form the tissues of stems, roots and leaves.

Two forms of micropropagation are callus culture and meristem culture (see Figure 12.7).

● In callus culture small pieces of tissue called explants are removed by cutting from leaves, stems or roots. These are surface sterilised to remove any microorganisms and then placed into or onto the culture medium. Differentiated cells then start dividing into a *callus*, which is a mass of undifferentiated cells produced in plants for wound healing. This can be subdivided to increase the number of plantlets produced. The callus is transferred to a medium with a different ratio of plant hormones to encourage differentiation and then the plantlets potted into soil in sterile and very moist conditions. Auxins stimulate root growth and cytokinins stimulate shoot growth. With equal concentrations of the two hormones growth of the callus is encouraged.

● In meristem culture meristems cut from the stem apex and from axillary buds are placed onto an agar medium in the same way as for callus culture. The advantage of using meristematic tissue is that it is free of viruses. Meristems can also be subdivided to give plenty of virus-free plant stock.

Tissue culture must be carried out under sterile conditions otherwise bacteria and fungi from the atmosphere will contaminate the culture media (Figure 12.8). Aseptic technique is used to ensure that no contaminants ruin the stock of plants. This involves sterilising the media, containers and implements; bacteria and fungal spores are prevented from reaching the cultures by using a chamber that blows filtered air over the work surface. All staff must avoid contaminating the cultures by wearing appropriate sterilised clothing, masks and gloves.

Tip

There is more about aseptic technique as applied to the culturing of microorganisms later in this chapter.

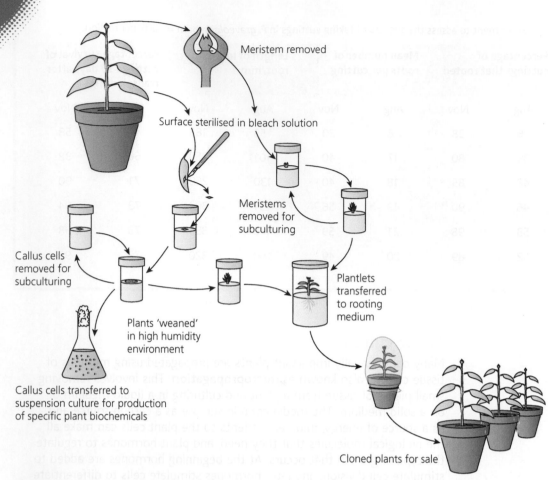

Meristem removed

Surface sterilised in bleach solution

Meristems removed for subculturing

Callus cells removed for subculturing

Plants 'weaned' in high humidity environment

Callus cells transferred to suspension culture for production of specific plant biochemicals

Plantlets transferred to rooting medium

Cloned plants for sale

Figure 12.7 Tissue culture is used to produce large numbers of plants with commercially important features.

Figure 12.8 Plantlets growing on agar. It has been estimated that one chrysanthemum apex placed in tissue culture could produce up to a million new plantlets in one year.

Advantages and disadvantages of vegetative propagation

There are many plants across the world that are edible, but only a small percentage of them are cultivated to form part of the human diet. Only 15 species of crop plants provide 90% of the global human food energy intake, with three cereal species – rice, maize and wheat – making up two-thirds of this. Species such as cereal crops are propagated by sexual reproduction, and so too are many that produce fruits, although some, such as apples and pears, are propagated vegetatively.

The advantages of vegetative propagation include the fact that the plants produced by cuttings or divisions are relatively easy to establish in the ground. Growth from seed is often more risky as the seeds may not be viable and fail to germinate. Seeds and especially young seedlings are very susceptible to diseases, pests and environmental factors such as cold, wind and drought.

Plants propagated vegetatively are clones and genetically uniform (Figure 12.9). This means that they have a uniform appearance, ensuring reliable quality control. Growers can provide a reliable supply to the consumer. Uniform size also aids harvesting and packing. Since all cloned plants will grow in a fixed period, it is easier to predict the time between planting and harvesting. These advantages help to keep production costs low.

Figure 12.9 Young banana plants that have been produced by tissue culture being grown in a protected environment before planting out.

Tip

In Chapter 10 of *OCR A level Biology 1 Student's Book* you read about Black sigatoka, a disease of bananas. Panama disease is another disease of bananas caused by the fungal pathogen *Fusarium oxysporum*.

Plants that are grown from seeds do not always 'breed true'. This is especially true of F_1 hybrids that are formed by the crossing of two varieties. The only way to ensure the sale of uniform plants is by repeatedly crossing the parent varieties or propagating the hybrids by asexual means. A cultivar that can be propagated vegetatively is therefore much cheaper to produce.

Facilities to propagate plants by tissue culture can be set up anywhere in the world as the growth of clones and plantlets is not dependent on the climate or weather. Plants can also be produced at all times of the year to meet demand from growers.

The main disadvantage of vegetative propagation is that all plants in a genetically uniform crop are at equal risk from pests and diseases for which they have no resistance. If there is an epidemic of a pest or disease there is a chance that all plants with the identical genotype will be wiped out. This happened with the Gros Michel cultivar of banana that was attacked by a fungal disease known as Panama disease in the 1950s. Another cultivar, Cavendish, is grown in plantations all over the tropics. A new strain of Panama disease (tropical race four) that infects the Cavendish variety has emerged in South-east Asia and may well spread to plantations in the Caribbean and Central America. If growers cultivated a range of cultivars that were resistant to different diseases then a world-wide epidemic and loss of income for many farmers would be less likely.

With many agricultural and horticultural crops being cloned on a wide scale there has been a loss of variation and genetic diversity in cultivated species. This puts these species at risk of not adapting to any future changes in environmental factors, such as global temperature. Mutation occurs in all plants, but without meiosis and fertilisation there is a much smaller chance that new mutations will be expressed and so their frequency in populations remains low.

Planting stock regenerated by cloning also frequently suffers from a build-up of pests and diseases. Most garlic cultivars, for example, carry some degree of virus infection that is passed on during asexual reproduction. This can only be countered by encouraging garlic to produce seeds and using these to breed virus-free stock.

Test yourself

1 Define the following terms: *vegetative propagation, tissue culture, explant, callus* and *meristem*.
2 Suggest the reasons why some plants are propagated vegetatively rather than by seed.
3 State the tissue that contains pluripotent stem cells in an adult plant.
4 Explain why a root hair cell and a guard cell in an individual plant are usually genetically identical, but differ in structure and function.
5 Explain why explants are surface sterilised before placing on agar.
6 Explain the advantage of culturing meristems rather than tissue from explants of stems or leaves.
7 Explain why plant hormones are used in tissue culture.

Cloning in animals

Figure 12.10 Identical twins are a valuable scientific resource! Any differences between them must be due to the influence of environmental factors or the interactions between those factors and the genome.

Figure 12.11 This flow chart shows artificial twinning of sheep by embryo splitting.

Natural cloning

Asexual reproduction is less common among animals than it is in plants. In general, animals with more complex bodies rarely reproduce asexually. Aphids are a pest of many crops. During the growing season female aphids produce diploid eggs by mitosis. These eggs develop inside the body of the female, hatch and emerge as miniature adults. These eggs are not fertilised by sperm to form zygotes so contain genetic material from the female. All the young produced by a female aphid in this way are a clone as they all have the same genomes. This natural form of cloning is a type of asexual reproduction called **parthenogenesis**.

Embryos, which result from sexual reproduction, can divide into two. The twins that result are identical twins because they have formed from the fusion of the same egg and sperm at fertilisation. They are not clones of a parent, but they are clones of each other and are always the same sex (see Figure 12.10). Non-identical twins form from two separate fertilised eggs and are not genetically identical and may therefore be the same sex or different sexes. They are not clones.

Artificial cloning

Embryo twinning is similar to the process that produces identical twins. Embryos are divided into two or more groups of cells and cultured in the lab to develop into duplicate embryos with identical DNA. After several days, the embryos are inserted in the uterus of one or more surrogate mothers to develop to full term (see Figure 12.11). The embryos are

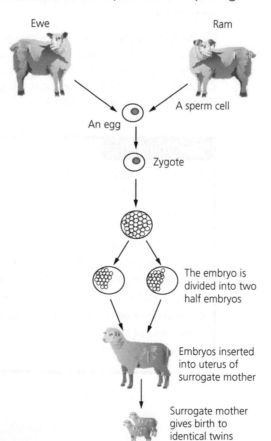

Ewe

Ram

A sperm cell

An egg

Zygote

The embryo is divided into two half embryos

Embryos inserted into uterus of surrogate mother

Surrogate mother gives birth to identical twins

usually derived from an elite female animal, whose life is not put at risk by implanting them in her to undergo the rigours of a pregnancy. Instead she is used to provide more embryos which may be formed by *in vitro* fertilisation. Embryo cloning has been a routine procedure to clone livestock, such as cattle, pigs and sheep, since the 1980s.

The disadvantage of embryo cloning is that it is not possible to predict exactly how productive the animals will be. They are, after all, the result of a sexual process involving meiosis and fertilisation. In 1958, John Gurdon in Oxford produced the first animal that was a clone of an adult. He did this by removing the nucleus from eggs of the South African clawed toad, *Xenopus laevis*, and inserting a nucleus from a somatic cell from the lining of the gut. This type of cloning was repeated with other species, but the first livestock animal to be cloned was Dolly the sheep in 1996 at the Roslin Institute near Edinburgh. Dolly was an example of reproductive cloning.

Reproductive cloning involves reproducing a known genome in a new animal. This is achieved by the technique of somatic cell nuclear transfer (SCNT). Figure 12.12 shows the procedure. The nucleus is removed from an egg and replaced with a nucleus from a somatic cell of an adult organism. Cells from the skin, lining of gut or udder are used. This can be done in two ways: removing the nucleus from a somatic cell with a needle and injecting it into the empty egg or using an electrical current that causes the entire somatic cell to fuse with the enucleated egg. The procedure is then the same as with embryo cloning: the egg containing the foreign nucleus is encouraged to divide and, once it has formed a small embryo, it is inserted into a surrogate mother.

> **Key terms**
>
> **Reproductive cloning** A method to clone an adult using nuclei from some of its somatic cells, for example skin, gut or udder cells.
>
> **Somatic cell nuclear transfer (SCNT)** The method in which a nucleus is removed from a somatic (body) cell of an adult and inserted into an egg from which the nucleus has been removed.

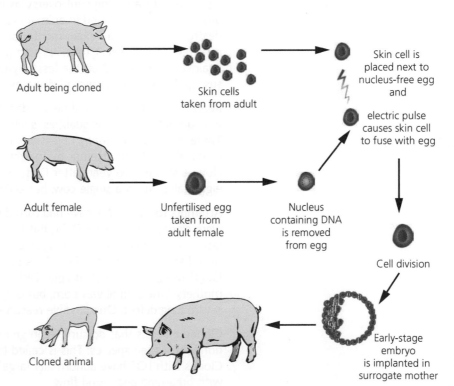

Figure 12.12 Artificial clones of adult animals can be formed by removal of the nucleus from an egg and insertion of a nucleus from a somatic cell from a suitable donor. This is somatic cell nuclear transfer (SCNT).

Besides cattle and sheep, other mammals that have been cloned from somatic cells include: cat, deer, dog, horse, mule, ox, rabbit and rat. SCNT is used in the effort to preserve rare breeds of livestock and to preserve endangered species. In the latter case there are rarely enough females of the same species that are suitable to receive the embryos, so they are implanted into a domesticated animal. This is interspecific SCNT or iSCNT.

Therapeutic cloning follows the same initial stages as reproductive cloning, but, once the embryo has been formed with nuclei derived from a somatic cell of an adult, its cells are removed and subdivided. These stem cells can grow into any type of cell, so researchers hope to harvest them and use them to replace ones that have become worn out or damaged in existing organisms. For more information about the use of stem cells derived from therapeutic cloning see Chapter 6 in *OCR A level Biology 1 Student's Book*.

The uses of animal cloning

Animal cloning is a powerful way to produce many individuals with the same genome. As in plants, this is a method that produces individuals that have the same productive features; for example, high milk yield. Although this is also possible with animal breeding, because that process involves sexual reproduction the performance and productivity of the offspring cannot be guaranteed. Reproductive cloning has been used to produce sheep genetically engineered to produce human proteins. The production of laboratory animals by cloning is useful in medical research as their use removes the genetic variability in response to drugs and other treatments that are being researched. This is also why human identical twins are so useful to medical researchers.

Arguments for and against artificial cloning in animals

When most people talk about animal cloning they are referring to reproductive cloning not embryo cloning, which has occurred for many years without arousing controversy as it is an extension of a natural process. It is widely used in animal breeding and, since no embryos are destroyed and human experimentation with embryos formed this way is rare, few people object to it. As with plants, there are the same arguments against cloning based on the loss of genetic diversity and increased susceptibility to epidemics of diseases and pests.

Reproductive cloning could be an ideal way to mass produce elite farm animals with particular qualities, such as dairy cows with a high milk yield. There are, however, many objections to reproductive cloning. Dolly was the result of the 277th attempt by the researchers at the Roslin Institute to clone a sheep. Two years after Dolly was born, researchers in Japan cloned eight calves from a single cow, but only four of them survived.

Attempts to use iSCNT have had mixed success. Several species of wild cattle were cloned shortly after Dolly, but the calves died soon after birth. The last surviving Pyrenean ibex (a type of wild goat) died in 2000. Some cells from her were stored and in 2009 an attempt was made to clone her with iSCNT using domestic goat eggs. Different species were used as the surrogate mothers. One animal was born, but only lived for a few hours as she had a severe lung defect. One possible reason is that freezing damaged the DNA.

Cloned animals that do survive pregnancy tend to be larger at birth than normal for their species. This is called the large offspring syndrome (LOS). Clones with LOS have abnormally large organs, which can lead to problems with breathing and blood flow.

Because LOS does not always occur, scientists cannot reliably predict whether it will happen in any given clone. Also, some clones without LOS have developed liver, kidney or brain malformations and impaired immune systems, which can cause problems later in life. Recent research suggests that such serious abnormalities may be due to the cloning process interfering with the precise regulation of genes during development.

The use of animal cloning could help preserve endangered species, especially if the young can be reared by domesticated animals. If eggs from another species are used, as in iSCNT, the resulting animals have an identical nuclear genome, but mitochondria are inherited from the domesticated animal. The method could also be used to bring back extinct animals as researchers tried to do with the Pyrenean ibex, which has now become extinct twice over.

Tip

People who oppose any form of alteration often say that scientists are 'playing God'. This is not a good way to answer a question about the ethical issues surrounding cloning. It is better to use ethical arguments and also arguments based on the science involved.

Some people think that the risks associated with cloning endangered animals outweigh the benefits of producing a few individuals, at great cost, that are not complete clones and that ultimately live in zoos and other *ex situ* conservation areas. Also many attempts have been unsuccessful because of problems at all stages from insertion of nuclei until the first few months after birth.

The use of SCNT to produce cells that are genetically identical to a patient offers a possible therapy for many medical conditions, although there are many technical problems to overcome. Therapeutic cloning of human cells might create perfect-match organs for transplant or cells to fight degenerative human conditions like Alzheimer's disease. Cloning for stem cells destroys the pre-embryos used, while reproductive cloning risks the lives of both mother and clone, with most attempts failing. As yet no stem cell therapy exists based on cells produced by SCNT.

Test yourself

8 Suggest why embryonic stem cells might divide more rapidly than the cells produced in SCNT.
9 Suggest why an animal formed by SCNT is not a complete clone of the adult organism that donated the nucleus.
10 Explain the difference between reproductive cloning and therapeutic cloning.
11 Make a flow chart diagram to show how therapeutic cloning is carried out.
12 Explain why it is beneficial that stem cells used in therapeutic treatments should be genetically identical to those of the patient.
13 Describe the use of stem cells in the treatment of human diseases. What category of human diseases may be treated with stem cells?
14 Explain why therapeutic cloning arouses ethical issues.

Using microorganisms in biotechnology

Key terms

Biotechnology The use of living organisms in the production of useful products, such as foods and drugs, or the carrying out of useful services, such as sewage disposal.

Primary metabolite Any compound produced by metabolism during normal growth of an organism, such as bacteria and fungi.

Secondary metabolite Any compound produced by metabolism after a microorganism has stopped growing.

Tip

This is a good place to revise the classification of microorganisms within the kingdoms Prokaryota, Fungi and Protoctista. Viruses are also used in biotechnology.

Biotechnology is the use of living organisms in the production of useful products, such as foods and drugs, or the carrying out of useful services, such as sewage disposal and the removal of toxic wastes. Biotechnological processes use either enzymes or whole microorganisms, such as bacteria, archaeans, yeasts, mould fungi and protoctists.

Microorganisms are useful in these processes because under optimum conditions for their growth they have a short life cycle, some reproducing as often as every 20 minutes. The microorganisms used in biotechnology usually have simple growth requirements. They use simple carbon and energy sources, such as glucose. These substances are often available very cheaply. One great advantage is that microorganisms do not produce unproductive cells and tissues. For example, they do not make anything like wood or bone, which crop plants and livestock produce. The products that they make for us include primary metabolites that are made as part of their normal growth and secondary metabolites that are made when their growth slows or stops. As we have seen in Chapter 11, it is relatively easy to genetically modify microorganisms to produce substances coded by foreign genes.

Microorganisms have specific uses, as shown in Table 12.2. Some of these processes are traditional biotechnologies that humans have carried out for centuries. Others are newer.

Table 12.2 The microorganisms used in selected biotechnological processes. Under 'Role', 1 = anaerobic fermentation, 2 = aerobic fermentation.

Biotechnological process	Microorganism(s)	Role
Baking	Yeast, *Saccharomyces cerevisiae*	Production of carbon dioxide (1 and 2)
Brewing	Yeasts: *Saccharomyces cerevisiae* and *Saccharomyces pastorianus*	Production of ethanol (1)
Cheese making	Bacteria, e.g. *Lactobacillus* and *Streptococcus* Mould fungi: *Penicillium roqueforti*	Bacteria produce lactic acid which curdles milk (1) Bacteria and fungi produce fatty acids and ketones that give flavour (1 and 2)
Yoghurt production	*Streptococcus lactis Lactobacillus bulgaricus*	Production of lactic acid (1)
Mycoprotein	*Fusarium venenatum*	Biomass is harvested to make Quorn™ (2)
Penicillin production	*Penicillium chrysogenum*	Production of penicillin (2)
Insulin production	*Escherichia coli* and *Saccharomyces cerevisiae*	Production of insulin (see Chapter 11) (2)
Production of vaccines	*Saccharomyces cerevisiae Pichia pastoris*	Hepatitis B vaccine (2)
Bioremediation (use of organisms to remove toxic materials from the environment)	*Desulfovibrio desulfuricans*	Removal of toxic materials from polluted sources (1)

Figure 12.13 Folding and kneading pizza dough to mix in air to provide oxygen for the yeast and to distribute the yeast throughout the dough.

Uses of microorganisms in producing human food and drink

Baking

The role of yeast in bread making is to produce carbon dioxide to cause bread dough to rise. Yeast is mixed with flour (made from wheat, rye or maize), sugar (sucrose), salt, ascorbic acid and water to make dough. The dough is folded and kneaded to distribute the yeast evenly and mix in some air (see Figure 12.13). The dough is then left in a humid atmosphere at 35 °C. Enzymes from yeast break down starch in the flour to maltose and break down the disaccharides (maltose and sucrose) to monosaccharides. Yeast respires aerobically using any oxygen in the dough but carries out alcoholic fermentation when there is no more oxygen left. The carbon dioxide released forms of pockets of gas in the dough causing it to rise. After the dough has risen sufficiently it is kneaded again and left to rise for a second time. The dough is then baked in a hot oven.

Brewing

Cereal grains, usually barley, are rich in starch and are the main raw material for brewing (see Figure 12.14). Yeast cannot respire starch, so the grains are soaked and allowed to germinate. The sugar is extracted and mixed with yeast to ferment and produce ethanol and carbon dioxide, both of which are required.

Figure 12.14 The fermentation room in a small brewery. The surface of the brew is covered with a scum of yeast and carbon dioxide.

The stages of brewing are listed below.

- Malting: the grains produce amylase to catalyse the hydrolysis of starch to maltose. Yeast can respire maltose. The germination is stopped by heating the grain to a temperature between 40 and 70°C to denature the enzymes and stop the hydrolysis.

- Milling: the grains are crushed to help remove the sugars.

- Mashing: hot water is poured over the grain to dissolve the sugars and other soluble compounds to form 'wort'.

- Boiling: the wort is boiled with hops, which provide flavour. The boiling concentrates the wort, which is then cooled.

- Fermentation: yeast is added to ferment the wort to produce ethanol and carbon dioxide (Figure 12.14). Yeasts of the sort typified by *Saccharomyces cerevisiae* are top fermenters that produce ales. Bottom fermenters of the type *Saccharomyces pastorianus* produce lager-type beers (*S. pastorianus* used to be called *S. carlsbergensis*). Temperatures between 10°C (lagers) and 18°C (ales) are used for the fermentation. Ethanol is a primary metabolite and is produced as soon as conditions become anaerobic.

When fermentation is complete the mixture is filtered. The beer is ready to be bottled or put into casks.

Yoghurt production

The production of all yoghurts starts with the addition of starter cultures of two bacteria: *Lactobacillus bulgaricus* and *Streptococcus thermophilus*. *L. bulgaricus* breaks down proteins into short chains of amino acids. *S. thermophilus* uses the chains of amino acids to make formic acid. *L. bulgaricus* uses lactose and formic acid to make lactic acid. If a food is to be marketed as 'yoghurt' then it has to be made using these two bacteria that can survive in milk, respire lactose and make lactic acid (see Figure 12.15). They ferment the lactose sugar in milk to give lactic acid, which coagulates the milk proteins, lowers the pH and gives the yoghurt a sour taste.

Cheese making

In making most cheeses, pasteurised milk is heated and a starter culture of *Streptococcus lactis* and *Streptococcus cremoris* is added. These respire lactose to produce lactic acid and the pH of the milk decreases. Rennet, containing the protein-digesting enzymes chymosin and pepsin, is added. These enzymes coagulate the milk proteins, so forming the semi-solid curd and liquid whey. The two are separated and curd is milled and pressed into moulds to make hard cheeses like cheddar. Bacteria and enzymes in the ripening cheese break down fats to fatty acids and glycerol, proteins to amino acids and lactose to lactic acid. The bacteria also release ketones and aldehydes that give cheeses their characteristic flavours. Blue cheeses are pierced with skewers to introduce air and spores of *Penicillium roqueforti* which grows throughout the cheese, giving their blue-veined appearance.

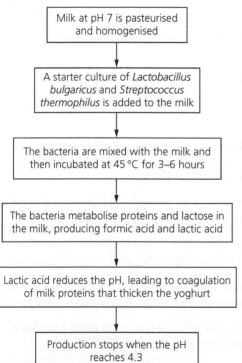

Figure 12.15 This flow chart shows the stages involved in producing yoghurt.

Mycoprotein

Since 1985 the wheat foot-rot fungus, *Fusarium venenatum*, has been used to make a meat substitute marketed as Quorn™. *Fusarium* is a filamentous fungus rather than a single-celled fungus like yeast. Mycoprotein is produced in an air lift fermenter (see Figure 12.16). The fungal hyphae are converted into meat-like products.

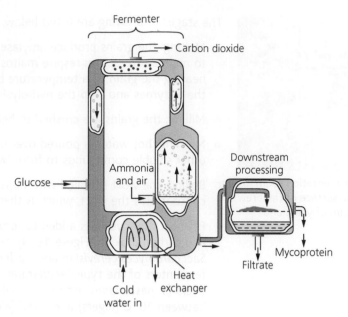

Fermenter

Carbon dioxide

Glucose

Ammonia and air

Cold water in

Heat exchanger

Downstream processing

Mycoprotein

Filtrate

Figure 12.16 Mycoprotein is produced in a continuous production process. The fungal hyphae are converted into meat-like products such as sausages and burgers.

Advantages and disadvantages of using microorganisms for making human food

Advantages of using microorganisms are

- growth is not dependent on seasons: production can occur throughout the year

- waste materials from other industries are used as substrates

- carbon and energy sources for microorganisms are relatively cheap, such as glucose made from corn (maize)

- microorganisms carry out multi-step processes using many enzymes (e.g. respiration to produce ethanol and carbon dioxide), which is cheaper than using a sequence of enzyme-catalysed stages

- factories containing fermenters are not as extensive as farms and can be established where the necessary infrastructure and raw materials exist

- it is much faster and easier to carry out selective breeding and genetic engineering on microorganisms than with domesticated plants and animals

- there are fewer ethical issues than with keeping livestock, although some people object to the use of genetically modified organisms in their food

- Quorn™ is very low in fat and is useful for people who want to reduce their intake of saturated fat from animal products

- using microorganisms is cheap, often cheaper than growing crops, which require large inputs of fossil fuels, and livestock, which are inefficient in converting biomass to human food (see Chapter 13).

Disadvantages include

- bacteria and fungi can be infected by viruses: if this happens then the production plant is shut down and sterilised

- if other bacteria enter the fermenter they may compete with the bacteria being cultured so yields are not as high: the product may need more purification treatment so costs increase

Tip

Most actinobacteria are found in soils; some have branching filaments so look very like mould fungi. The antibiotic streptomycin was discovered in *Streptomyces*.

Tip

Bacteria and yeasts are used in the production of human and animal proteins for use as drugs. The production of some human proteins including insulin is described in Chapter 11.

- the fungus extracted from the fermenter that is used to make Quorn™ is high in nucleic acids (DNA and RNA), which have to be removed before the food is safe. Purines in the nucleic acids are metabolised in the body to uric acid, which can cause gout. Processing costs may be high and reduce the profit margin for foods made from microorganisms.

Using microorganisms to produce drugs

Antibiotics, such as penicillin, are made by fungi, such as *Penicillium*, and actinobacteria, such as *Streptomyces*. The production of penicillin is a batch process similar to brewing and yoghurt making. A fermenter is filled with materials and an inoculum of *Penicillium chrysogenum* and the fermentation continues for about 140 hours. During this time conditions in the fermenter are monitored constantly. Penicillin is a secondary metabolite.

Penicillin production is a fed-batch process in which glucose and lactose are added at the beginning and corn steep liquor prepared from maize is added every 30 minutes during the fermentation. This addition supplements the nutrients that are being depleted. *Penicillium* grows well on glucose, but is stimulated to produce more penicillin when given lactose and corn steep liquor (see Figure 12.17).

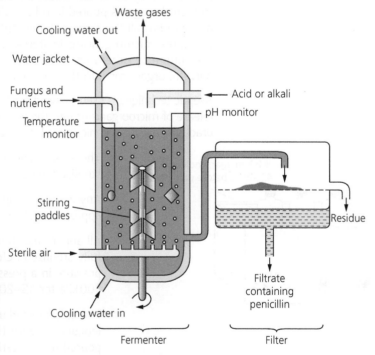

Figure 12.17 The production of penicillin is a fed-batch process carried out in a fermenter that is constantly monitored for changes in temperature and pH.

Bioremediation

Bioremediation is the use of organisms to remove toxic materials from the environment. Such materials may be heavy metals, such as mercury, lead, cadmium, zinc, chromium or nickel (see Figure 12.18). They can accumulate in the environment and in organisms, increasing in concentration in the trophic levels of food chains.

Hydrocarbons in petroleum are relatively simple for microorganisms to break down. The addition of nutrients to areas affected by oil spills encourages growth of bacteria that metabolise the hydrocarbons in crude

Figure 12.18 Wastewater from industrial, domestic and agricultural sources is often a source of pollution by heavy metals. Waste from industrial plants should pass through a bioreactor to remove any toxic waste before it enters the environment.

233

oil. Also added are sources of nitrogen and phosphorus, and trace elements that microorganisms require as cofactors. Bacterial fertilisers are also added to boost growth and decomposition of petroleum. There are few adverse ecological effects because the end products are carbon dioxide, water and biomass.

Culturing microorganisms

Microorganisms can be cultured on solid media, which is usually prepared using agar modified from a polysaccharide extracted from seaweed, and in liquid media that are usually called broths.

Any medium for culturing microorganisms should contain

- an energy source
- a carbon source
- a nitrogen source
- mineral salts
- growth factors
- water.

To prepare agar plates, the agar powder is used to make a 1 or 2% solution and boiled. This is cooled until it is at a temperature of about 42–45 °C. Nutrients may be added unless a ready-made agar is used. While still molten the agar is poured into Petri dishes (see Figure 12.19). A liquid medium has all the same requirements but without the agar. It is usual to culture microorganisms in sterile conditions to avoid contamination of the culture by other microorganisms and to prevent release of potentially harmful organisms into the environment.

Aseptic technique is used to avoid contamination of culture media and the release of microorganisms into the environment. Some of the procedures used in practising aseptic technique are listed here.

- The area surrounding the experiment must be carefully sterilised, by using a disinfectant, for example.

- Anyone preparing cultures must wear gloves and appropriate clean protective clothing.

- All apparatus, glassware and collecting loops must be fully sterilised before use; this is done by steam under pressure in a pressure cooker or autoclave at 121 °C and 100 kPa for 15–20 minutes.

- The culture medium must be sterilised before inoculation with the microorganism and it must be poured under sterile conditions (see Figure 12.19).

- When sampling and transferring microorganisms, any collecting loop should be flamed in a Bunsen burner and cooled (see Figure 12.20).

- All disposable items, such as syringes and pipette tips, should be put into a container for such items.

- After use, all non-disposable apparatus, such as glass bottles, must be cleaned and sterilised, preferably by placing into a pressure cooker or autoclave.

- All plastic Petri dishes should be wrapped in a plastic bag, tied and sterilised in an autoclave. The sterilised dishes can then be disposed of with other non-harmful solid waste.

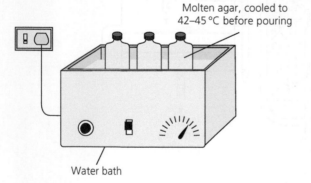

Molten agar, cooled to 42–45 °C before pouring

Water bath

Thumb and index finger used to lift lid

Cloth wrapped around bottle

Lid lifted at angle of 30–45°

Agar cooled to 42–45 °C

Bottle neck held close to Petri dish

Petri dish rested on bench surface

Figure 12.19 Pouring an agar plate. The surrounding bench must be cleaned first with disinfectant. The Petri dish has been sterilised by exposure to ultraviolet light by the manufacturer.

Key terms

Direct counting Samples are removed at intervals and examined in a counting chamber. If there are too many cells to count then the sample must be diluted with water. Alternatively an automated cell counter is used. A disadvantage is that it gives total counts including both dead and living microorganisms.

Viable counting Samples are diluted and plated onto agar. Each bacterium or yeast grows into a colony. Colonies are counted by eye or by using a colony counter. This method only counts the cells that were alive when the samples were taken.

Turbidimetry Samples are placed into a colorimeter and readings taken. This method too includes living and dead cells. It is also necessary to calibrate the turbidity readings by comparing them against total counts or viable counts.

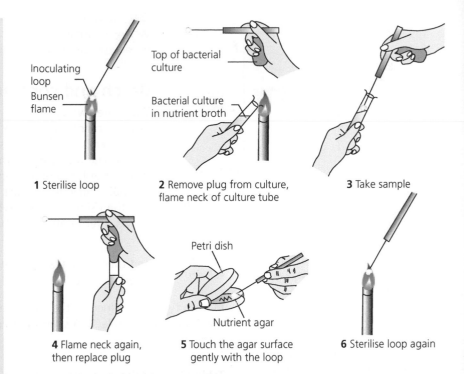

1 Sterilise loop

2 Remove plug from culture, flame neck of culture tube

3 Take sample

4 Flame neck again, then replace plug

5 Touch the agar surface gently with the loop

6 Sterilise loop again

Figure 12.20 This is the method to follow when culturing bacteria on a streak plate. A Bunsen burner should be nearby to create an updraft of air carrying any bacteria or fungal spores away from the culture medium.

Growing microorganisms

The growth of a population of microorganisms can be investigated in a closed culture using a simple fermenter as in Figure 12.21.

The growth of a population can be measured in three ways

1 direct counting

2 viable counting

3 turbidimetry using a colorimeter.

Figure 12.22 shows the change in numbers of bacteria in a population in a closed vessel over a period of time. Results were taken with both direct counting and viable counting.

The factors that affect the growth of a population of bacteria or yeast can be investigated in the following ways.

- Temperature: the water that flows through the water jacket is set at different temperatures or the fermenters are placed into temperature-controlled chambers or water baths.

- pH: the culture medium containing the nutrients required is made up in different buffer solutions.

- Type of substrate: different carbon and energy sources are provided, such as glucose, sucrose and starch.

- Concentration of the carbon source: the carbon source, such as glucose, can be made up at different concentrations, such as 0.05, 0.5 and $0.5\,mol\,dm^{-3}$.

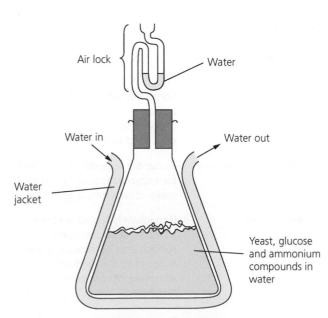

Figure 12.21 Yeast and other microorganisms can be cultured in simple fermenters like this one. Cell counts are made on samples taken at intervals and the results used to estimate the population in a certain volume, for example $1\,mm^3$.

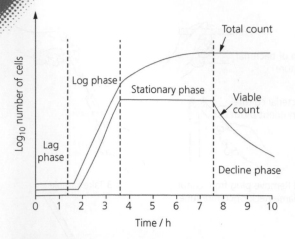

Figure 12.22 A graph showing the growth of a bacterial population in a closed fermenter. You will learn more about why the graph is this shape in chapter 14. Refer also to question 3 on page 237.

Some microorganisms, such as cyanobacteria, are photosynthetic. When culturing these microorganisms, the effects of carbon dioxide concentration and light intensity can also be investigated.

Batch and continuous fermentation

There are two types of fermentation: batch and continuous. Most fermentations are of the batch variety in which the fermenter is filled with a suitable medium and an inoculum is added. The microorganism grows and when the fermentation is over the fermenter is emptied, sterilised and filled again. Examples include yoghurt production, penicillin production and the rising of dough when making bread. Figure 12.17 shows a fermenter for a batch process.

A continuous fermentation involves setting up a fermenter and then continually adding nutrients and removing products (see Figure 12.16).

In both types of fermenter the conditions must be controlled to ensure the maximum yield. Table 12.3 shows the conditions that are controlled during fermentation.

Table 12.3 Conditions that are controlled during fermentation.

Condition		Reasons
pH	Within 0.5 of the optimum	Microorganisms and enzymes are sensitive to changes in pH.
Temperature	Within 2–3 °C of optimum	Microorganisms function best at certain temperatures; at high temperatures enzymes are denatured.
Carbon source	Monosaccharide (e.g. glucose), disaccharide (e.g. sucrose), polysaccharide (e.g. starch) if organism is heterotrophic CO_2 if microorganism is autotrophic	The type of carbon source depends on the hydrolytic enzymes produced by the microorganism.
Nitrogen source	Ammonia, amino acids, peptides	Some microorganisms can utilise ammonia to make amino acids; if not, then some may need to be supplied.
Minor nutrients	Ions, vitamins, growth factors	These are required as cofactors for enzymes (e.g. zinc) and to make coenzymes (e.g. NAD).
Turbidity	Below a certain level	If medium is too turbid then there is not an adequate surface area of microorganisms exposed to the medium so uptake of raw materials is reduced.

Tip

Each sample taken from a liquid culture must be representative of the culture as a whole, otherwise results will be inaccurate. The culture vessel must be shaken or stirred to distribute the bacteria evenly throughout the vessel. Samples are removed with a sterile pipette placed halfway down the vessel. Several samples, at least three, should be taken to ensure results are repeatable.

Example

Bacteria were grown in nutrient broth in a fermenter at 25 °C and samples were taken at regular intervals. A 10-fold serial dilution was carried out with each sample as shown in Figure 12.23. A volume of 0.1 cm³ from each tube was mixed with molten nutrient agar medium. The agar was poured into Petri dishes and incubated for 24 hours with the results shown in Figure 12.23.

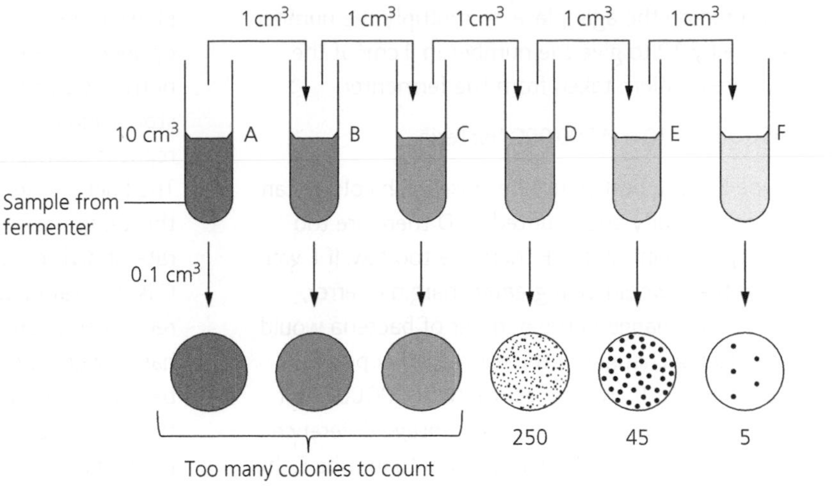

Figure 12.23 A diagram showing dilution series and colonies on each Petri dish of agar.

1 a) Describe the steps involved in making dilutions B and C.

b) Describe three precautions that should be taken when preparing the agar plates with samples of the bacterial culture.

c) Explain the disadvantage of using direct counting or turbidimetry to obtain results in an investigation to measure bacterial population growth.

2 a) Use the results in tube E to estimate the population of bacteria in 1.0 cm³ of the nutrient broth. Show your calculation.

b) Explain why tube E is the best to use for estimating the population.

3 Figure 12.24 shows the results for the whole investigation. Describe and explain the change in the population of bacteria during the course of the investigation.

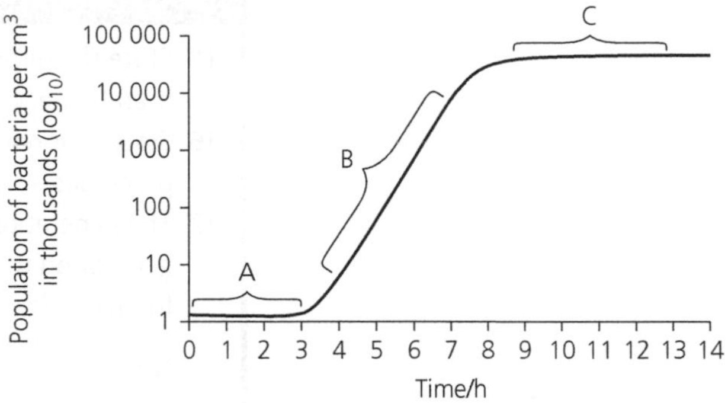

Figure 12.24 The bacterial growth curve.

Answers

1 a) Stir, shake or invert the test tube with the sample taken from the fermenter. Then use a pipette or syringe to take 1 cm³ of the sample and add to 9 cm³ water in test tube B. Shake, stir or invert the test tube to mix the sample with the water. Now take 1 cm³ from tube B and add to 9 cm³ water in test tube C. Stir, shake or invert to mix.

b) The surface of the bench should be cleaned first with disinfectant. A Bunsen burner should be lit and kept nearby. The lid of the Petri dish should only be opened a little and closed immediately the agar is poured in.

c) Both of these procedures include dead bacteria so give a total count rather than a count of the living bacteria in the population.

2 a) This is a 10-fold dilution, so in E the original sample is diluted by ×10 000. Each colony on the agar plate is formed by a single bacterium. To calculate the number of bacteria in the sample, multiply the number of colonies by the dilution factor

= 45 × 10 000

= 450 000

But only 0.1 cm³ was taken from test tube E and cultured on the agar plate, so multiply the number above by 10 to give the number in 1 cm³ of the culture medium taken from the fermenter

$450\,000 \times 10 = 4\,500\,000$ bacteria

b) Tube E is the best to use because each colony can be seen clearly and counted. In D there are too many to count and in F there are too few. If F were used there would be a greater margin of error, since any change in the number of bacteria would lead to big changes in the estimated population. For example, 5 gives an estimate of 500 000, but 6 gives 600 000, which is a percentage difference of 20% whereas if there had been 46 in tube E the result would be 4 600 000, a percentage difference of 2% from the estimated result.

3 Region A on the graph is the lag phase. During this phase there is no increase in the number of bacteria or only a small increase. Bacteria are absorbing nutrients, synthesising enzymes, building up energy stores and carrying out protein synthesis and DNA replication. Region B is the exponential (log) phase. The bacteria are dividing at their maximum rate for this temperature as there are no limiting factors. The rate of division is greater than the death rate. Region C is the stationary phase when the number of bacteria remain constant. The rate at which the bacteria divide has slowed down and the death rate has increased because limiting factors are now having an effect: there is less of the carbon/energy source and waste products are increasing. The pH may be changing so that it is less favourable for the bacteria.

Test yourself

15 State the difference between a primary metabolite and a secondary metabolite. Give an example of each.
16 State four reasons why microorganisms are useful in biotechnological processes.
17 The generation time (G) is defined as the length of time from one generation to the next. The mean generation time is calculated using the following formula

$G = t/n$

where t = time and n = number of generations.
The table shows the rate of division of some species of bacteria under optimum conditions.

Bacterium	Number of divisions in 24 hours
Streptococcus lactis	55
Escherichia coli	72
Staphylococcus aureus	48
Mycobacterium tuberculosis	18

Calculate the generation time for each bacterium in minutes.

18 Describe the growth of the population of bacteria shown in Figure 12.24.
19 Explain the difference between the results for direct counting and viable counting.

Immobilised enzymes

The pads at the ends of the test strips that detect glucose contain immobilised enzymes. These are also in the test strips in biosensors, such as those used by people with diabetes to monitor their blood glucose concentration.

Key term

Immobilised enzyme Any enzyme that is fixed to an inert substance or encapsulated in a bead or kept within a partially permeable membrane so it is not free in solution.

When you have investigated the activities of enzymes you have used enzymes in solution. Many enzymes used in medicine and in industrial processes are in solution. They are quite cheap to produce but are usually used only once as it is costly to recover them for reuse. The high costs of enzymes prompted scientists to find a way in which they could be reused. Enzymes for use in industrial processes are extracted from microorganisms that are grown in culture. Enzymes that the microorganisms secrete into the medium are simply extracted from the medium; those that are not secreted must be obtained by breaking open the microorganisms to release them. It is these that are more expensive to produce.

The solution to this problem was to immobilise enzymes so they are attached to an inert substance, such as glass beads, resin beads and collagen fibres. Enzymes are also immobilised by trapping them in microcapsules of polyacrylamide or alginate or they are kept within partially permeable membranes. Enzymes that are attached to an inert substance are linked by bonds to that substance so that substrate molecules can easily reach the active sites. Table 12.4 lists six examples of immobilised enzymes.

Table 12.4 Six immobilised enzymes, the microorganisms that make them and their roles in industrial processes.

Enzyme	Source of enzyme	Role in industrial process
Glucose isomerase	GM *Streptomyces*	Conversion of glucose to fructose in manufacture of high-fructose corn syrup
Penicillin acylase	*Escherichia coli*	Manufacture of semi-synthetic penicillins
Lactase	Yeasts and mould fungi, e.g. *Aspergillus* spp.	Hydrolysis of lactose to glucose and galactose
Aminoacylase	*Saccharomyces cerevisiae, Bacillus, Streptomyces* and *Aspergillus oryzae*	Production of pure samples of L-amino acids, e.g. to produce amino acids in animal feeds
Glucoamylase	*Aspergillus* spp.	Conversion of dextrins (short-chain carbohydrates) to glucose
Nitrile hydratase*	*Rhodococcus*	Conversion of acrylonitrile to acrylamide for use in the plastics industry

*This enzyme is often called nitrilase, although nitrile hydratase is the correct name.

Using immobilised enzymes

One method of immobilising cells and enzymes is to encapsulate them in small beads of calcium alginate, which is a very viscous substance. Immobilised enzymes are often slower in action as substrates have to diffuse through the gel to reach the active sites of the enzymes, but they can withstand higher temperatures and a wider range of pH and can be put into columns (see Figure 12.25) or filtered easily from the product.

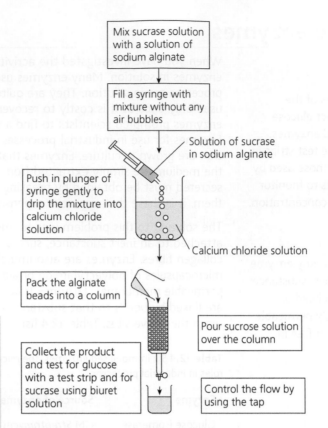

Figure 12.25 This flow chart shows how to prepare and use immobilised sucrase to hydrolyse sucrose and test the product for reducing sugars. The product is a mixture of sucrose, glucose and fructose.

Glucose isomerase is an intracellular enzyme and is therefore expensive. High-fructose corn syrup (HFCS) is produced by processing corn (maize) starch to yield glucose and then processing the glucose to produce a high percentage of fructose. The process starts with the breakdown of starch catalysed by soluble amylase. This produces short chains of sugars that are broken down by glucoamylase, again in solution. The product of these two reactions is poured through a column containing beads with glucose isomerase held on the surface. It converts glucose to a mixture of about 50% fructose and 50% glucose. The product (HFCS) collects at the base of the column and is free of contamination by the enzyme. The enzyme remains active for up to 100 days. The production of HFCS in the USA is over 9 million tonnes per annum. Most of the HFCS produced is used by the food-processing industry as a sweetener. Fructose is sweeter than sucrose so smaller quantities of HFCS can be added to sweeten foods and drinks achieving the same effect as larger quantities of sucrose, which is a more expensive raw material than corn starch. HFCS is present in a very wide range of products.

The advantages of immobilising enzymes are that the enzyme can be reused easily and does not contaminate the product. As you can see from Figure 12.25, when packed into columns enzymes can be used in continuous processes rather than batch processes. These enzymes are also more stable than soluble enzymes at extremes of temperature and pH.

Activity

Immobilised lactase

Lactose is milk sugar. Lactase catalyses the hydrolysis of lactose. A student investigated the effect of temperature on the activity of immobilised lactase and lactase 'free' in solution. The student followed this procedure.

1 Prepare a beaker as a water bath at 15 °C using cold water and ice.
2 Prepare at least 50 cm³ of a solution of 1% lactase in distilled water.
3 Mix 25 cm³ of the 1% lactase solution with the same volume of a sodium alginate solution.
4 Put the lactase alginate mixture in a pipette. Use the pipette to put drops of the mixture into a calcium chloride solution (see Figure 12.25). This will immobilise the enzyme in calcium alginate beads.
5 Put 10 cm³ milk in test tube A and 1 cm³ of the 1% lactase solution into test tube B. Put both test tubes in the water bath and leave for 3 minutes to equilibrate.
6 After 3 minutes mix the lactase solution and the milk together in test tube A and replace the test tube in the water bath.

7 Immediately take a sample of the reaction mixture and test for glucose using a test strip.
8 Continue taking samples and testing them until there is no further change in the colour of the test strips.
9 Repeat steps 5–8 a further seven times to give a total of eight results for 15 °C.
10 Repeat steps 5–9 using 10 cm³ milk and 1 cm³ of the beads of immobilised lactase.
11 Repeat steps 5–10, increasing the temperature up to 55 °C at intervals of 5 °C for both soluble lactase and immobilised lactase.

The student recorded the time, t, taken for each reaction mixture to reach a glucose concentration of $1\,g\,100\,cm^{-3}$. The rate of reaction for each reaction mixture was calculated as $1/t$. The student calculated the relative enzyme activity of each reaction mixture as the percentage of the highest rate achieved in the investigation, as shown in Table 12.5 (which continues on the next page).

Table 12.5 Results of the experiment.

Temperature/°C	Relative enzyme activity/percentage of maximum activity							
	Immobilised lactase							
	1	2	3	4	5	6	7	8
15	16	18	18	15	17	18	19	17
20	21	22	25	23	22	21	25	21
25	33	35	29	34	28	29	33	32
30	47	45	43	45	46	45	49	46
35	65	59	58	67	68	59	60	63
40	76	77	75	78	75	74	70	79
45	100	100	99	100	100	98	97	100
50	69	70	65	70	73	67	68	65
55	29	33	31	30	27	23	31	32

Temperature/°C	Relative enzyme activity/percentage of maximum activity							
	Soluble lactase							
15	29	25	27	24	29	27	28	30
20	36	34	33	35	32	38	31	34
25	51	53	46	50	48	47	51	52
30	67	63	65	67	68	60	59	61
35	92	88	90	87	89	93	95	90
40	100	100	100	100	99	98	99	100
45	96	93	96	95	98	93	96	97
50	23	27	28	26	27	23	29	31
55	12	18	14	11	14	13	20	10

Questions

1 Calculate the mean and standard deviation for each temperature for both the immobilised lactase and the soluble lactase and present your answers in a table.
2 Use one pair of axes to draw a graph to show the effect of temperature on the activity of immobilised and soluble lactase.
3 State and explain the effect of increasing temperature on the activity of the soluble lactase.
4 Suggest why the activity of immobilised lactase at 40 °C is less than the activity of the soluble lactase at 40 °C.
5 Explain why the activity of immobilised lactase is higher at 50 °C than soluble lactase at 50 °C.

6 Describe two controls that should have been included in the procedure and explain why they should have been included.
7 The student included replicates in the investigation. Explain how the inclusion of replicates improves the validity of any conclusions that are made from the results.
8 State two limitations of the method described above that may have affected the quality of the results obtained. For each limitation
   • explain how it influenced the quality of the results
   • describe how you would modify the procedure to overcome the limitation.

Tip

Immobilised lactase can be used to break down the lactose in milk for people who are lactose intolerant. These people cannot digest lactose and this makes them unwell with symptoms such as diarrhoea.

Test yourself

20 Describe two methods of immobilising enzymes.
21 State the advantages of using immobilised enzymes in industrial processes.
22 State the roles of immobilised glucose isomerase and immobilised lactase.

Exam practice questions

1 In tissue culture, whole plants can be formed from callus tissue. Which property of callus cells makes this possible?

- **A** Homozygosity
- **B** Multicellularity
- **C** Polymorphism
- **D** Totipotency *(1)*

2 Which type of roots grow on cuttings taken from stems and leaves?

- **A** Adventitious roots
- **B** Fibrous roots
- **C** Lateral roots
- **D** Tap roots *(1)*

3 *Desulfovibrio desulfuricans* is a bacterium used in bioremediation. What is the role of sulfate in the metabolism of *D. desulfuricans*?

- **A** Carbon source
- **B** Electron acceptor
- **C** Electron donor
- **D** Sulfur source *(1)*

4 Tissue culture is a method of artificial propagation. One technique for carrying out tissue culture is to remove meristems from shoot tips and culture them.

- **a)** Describe the appearance of cells taken from a plant meristem. *(3)*
- **b)** Suggest an advantage of using meristems in plant tissue culture. *(1)*
- **c)** Outline two advantages and two disadvantages of using tissue culture to produce ornamental and crop plants. *(4)*
- **d)** Suggest why it is easy to genetically engineer plants using callus tissue. *(3)*

5 a) State the term that applies to each of the following descriptions.

- **i)** The chemical produced by *Lactobacillus bulgaricus* and *Streptococcus thermophilus* that causes coagulation of protein in yoghurt manufacture. *(1)*
- **ii)** The procedures to follow when culturing microorganisms on sterile agar. *(1)*
- **iii)** An enzyme added to milk to make cheese. *(1)*
- **iv)** The gel used to form beads when immobilising enzymes. *(1)*
- **v)** The type of fermentation used to culture the fungus *Fusarium venenatum* in the manufacture of mycoprotein. *(1)*

b) Outline the reasons for using microorganisms in biotechnological processes. *(5)*

Among the many enzymes that are immobilised for use in industrial processes are aminoacylase, lactase, glucose isomerase and penicillin acylase.

c) Aminoacylase requires zinc. Suggest a reason for this. *(2)*

d) Lactase is inhibited by galactose. Suggest how this inhibition might occur. *(3)*

e) Glucose isomerase is used to convert glucose to fructose. Explain why Benedict's solution could not be used to detect the progress of the reaction. *(2)*

f) The graph shows the effect of pH on the activity of free (soluble) and immobilised penicillin acylase at 37 °C. Describe the effect of pH on the activity of the immobilised enzyme compared with the soluble penicillin acylase. *(4)*

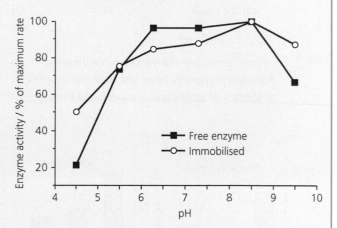

6 An investigation was carried out into the effects of plant hormones on callus tissue. The callus tissue was prepared from leaves of tobacco, *Nicotiana tabacum*, and cultured on solid media containing different concentrations of two plant hormones, the auxin IAA and kinetin, a cytokinin. The results are shown in the table.

Treatment	Concentrations of plant hormones/ mg dm⁻³		Effect of plant hormones on growth of callus tissue
	IAA	Kinetin	
1	2.00	0.00	Little or no growth
2	2.00	0.02	Growth of roots
3	2.00	0.20	Increased growth of callus with no differentiation
4	2.00	0.50	Growth of shoots
5	0.00	0.20	Little or no growth

a) State three precautions that should be taken to ensure valid comparisons can be made between the treatments. (3)

b) Summarise the effects of the plant hormones on the growth of callus tissue of *N. tabacum*. (4)

c) Use the information in the table to explain how small quantities of callus tissue can be used to produce large numbers of plantlets of *N. tabacum*. (3)

d) Make three criticisms of the investigation as described above. Suggest one improvement for each criticism and explain how each improves the quality of the conclusion that can be made. (9)

Stretch and challenge

7 Hybrid clones are plants which combine the desirable properties from one or more varieties or species of plants and are produced through tree-breeding programmes. Hybrid clones of *Eucalyptus* bring together such properties as fast and uniform growth, high productivity, small crowns, resistance to pests and diseases, and straight stems which lead to a good-quality, highly marketable product for the timber and wood pulp industries.

A researcher investigated the effect of water stress on young plants of two different clones, A and B, of *Eucalyptus globulus* grown in a greenhouse. Plants in the two clones were exposed two different forms of water stress: one group received 25% of the usual watering regime and the other received 18%. After three weeks the plants were given a one-week recovery period when they were watered normally. One group of clones was well watered throughout the four-week period. The total dry mass of the plants was determined after the three weeks of water stress and at the end of the recovery week by taking samples from the three groups within each clone. The results are shown in the charts below.

a) Explain why the groups of well-watered plants were included in the study.

b) Discuss which of the two clones is better adapted to water stress.

c) The investigation was carried out with the plants growing in pots in a greenhouse. Explain how you would carry out a field investigation to find out which clone to recommend for planting in dry upland areas.

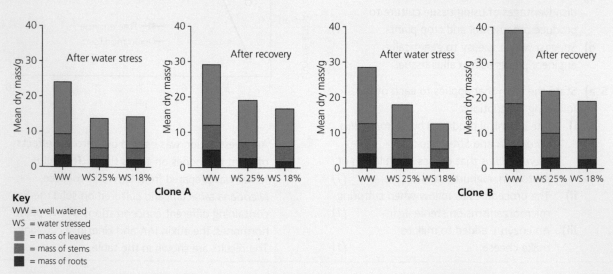

Key
WW = well watered
WS = water stressed
■ = mass of leaves
■ = mass of stems
■ = mass of roots

Four bar charts showing the biomass of the two clones at the end of the three-week period of water stress and the one-week recovery period.

8 A culture of the bacterium *Escherichia coli* was grown for 8 hours in a liquid medium containing glucose. The culture was kept at 37 °C and was aerated. A small sample was removed into a fresh medium containing limited quantities of glucose and lactose and kept under identical conditions. Samples were taken at regular intervals and the population of the bacteria determined.

The table shows the results.

Time/min	Bacteria population/ numbers $\times 10^8$ cm^{-3}
0	1
20	1
40	1
60	1.5
80	3
100	6
120	12
140	24
160	48
180	96
200	192
220	384
240	440
260	440
280	440
300	440
320	600
340	1200
360	2400
380	4800
400	9600
420	9700

a) Suggest how the population of the bacteria was determined at each sampling time.

b) i) Enter the data on a spreadsheet and use the spreadsheet facilities to plot a graph of the data in the table.

ii) Use the spreadsheet to calculate the $\log_{10}$ of each figure in the table and use the results to plot a semi-log graph ($\log_{10}$ population numbers against time).

c) i) With reference to your graphs, explain the advantage of converting the population figures to $\log_{10}$.

ii) Describe the growth of the population.

iii) Explain why the culture was kept at 37 °C and why it was aerated.

d) Explain the shape of the growth curve. During the first period of growth, the concentration of lactose in the medium remained constant. A test for glucose at 260 minutes was negative.

e) Describe how a test for glucose might have been carried out.

f) i) Explain fully why the concentration of lactose remained constant in the medium during the first period of growth.

ii) Explain how the bacteria continue to grow when there was no glucose in the medium.

g) In a replicate of the procedure uracil was added to the growth medium at time 320 minutes. The concentration of uracil in the medium decreased very rapidly. Explain why this was so.

h) *E. coli* lives in the large intestine of many mammals and can infect the udders of cattle and sheep. Use the information in this question to discuss the adaptive features of *E. coli*.

Chapter 13

Ecosystems

Prior knowledge

- The principles of sampling habitats using random and non-random sampling methods.
- Different methods of sampling habitats, such as using frame quadrats, pitfall traps and line and belt transects.
- Using mark–release–recapture and the Lincoln index to estimate the population of a motile animal.
- The definition of the term *biodiversity* and the reasons for maintaining it.
- The use of the Simpson's index of diversity to assess the biodiversity of habitats.
- Definitions of the terms *ecosystem, habitat, population* and *biospecies*.
- The differences between autotrophic and heterotrophic nutrition.
- The use of food chains and food webs to show the flow of energy through trophic levels.
- Carboxylation (carbon fixation) in photosynthesis and decarboxylation in respiration.
- The importance of biological molecules that contain nitrogen.
- Deamination is the removal of amino groups from amino acids.
- Plants absorb nitrate ions from their surroundings and use them in the synthesis of amino acids and other nitrogenous compounds.

Test yourself on prior knowledge

1 Distinguish between a habitat and an ecosystem.
2 Opportunistic, systematic and stratified sampling are different types of non-random sampling. Explain the main features of each type.
3 State the three aspects that are covered in the definition of biodiversity.
4 Define the term *biospecies*.
5 State how the following are determined: species density, species frequency and percentage cover.
6 Explain the terms species richness and species evenness.
7 State the formula for Simpson's index of diversity and explain what it shows about a particular habitat.
8 The food chain below has five trophic levels. Name the trophic level for each organism in the food chain.

phytoplankton → small crustacean → herring → common seal → killer whale

Tip

This is a good opportunity to review Chapter 12 in *OCR A level Biology 1 Student's Book* on distribution and abundance of species as well as sampling.

Sampling the marine environment

Around the coast of the British Isles there are areas that are environmentally important. Biologists study these and their biodiversity just as they do with terrestrial habitats. Sampling the sea bed can be difficult but it is still done in much the same way as on land; for example, by using random quadrats

Figure 13.1 This diver is sampling the sea bed.

over areas of uniform habitat or along transects if there is a gradient of one or more physical factors, such as the type of substrate. Divers working in pairs or in teams use these methods to sample habitats, identify organisms and record where different species are found (their distribution) and how many there are (their abundance).

Quadrat sampling underwater may involve searching each area by hand, identifying and recording the data or taking photographs for analysis at a later time (see Figure 13.1). This gives an indication of the abundance and distribution of organisms in the different habitats and the biodiversity. If sampling is repeated at intervals, changes in the communities can be tracked.

Recording biodiversity of marine habitats is a way of checking on the effect of human influences such as pollution and fishing on communities. In many places around the UK there are marine conservation areas to protect the rich diversity that exists in the seas around our coasts.

Ecosystems are dynamic

Key terms

Ecosystem A community of different species that are dependent on each other together with the non-living environment of a relatively self-contained area.

Population All the members of the same species living in a particular area at the same time. In species which reproduce sexually there is interbreeding between the males and females.

Community All the organisms of all the populations that live in a particular area at the same time.

Tip

These three key terms have a variety of meanings depending on the context. For example, population often refers to the *human* population, but here it refers to any species including prokaryotic organisms. You should know and use these three terms in their biological meanings.

Ecosystems are places with characteristic physical and geographical features with the different species that live there permanently or temporarily. Within an ecosystem populations of each species interact with each other to form a community and the community interacts with the environment.

Any fairly self-contained environment where there are habitats for organisms that interact with each other represents an ecosystem. In this place there is a supply of energy, usually in the form of light, and organisms that are able to absorb and use the light to fix carbon and make biological molecules such as carbohydrates, proteins, lipids and nucleic acids. In an ecosystem there is a flow of biomass and the energy it contains through organisms that are at different trophic (feeding) levels. Few if any species can exist in a place entirely on their own and in most places there are many different organisms. Some places teem with life. Examples are those areas with very high biodiversity, such as coral reefs and tropical rainforests.

Ecosystems vary considerably in size. Ecologists usually demarcate areas that have specific characteristics and features both of the physical environment and of the community of organisms that inhabits that area. Ecosystems can be as large as the open ocean and as small as a garden pond. Extensive ecosystems such as the East African savanna are known as biomes. Most ecosystems are described as open ecosystems as some species in the community move between different ecosystems. This happens with many species of fish that visit rocky shores at high tide to feed, but retreat to deeper water when the tide goes out. No ecosystem is completely isolated, but there are those that have few interactions with other ecosystems. Examples of these are oceanic islands, such as the Galápagos Islands (see Chapter 10), and undersea vent communities.

While it is possible to study the biology of one species and the ways in which it interacts with different aspects of its environment, other aspects can only be studied at the ecosystem level. Such aspects include biodiversity, biomass flow, energy flow and the recycling of biologically important elements.

Ecosystems are dynamic because energy is constantly flowing through organisms; numbers of species are increasing and decreasing over time. Also, ecosystems respond to external and internal changes, such as natural catastrophes and human influences such as pollution.

The species in an ecosystem can be divided into those that are **autotrophic** and those that are **heterotrophic**. Autotrophic species can use light or chemical energy to fix carbon. Those that are heterotrophic cannot fix carbon but have to take in complex carbon compounds as their source of energy. Organisms produce waste and die and there are decomposers that recycle the components of this waste and dead matter. They are an integral part of how the ecosystem functions as without them there would be a decrease in the finite quantities of the elements that plants need to build complex organic compounds.

Methods of sampling for distribution and abundance

Chapter 12 in *OCR A level Biology 1 Student's Book* dealt with the different methods that you are likely to use to sample ecosystems. This section gives some examples of ecosystems that you might investigate and the techniques of sampling, recording and data presentation that you might use.

Studies of the abundance of different species rely on taking samples from a habitat, often by using a quadrat. The abundance may be measured directly by counting individuals and calculating species density. This is sometimes difficult to do with plants and some animals (such as barnacles; see Figure 13.2) and percentage cover is used instead by estimating what percentage of a quadrat is occupied by an organism. When something quicker is needed then an abundance scale might be used, such as

- DAFOR: D = dominant; A = abundant; F = frequent; O = occasional; R = rare

- ACFOR: A = abundant; C = common; F = frequent; O = occasional; R = rare

- Braun-Blanquet scale for plant cover (see Table 13.1).

Table 13.1 The Braun-Blanquet scale for recording vegetation within quadrats.

Description	Rating
Very few plants, cover is less than 1%	+
Many plants, but cover is between 1 and 5%	1
Very many plants or cover = 6–25%	2
Any number of plants; cover = 26–50%	3
Any number of plants; cover = 51–75%	4

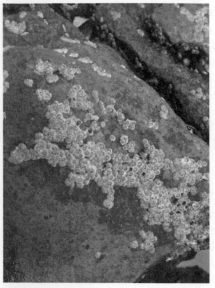

Figure 13.2 An abundance scale is useful for recording the numbers of very small organisms like these barnacles.

Figure 13.3 An Arctic tern, which migrates between breeding grounds around the Arctic, including the UK, to feeding grounds around the Antarctic coast. The birds travel up to 90 000 km during each round trip each year, exploiting ecosystems across the globe. The birds may live for up to 30 years.

Surveys of small mammals give data on population trends and are useful for studies of selection (for example, see the Exam practice question about mice on the island of Skokholm in Chapter 10). To assess their population size, they can be caught, marked, released and then recaptured. Longworth mammal traps can be used to catch them (see Figure 13.4). Only people who hold a licence to use these traps may set them. You may take part in a survey of a small mammal population if you carry out your fieldwork at a field centre run by organisations like the Field Studies Council.

(a) The trunk and branches of large trees provide habitats for mosses, lichens and many insects that live in the bark and bore into the wood. The insects provide food for specialist feeders such as tree creepers and nuthatches.

(b) Lawns and playing fields are usually uniform ecosystems suitable for random sampling. The effect of trampling can be assessed by using belt transects.

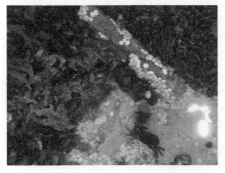

(c) This rock pool supports a community of organisms including crabs, sea anemones, barnacles and limpets. Around the edge are some green and brown algae.

(d) The margins of ponds are places where an open ecosystem interacts with surrounding ecosystems.

Figure 13.5 Four 'mini-ecosystems'.

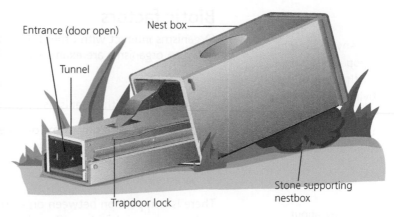

Figure 13.4 A Longworth mammal trap. It is filled with bedding and a suitable supply of food. The door closes behind the animal as it enters the trap. The traps are laid where it is likely that small mammals, such as mice and voles, will visit. The traps are checked frequently to avoid fatalities through cold or starvation.

Three 'mini-ecosystems' which you may study are large trees, playing fields and rock pools. Others are ponds and old walls that have been colonised by plants. All of these can be studied using the techniques described in Chapter 12 in *OCR A level Biology 1 Student's Book*.

Study the photographs in Figure 13.5 and think about appropriate methods to investigate the distribution, abundance and diversity of organisms in each mini-ecosystem. Then try the Test yourself questions.

Test yourself

1 State the term that describes each of the following
   a) all the organisms that live in an ecosystem
   b) all the barnacles of the species *Semibalanus balanoides* on a rocky shore
   c) the movement of a species between habitats to make best use of resources that become available at different times of the year
   d) sampling the most accessible parts of a rocky shore.
2 Explain what is meant by an abundance scale and give an example.
3 Explain how you would investigate the abundance of lichens and mosses on large trees in different ecosystems (see Figure 13.5a). How would you display your results?
4 Outline a method that could be used to investigate the distribution of animal species in rock pools at different positions on a shore. How would the results be displayed?
5 Predict the changes in temperature, pH and oxygen concentration in a rock pool that has a large percentage cover of seaweeds over a 24 hour period during a hot summer.
6 List four assumptions that are made when traps are used to assess the population of a small mammals by using mark–release–recapture.

The communities in each of these mini-ecosystems are influenced by their interactions with other organisms and with the physical and chemical features of the environment. **Biotic factors** are all the influences of organisms of all populations. **Abiotic factors** are all physical and chemical factors that influence populations in a community.

Figure 13.6 Lichens are a mutualism between algae and fungi. The fungi provide a habitat for the algae and the fungi gain sugars and other nutrients.

Biotic factors

Organisms interact with other organisms in an ecosystem. The interactions between organisms are examples of biotic factors. Five examples are

- competition
- cooperation with organisms of the same species
- cooperation with organisms of other species
- predation
- parasites and disease.

There is competition between organisms for resources, such as space, water, energy, nutrients, mates, shelter and light. This competition can be either

- interspecific: competition between different species
- intraspecific: competition between members of the same species.

Cooperation exists at different levels. The best example of cooperation within a species is seen in social insects such as ants, termites and honey bees. Individuals work together for the benefit of the colony even when that means those individuals do not get a chance to breed, as is the case with soldier ants and worker bees. Mole rats are an example of a social species of mammal. This cooperative behaviour is known as altruism.

Any association between two or more different species in which all partners benefit is an example of mutualism. The polyps of many coral species contain single-celled algae called zooxanthellae which photosynthesise. The algae gain protection, carbon dioxide and nitrogenous waste from the polyps and provide carbohydrate in the form of sugars to the polyps. Coral grows near the surface of the water, so providing the algae with sufficient light.

Many, possibly most, plants have associations with fungi. The fungi grow into the roots and extend outwards into the soil; fungi absorb ions for the plants, which provide them with sugars. This type of association is a **mycorrhiza** (literally 'fungus' and 'root').

Predators have a variety of adaptations for finding and catching their prey (predation). Populations of carnivores depend on the presence of prey species (Figure 13.7). They tend to take the young, sick and old from the populations of their prey. Most predators rarely have an effect in controlling populations of their prey, more usually it is the other way around (see Chapter 14). Grazing is a form of predation. On a coral reef, butterflyfish and parrotfish are 'predators' of algae that grow on bare rock. Left ungrazed, algae will grow to occupy much of the space.

Figure 13.8 A honey fungus, *Armillaria* sp., which infects living trees, kills them and then feeds as a decomposer on their dead bodies. Honey fungus is a plant pathogen.

Figure 13.7 Life at the top of the food chain can be tough. Predators like this cheetah use much energy when catching prey and often the chances to catch prey are few and far between.

Almost all plant and animal species have parasites. The host gains no benefit from the association, but the parasites do not cause enough harm to kill their hosts. Parasites and their hosts often have a relationship in which the host tolerates the presence of parasites. Parasites that cause harm are pathogens (Figure 13.8). The diseases that pathogens cause may spread through populations and have greater effects when host populations increase to a high density (see Chapter 14). Parasitoids are animals that lay eggs inside their host. Some species of ichneumon wasp lay their eggs inside other insects. The eggs hatch and develop into larvae. These larvae eat away at the host's tissues from the inside, ultimately killing the host. These species make effective biological control agents for use in protected environments such as glasshouses.

Abiotic factors

Abiotic factors are all the factors that influence the organisms within ecosystems that involve the non-living aspects of the environment. There are three groups

- climatic factors, such as temperature range (maximum and minimum), precipitation and exposure to wind

- edaphic factors, all the features of soil, such as depth, air content, pH, texture, humus content and mineral ion content

- physiographic factors, such as altitude, topography (shape of the land), aspect (for example, facing north or south), gradient, degree of erosion and drainage.

These factors, either individually or collectively, determine the distribution of species (Figure 13.9). For example, plant species that live at high altitude are able to survive the cold, but tend to grow very slowly. At lower altitudes they are outcompeted by species adapted to warmer temperatures.

Figure 13.9 In areas like this, abiotic factors are often more important than biotic factors in determining distribution and abundance of species. These students are laying down a transect in the Sierra de Gredos in Spain at over 1600 m above sea level. High altitude and exposure make wind speed, temperature extremes and thin soils important abiotic factors here.

Organisms affect the abiotic factors that influence ecosystems; forests, for example, modify the environment considerably. The wind speed is reduced; the effect of rainfall hitting the ground is much reduced compared with open ground. The canopy of most forests absorbs so much light that little is available for plants that grow on the forest floor. In tropical ecosystems, many plant species are adapted to grow on the trunks and branches of trees to gain enough light. Examples of these epiphytes are bromeliads and orchids.

There are many abiotic factors that influence terrestrial and aquatic ecosystems. Table 13.2 lists some abiotic factors that influence communities in freshwater streams.

Table 13.2 A summary of some of the abiotic factors that directly influence communities in freshwater streams.

Abiotic factor	Effects on abundance and distribution	Ways to take measurements
Temperature	Temperature range determines the species which can survive	Thermometer or a temperature probe
pH	Many species cannot survive in waters of low pH	Universal indicator paper or pH probe
Depth of water	This determines the size of fish that can survive in the stream	Metre rules
Flow rate	Some species prefer high flow rates; others cannot survive and are swept downstream	Flow meter or timing how long a floating object takes to travel a set distance
Oxygen concentration	Few species can survive low oxygen concentrations	Oxygen probe
Turbidity	Makes vision difficult for predatory fish	Colorimeter or arbitrary scale (0–10)
Dissolved solids (concentration of ions, e.g. Ca^{2+})	Provide ions for various physiological processes (see Chapter 2 in *OCR A level Biology 1 Student's Book*)	Conductivity meter
Light intensity	Light intensity influences the rate of photosynthesis of submerged plants	Light meter
Type of substrate	Some species burrow into mud and sand; others have flattened bodies to survive under stones	Visual description, e.g. rock, sand, mud

Most field work involves studying an area on one specific day and taking results for vegetation, sometimes animal life and maybe some abiotic factors. Single readings of most abiotic factors are meaningless as almost all of them change on an hourly basis and over the seasons. Measuring features of the soil, edaphic factors (see Table 13.3), tells us much more about the habitat than wind speeds and air temperatures at a certain time on one day. It is better to take readings over a period of time; for example, the changes in temperature and pH in a rock pool from the time when it is uncovered by the tide to when it is covered again at high tide.

Table 13.3 Some edaphic factors.

Edaphic factor	Ways of measuring
Soil texture	Pass dry soil through sieves of different mesh to find composition of gravel, sand and clay
Soil moisture content	Drying soil to constant mass
Mineral ion content	Conductivity of a soil solution
Humus content	Mass lost by heating dry soil to burn off organic matter
Mineral matter content	Mass left after organic matter is burnt off dry soil
Soil pH	pH of soil water; using a pH meter or universal indicator paper or solution
Temperature of soil	Temperature probe
Air content	Adding known volume of soil to water and stirring to drive off the air. The air content is the expected volume of soil plus water minus the actual volume of soil plus water
Depth of soil	*Either* cut a soil profile with a spade and measure the depth of topsoil and subsoil *or* use a soil auger to take a core of soil from the ground

Activity

Biodiversity on playing fields

A group of students investigated the abundance of plant species on a well-managed playing field and also on an abandoned playing field at a school that had closed four years previously. They did a timed search at each site and collected single leaves of the different plants that they could find. They then used the Field Studies Council's key to playing field plants to identify what they had collected. They photographed a leaf of each species that they had found and recorded the way in which each plant grew, known as its growth habit.

The students then carried out a random survey of the two playing fields using the ACFOR abundance scale to record their results, as shown in Table 13.4.

1 Explain how the students would do a timed search.
2 Explain how they collected the results shown in Table 13.4.
3 State three factors that influence plants on a playing field that is used frequently and would not affect those growing on an abandoned playing field.
4 Explain why rosette plants, such as *Plantago major*, are well adapted to habitats like playing fields.
5 Criticise the results shown in the table and explain how they could be improved.
6 Explain the differences between the species richness of the two playing fields.

Table 13.4 The abundance of selected plant species on a playing field in constant use and on a playing field abandoned for four years. Results were taken using the ACFOR abundance scale.

Species	Common name	Growth habit	Playing field in frequent use	Abandoned playing field
Plantago major	Greater plantain	Rosette	O	O
Plantago lanceolata	Ribwort plantain	Rosette	R	F
Trifolium repens	White clover	Spreading	C	A
Senecio vulgaris	Groundsel	Herb	–	R
Agrostis capillaris	Common bent	Grass	C	A
Taraxacum officinale	Dandelion	Rosette	R	O
Achillea millefolium	Yarrow	Rosette	R	O
Senecio jacobaea	Ragwort	Rosette	R	O
Anthriscus sylvestris	Cow parsley	Herb	–	A
Bellis perennis	Daisy	Rosette	A	A
Lolium perenne	Perennial rye grass	Grass	A	A
Festuca rubra	Red fescue	Grass	A	A
Anacamptis pyramidalis	Pyramid orchid	Herb	–	R
Brassica napus	Oil seed rape	Herb	–	R
Epilobium hirsutum	Great willowherb	Herb	–	R

Test yourself

7 Explain the difference between biotic factors and abiotic factors.
8 List three abiotic factors that influence plant communities.
9 List three biotic factors that influence small mammals living in woodlands.

The transfer of biomass and energy in ecosystems

Biomass is the mass of biological material; it is measured either as wet mass or as dry mass. The mass of water in animal matter tends to remain fairly constant, but in plants it can vary considerably as a result of fluctuations in water uptake and transpiration. Most often biomass is expressed as dry mass. One reason for this is that biomass represents the energy available to animals in ecosystems and water does not provide any energy. The flow of biomass through the trophic (feeding) levels of an ecosystem is usually considered as the flow of energy since the energy content of biological material varies considerably. Animals that feed on wood obtain much less energy per gram than those that consume seeds or meat. Calculating the biomass produced by a plant and eaten by an animal tells us very little about the efficiency of an ecosystem. For the rest of this section and in Chapter 14 we will deal with energy flow, not biomass flow.

Green plants, some protoctists and some prokaryotes absorb light energy and convert it into chemical energy in ATP and reduced hydrogen carriers to drive the reactions of carbon fixation. These organisms are phototrophic and many can obtain their carbon in the form of carbon dioxide and are autotrophic. All organisms that use photosynthesis are photoautotrophs. Photoautotrophs are responsible for making much of the biomass that enters food chains.

Tip

The process by which bacteria use chemical reactions to generate the energy they need to fix carbon is often called chemosynthesis. Nitrifying bacteria use the oxidation of forms of nitrogen (see later in this chapter).

Tip

This is a good place to revise the ways in which organisms are classified. Table 13.5 uses the five-kingdom classification system: prokaryotes, protoctists, fungi, animals and plants.

Tip

Energy is an important concept in biology. Here you need to think of energy being transferred through food chains and food webs. Remember that in these diagrams of energy flow, as in food chains and food webs, the arrows show the direction in which energy flows.

Organisms that cannot use light energy, but obtain energy from organic or inorganic chemical compounds, are chemotrophic. Many, like us, obtain carbon from complex carbon compounds and are chemoheterotrophic. However, some use energy released by oxidation reactions using inorganic materials to generate energy; they do not use light energy. These are chemoautotrophic organisms.

Table 13.5 summarises these different ways to obtain energy and carbon.

Table 13.5 All organisms can be divided into four groups according to their source of energy and source of carbon.

		Source of carbon	
		Carbon dioxide (autotrophic)	**Complex carbon compounds (heterotrophic)**
Source of energy	**Light (phototrophic)**	Photoautotrophic: photosynthetic bacteria, some protoctists including algae, plants	Photoheterotrophic: purple non-sulfur bacteria
	Chemical reactions (chemotrophic)	Chemoautotrophic: nitrifying bacteria (see later in this chapter)	Chemoheterotrophic: many bacteria, many protoctists, all fungi, all animals

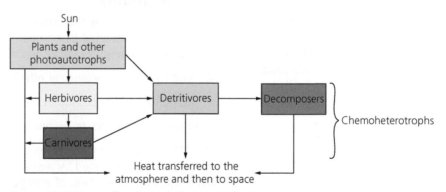

Figure 13.10 Energy flow through the trophic levels of an ecosystem.

Some of the energy captured by plants and present in plant biomass is consumed and converted into animal tissue by herbivores. These herbivores may be eaten by other consumers at higher trophic levels; that is, carnivores and humans. This pattern of energy flow is the grazing food chain. Waste organic material in the form of dead leaves, roots and branches together with animal waste in faeces and dead bodies forms detritus. This dead, waste material is food for many large and small animals known as detritivores. Dead material and the faeces of detritivores provide the energy source for decomposers, such as fungi and bacteria. This pattern of energy flow is the detritus food chain. Energy is transferred from the grazing to the decomposer food chain in the form of faeces and animal tissues after death (Figure 13.10).

Energy is not recycled; instead it leaves ecosystems as infrared radiation that warms the atmosphere. Food chains tend to be short. Much of the energy that enters the organisms at one trophic level is used by those organisms and is not available to be consumed by the next trophic level, as explained below. Much energy is transferred to heat during respiration and is 'lost' from organisms to heat the surroundings. This means that only a small percentage of the energy that enters a trophic level becomes stored in the bodies of the organisms in that trophic level to be eaten by those of the next.

Also note that biomass is not recycled either. Plants do not absorb complex biological compounds that constitute biomass, but simple inorganic compounds (carbon dioxide, water, mineral ions) that are the end products of respiration, decomposition and the metabolism of some microorganisms.

There are three ways in which biomass and energy transfers in ecosystems can be determined

1 measure the dry mass of biological material that is produced by autotrophs and consumed by herbivores, carnivores and decomposers; suitable units are $kg\,m^{-2}\,year^{-1}$

2 determine the energy content of the biomass that is produced and consumed; suitable units are $kJ\,m^{-2}\,year^{-1}$

3 determine the mass of carbon in the biomass produced by autotrophs and consumed by animals and decomposers; suitable units are $kg\,C\,m^{-2}\,year^{-1}$.

Efficiency of energy transfers between trophic levels

Ecological efficiency is the efficiency of energy transfer between trophic levels. This is calculated by comparing the energy available to a trophic level with the energy available to the next trophic level.

We will look at the ecological efficiency of two transfers

1 plants to primary consumers

2 primary consumers to secondary consumers.

The efficiency of energy transfer between producers and primary consumers is determined by assessing the net primary productivity of autotrophs and the intake of energy by primary consumers.

To find the biomass produced by plants it is possible to measure the uptake of carbon dioxide. This represents the gross productivity of the plant since some of the carbohydrate produced is used in respiration. The output of carbon dioxide from respiration can also be determined so that the net productivity of plants is calculated as follows:

net primary productivity = gross primary productivity − respiration

An alternative method is to take samples of leaves during a 24 hour period, dry them and find the change in dry mass.

Very little of the light energy that strikes plants is used in photosynthesis. Some of the reasons for this are

● plants do not have the necessary pigments to absorb visible light of all wavelengths

● light is reflected from the surfaces of leaves

● light passes straight through leaves.

Much of the light energy that strikes plants is wasted because photosynthesis cannot make use of all the light energy. Another reason may be because it is too cold for the chloroplast enzymes to function efficiently or because carbon dioxide is in short supply. At best, our crop plants may pass on to us 5% of the energy that strikes their leaves. In natural ecosystems, the percentage is even lower than this, often as little as 1%. In calculations of energy flow it is best to determine the light energy that plants can use that strikes their leaves. Photosynthetically active radiation (PAR) is the energy that plants can use in photosynthesis.

Not all of the net productivity in plants reaches the primary consumers as some plant matter is not eaten, some cannot be digested so it is not assimilated and much will die and decay rather than be eaten by consumers that graze on plants. Dead plant material will be eaten by detritivores, such as earthworms, and pass to the detritus food chain. The energy transfer from plants to primary consumers is quite variable and on average may be about 10% of the net productivity of producers.

Herbivores are preyed on by predators. The energy in herbivores that is eaten by predators is the energy transferred to the next trophic level. Much of the energy that was consumed by herbivores is not available to their predators because

- herbivores use energy as they move about in search of food

- herbivores 'lose' heat during the digestion of their food

- herbivores 'lose' heat in respiration

- herbivores use energy in reproduction

- predators do not eat all of the bodies of herbivores

- predators do not digest all of the bodies of herbivores.

The only energy transferred to the next trophic level is the energy in new herbivore flesh that is eaten by the predators. As a percentage of the energy input from the producers to herbivores that is very small.

Figure 13.11 shows energy flow through the following food chain.

Meadow grass, *Poa compressa* → meadow vole, *Microtus pennsylvanicus* → weasel, *Mustela rixosa*

The efficiency of biomass and energy transfer may be calculated in a variety of ways. One of the most common is calculating the net productivity of one trophic level as a percentage of the net productivity of the preceding trophic level. For example, the efficiency of energy transfer between primary consumers and secondary consumers is calculated as

$$\text{efficiency of energy transfer} = \frac{\text{net productivity of secondary consumers}}{\text{net productivity of primary consumers}} \times 100\%$$

- The efficiency is variable between producers and consumers and between consumers; it may be as high as 20%, but more often it is much lower than this.

- Energy is used by organisms at all trophic levels for their body maintenance and movement.

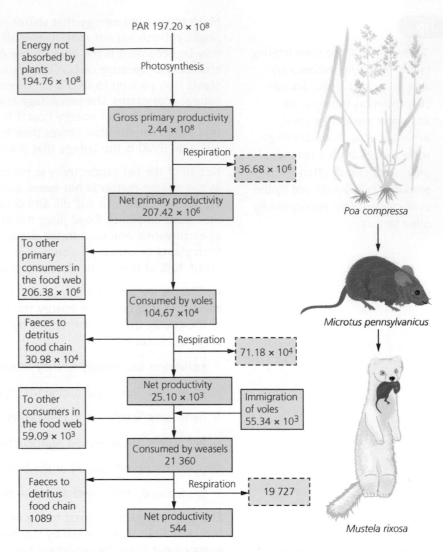

Figure 13.11 Energy flow through a food chain in an old field ecosystem in Michigan, USA. All figures are in kJ m^{-2} year^{-1}. PAR (photosynthetically active radiation) is the energy that plants can use in photosynthesis.

- Only the energy in new growth and new individuals (reproduction) is available from one trophic level to the next.

- Energy is transferred to the surroundings as heat when organisms respire and move.

- Energy is transferred to detritus food chains at all trophic levels.

- Energy transferred to detritus food chains and to the surroundings as heat is not available to consumers, so limiting their numbers and biomass.

- In open ecosystems, organisms or their wastes are lost to other ecosystems, so limiting energy available to consumers (for example, detritus in river ecosystems is carried downstream to the sea; in the open ocean dead bodies sink to the bottom).

Arable farmers and growers attempt to make net primary productivity as high as possible. Livestock farmers do the same for secondary productivity. As you can see from Table 13.6 human activities can alter the efficiency of biomass and energy flow.

Table 13.6 Some of the ways in which humans manipulate energy flow in artificial ecosystems.

Method	Crop plants (producers)	Livestock (primary consumers)
Maximise energy input	Optimum planting distances between crop plants; provide light for greenhouse crops on overcast days	Provide good-quality feed
Maximise growth	Provide water (irrigation); fertilisers (containing NPK and other elements, e.g. S); selective breeding for fast growth	Provide food supplements, e.g. vitamins and minerals; selective breeding for fast growth
Control disease	Fungicides	Antibiotics and vaccines
Control predation	Fencing to exclude grazers, e.g. rabbits, deer; use pesticides to kill insect pests, nematodes, slugs, snails, etc.	Extensive systems (ranching): control predators such as wolves and foxes; intensive systems: keep animals in sheds protected from predation
Reduce competition	Ploughing and herbicides kill weeds	Control competitors such as rabbits and deer
Reduce energy loss	Breed plants that maximise energy storage in edible products, e.g. seeds, fruits and tubers	Keep animals in sheds: less energy lost in movement and maintaining body temperature

Tip

You will find ecological efficiencies of 10% given quite often between consumers. This figure was derived from original research done in the 1940s and involved a miscalculation of the data. A study carried out in the 1960s found energy transfer between primary consumers and secondary consumers to be about 1%!

Test yourself

10 State three ways in which energy flow through ecosystems can be determined.
11 Using examples from this chapter write out a grazing food chain and show how it is linked to a detritus food chain.
12 Explain the difference between the grazing food chain and the detritus food chain.
13 Calculate the efficiency of energy transfer by the following in the old field ecosystem in Figure 13.11.
   a) The producers
   b) The voles
   c) The weasels

Recycling in ecosystems

Tip

In various places on Earth there are reservoirs of each of these elements; for example, the air is a reservoir of nitrogen and carbon (as carbon dioxide). The flow of an element between reservoirs is a **flux**. Long-term reservoirs where the recycling of the elements is very slow are known as **sinks**. Fossils fuels have been sinks for carbon for millions of years, until recently when they began to be exploited by us (Figure 13.12).

Tip

Do not confuse the recycling of elements in the biosphere with energy flow through the biosphere. Energy cannot be recycled and so, although you may read about 'the energy cycle', no such thing exists.

Tip

This is a good opportunity to review Chapter 7 on photosynthesis and also Chapters 2 and 3 in *OCR A level Biology 1 Student's Book* on biochemistry. Remember that carbon is *fixed* in photosynthesis to form carbohydrate molecules which are then used to make the great variety of biological molecules in plants.

The elements carbon, hydrogen, oxygen, nitrogen, sulfur and phosphorus are some of the building blocks of organic molecules. Autotrophic organisms absorb these elements in simple, inorganic forms. They absorb carbon, oxygen and hydrogen as carbon dioxide and water, the raw materials for photosynthesis; nitrogen is absorbed in the form of nitrate ions. Sulfur is absorbed as sulfate and phosphorus as phosphate.

Unlike energy, there is a finite quantity of these elements available in the biosphere to organisms. If they are not recycled, then life will come to a complete stand still. We will consider how two elements – carbon and nitrogen – are recycled by organisms so that there is a continuous supply of them for autotrophs and hence for consumers and decomposers.

Recycling of carbon

Figure 13.12 How much of the carbon cycle can you see in this view of a coal-fired power station? The gases from the power station are part of the flux of carbon from coal to the reservoir of carbon dioxide in the air.

Figure 13.13 shows how carbon atoms are continuously taken up, converted into compounds and recycled. Some carbon atoms may remain in places in the cycle for thousands or millions of years. Think of the carbon atoms in human bodies preserved for a thousand years in the peat in Denmark or the coal that has been in the ground for 300 million years.

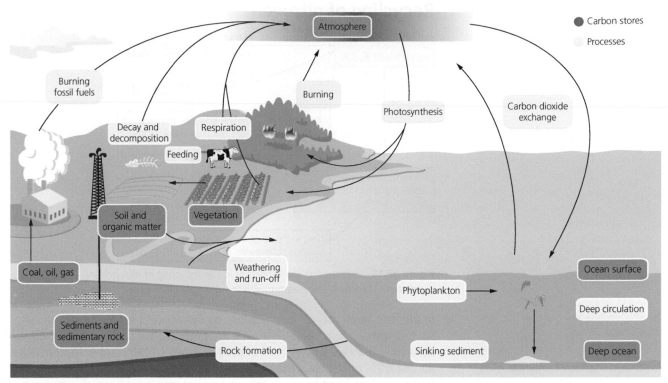

Figure 13.13 The cycling of carbon.

Even though the concentration of carbon dioxide in the atmosphere is 400 ppm (0.04%) it forms a huge reservoir estimated to be about 750 000 million tonnes of carbon. There is an even larger reservoir of dissolved carbon dioxide in the oceans, about 40 trillion tonnes. Carbon dioxide is taken from the atmosphere and from waters by autotrophs. Carbon dioxide is fixed in the Calvin cycle to form compounds of carbon, such as carbohydrates, fats and proteins, which pass along grazing and detritus food chains.

Carbon dioxide is added to the atmosphere and to water by decarboxylation, which occurs in respiration in all organisms. It is also added to the atmosphere by the combustion of wood and fossil fuels.

Some carbon compounds are not decomposed or burnt; instead they accumulate in carbon sinks, for example in peat bogs (see Figure 13.14). If undisturbed for millions of years, these undecomposed carbon compounds form fossil fuels. Peatland forming a massive carbon sink the size of England was discovered in 2014 in The Republic of the Congo in Central Africa.

Figure 13.14 Carbon sinks like this peat bog are very important for holding carbon that might otherwise become carbon dioxide in the atmosphere. The peat from this bog has been cut for use as a fuel.

Carbon atoms that were absorbed from the atmosphere 350 million years ago are still 'locked up' in fossil fuels. They are also 'locked up' in the rocks formed from the shells of tiny marine organisms. Limestone and chalk are sedimentary rocks formed over millions of years from deposits of shells made of calcium carbonate. These carbonates form a reservoir of 100 000 trillion tonnes.

Recycling of nitrogen

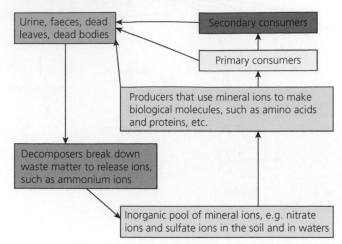

Figure 13.15 Decomposers are important in the recycling of nitrogen in ecosystems. Compare this diagram with the energy flow diagram in Figure 13.8.

Key term

Nitrogen (N) A chemical element that is in many biological molecules such as proteins and DNA; nitrogen gas (dinitrogen, N_2) makes up most of the atmosphere and is not very reactive so most organisms cannot use it.

Tip

Take care when writing about nitrogen. In this book we refer to nitrogen the element (N), dinitrogen the unreactive gas (N_2), and fixed nitrogen (N combined with other atoms as in NH_3 or amino acids, proteins, nucleotides and nucleic acids). If you just write 'nitrogen' it may not be clear what you mean.

In just the same way that carbon has to be recycled for life to continue, so it is with nitrogen (see Figure 13.15). Nitrogen is found in many organic molecules, particularly proteins and nucleic acids (DNA and RNA). Although almost 80% of the air is made up of dinitrogen (N_2, N≡N), as it has a triple covalent bond between the nitrogen atoms it is unreactive and not available to most organisms. Compare this with oxygen and carbon dioxide which are readily used by organisms in respiration and in photosynthesis. Most organisms cannot make use of it directly because the two nitrogen atoms are bonded together very firmly. It takes much energy to break them apart so that nitrogen can combine with atoms of other elements, for example hydrogen, oxygen and carbon.

Nitrogen that is bonded to other atoms, such as oxygen, hydrogen or carbon, is called fixed nitrogen to distinguish it from dinitrogen (N_2). Most nitrogen enters communities as fixed nitrogen in the form of nitrate ions (NO_3^-), absorbed by autotrophic organisms and used to make amino acids. Heterotrophs need their fixed nitrogen in the form of amino acids.

Autotrophs use nitrate ions to make amino acids. They reduce nitrate ions to nitrite ions (NO_2^-) and then to ammonium ions (NH_4^+) in energy-consuming reactions that occur mostly in chloroplasts. Fixed nitrogen in the form of NH_4^+ is combined with products of the Calvin cycle to make amino acids in the process of **amination**. These amino acids are exported from chloroplasts to be used by the rest of the plant. Autotrophs use amino acids to make proteins; they also use the amine group from amino acids for the biosynthesis of purines and pyrimidines.

Fixed nitrogen as part of biomass enters food chains. Primary consumers eat and digest proteins to amino acids. They absorb them and use them in making their own proteins. The process is repeated in secondary consumers that feed on primary consumers. The processes of feeding, digestion and biosynthesis of proteins continue along grazing and detritus food chains.

Consumers cannot store amino acids or proteins. Carnivores, for example, gain most of their energy from proteins and therefore break down amino acid molecules that they do not need for biosynthesis. This breakdown releases ammonia in the process of **deamination**. The rest of each molecule of amino acid may be respired in the Krebs cycle or converted to glucose and stored as glycogen. Many aquatic animals excrete ammonia; mammals convert ammonia to urea and birds convert it into uric acid.

Decomposers break down all the materials that are excreted and egested by animals. They also break down the dead bodies of plants and animals. They digest proteins to amino acids, absorb them and use them in biosynthesis. They also deaminate excess amino acids and excrete ammonia. Some bacteria use urea as a source of energy and convert it to ammonia. The production of ammonia by these microorganisms is **ammonification**. Ammonia does not remain in the environment very long. Some bacteria use it in their energy-transfer reactions, oxidising it to nitrite ions which they excrete.

$$2NH_3 + 3O_2 \rightarrow 2NO_2^- + 2H^+ + 2H_2O$$

Other bacteria oxidise nitrite ions in similar reactions and excrete nitrate ions.

$$2NO_2^- + O_2 \rightarrow 2NO_3^-$$

This process of conversion of ammonia to nitrate ions is nitrification and the bacteria involved are nitrifying bacteria (see Figure 13.16).

Key term

Nitrification The process in which ammonium ions are oxidised to nitrate ions by the bacteria *Nitrosomonas* and *Nitrobacter*.

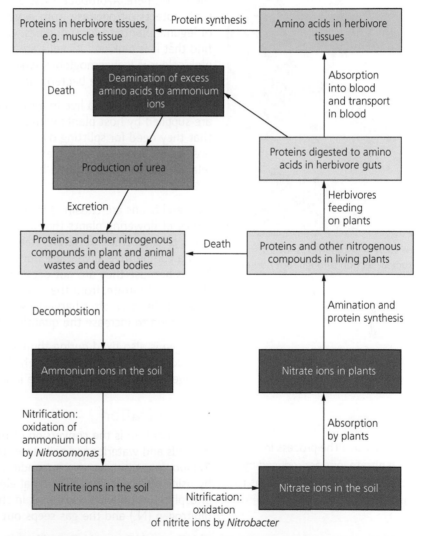

Figure 13.16 The recycling of fixed nitrogen from proteins in dead plant material to nitrate ions that are absorbed by plants to make amino acids, proteins and other nitrogenous compounds.

Tip

Notice that carnivores have been omitted from this flow chart. They also produce urea and provide protein in wastes and dead bodies.

Tip

Nitrosomonas and *Nitrobacter* are the nitrifying bacteria. They have a special form of nutrition in which they gain energy from these chemical reactions.

Fixed nitrogen has now been recycled to nitrate, which is where we started. This cycling of nitrogen is important to maintain the growth of producers in ecosystems. Nitrate ions are an important limiting factor for growth of producers, which is why farmers add fertilisers containing fixed nitrogen in the form of compounds such as ammonium nitrate. These fertilisers are mass produced by the Haber process which combines atmospheric nitrogen with hydrogen to make ammonia (often using energy from fossil fuels).

Nitrogen fixation

There are some organisms that can break molecules of N_2 apart and combine nitrogen with other elements. This is nitrogen fixation: remember, nitrogen that has been chemically combined in this way is said to be 'fixed'. Unlike the fixing of carbon in photosynthesis, which is carried out by many organisms, nitrogen fixation is only possible for the few specialist bacteria that are able to break the molecules of N_2. All nitrogen combined with hydrogen is catalysed by the enzyme nitrogenase. This reaction needs anaerobic conditions as oxygen can occupy the active site of nitrogenase. *Azotobacter* and *Rhizobium* are two types of bacteria that fix nitrogen.

Azotobacter lives in soils around the roots of plants: the area known as the rhizosphere. *Azotobacter* has a special nitrogenase-protective protein that protects nitrogenase from oxygen. *Rhizobium* stimulates the roots of legume plants to form nodules. Cut open a root nodule and you will find that it is pink and its cells are full of bacteria (see Figure 13.17). The pink colour is leghaemoglobin, which absorbs oxygen, thus keeping the conditions around the bacteria anaerobic.

Both types of bacteria live in a mutualism with their hosts. Both types are supplied by host plants with an energy source in the form of sugars that they need for splitting dinitrogen. In return the bacteria provide the host plants with fixed nitrogen as ammonium ions or as amino acids. The relationship between legume hosts and *Rhizobium* is the more complex of the two.

Peas and beans are classified in the Leguminoseae (or Fabaceae), a large family of flowering plants that includes species ranging in size from small herbs to large trees. In natural habitats, legumes, as they are usually known, gain a competitive advantage from having these bacteria in their roots. This is because they are not dependent on a supply of fixed nitrogen, in the form of nitrate, from the soil. Cultivated legumes are grown in crop rotations because after harvest the remains of the plants are ploughed into the ground to increase the quantity of fixed nitrogen in the soil.

Nitrogen is also fixed during thunderstorms. Lightning causes nitrogen and oxygen to react together at high temperatures with the formation of nitrogen oxides that form nitrate ions in the soil.

Denitrification

Denitrification is the conversion of nitrate ions to dinitrogen. This happens in muds and waterlogged soils that are anaerobic. Bacteria, such as *Pseudomonas*, can survive in conditions of very low oxygen concentration by using nitrate ions as a terminal electron acceptor in oxidative phosphorylation when oxygen is in short supply. Nitrate is reduced to dinitrogen (N_2) and the gas seeps out of the mud into the atmosphere:

$$NO_3^- \rightarrow NO_2^- \rightarrow N_2O \rightarrow N_2$$

Denitrification depletes soils of fixed nitrogen so that there are fewer nitrate ions to support plant growth.

The chemistry of nitrogen cycling

Nitrogen exists in nine different oxidation states. Four of them are relevant to the cycling of nitrogen in the biosphere.

> **Key term**
>
> Nitrogen fixation The process in which nitrogen gas is changed into ammonium ions by nitrogen-fixing organisms, such as *Azotobacter*, *Rhizobium* and some cyanobacteria. Only some prokaryotes are able to do this.

Figure 13.17 A photograph showing a longitudinal section through root nodules.

> **Key term**
>
> Denitrification The process in which nitrate ions are reduced by bacteria in anaerobic conditions to dinitrogen (N_2).

- In ammonia (NH_3) and ammonium ions (NH_4^+) nitrogen has a valency of −3.
- The valency of nitrogen in dinitrogen (N_2 or $N\equiv N$) is 0.
- In nitrous acid and nitrite ions (NO_2^-) it is +3.
- In nitric acid and nitrate ions (NO_3^-) it is +5.

NH_3 and NO_2^- are oxidised aerobically by nitrifying bacteria. Electrons from ammonia and nitrite ions enter the electron transfer chain to establish a proton gradient to synthesise ATP by chemiosmosis. This is how *Nitrosomonas* and *Nitrobacter* obtain the energy that they need to fix carbon (see Table 13.5). Nitrogen fixation requires energy as the bonds between the nitrogen atoms are very stable. Thus nitrogen fixers require a plentiful source of carbohydrate, which *Azotobacter* gets from the rhizosphere and *Rhizobium* achieves by living inside root nodules.

Test yourself

14 In recycling what is meant by a flux?
15 Suggest two ways in which deforestation leads to an increase in carbon dioxide concentration of the atmosphere.
16 Explain how leaf litter and soils contribute to the carbon dioxide concentration in the atmosphere.
17 State the role of peat bogs, limestone rocks and oil shale in the carbon cycle. Explain why they are important.
18 Explain the difference between nitrification and denitrification.
19 Explain the importance of *Nitrosomonas* and *Nitrobacter* in ecosystems.
20 Name two genera of nitrogen-fixing bacteria.
21 Explain the difference between the rhizosphere and a root nodule.

Succession

Key terms

Succession The change in communities that occurs when new land becomes available to colonise (primary succession) or when the vegetation is cleared (secondary succession).
Primary succession The colonisation of land that has never been colonised before followed by the progression of changes until a climax community is established.
Seral stage One of the communities during a succession, e.g. pioneer community.

Abiotic factors strongly influence organisms that live in exposed conditions, such as at high altitude, on exposed cliffs and on wave-battered shores. Where new land has recently become exposed and the bare rock or sand is colonised abiotic factors strongly influence the species that become established. With time, organisms gradually change the environment and more colonisers become established. A community develops and biotic factors become more important. Over a much longer period of time communities are replaced until eventually the ecosystem has the features of others locally that experience the same climatic conditions. This series of changes is called succession and occurs on land which has never been colonised and also to ecosystems following catastrophes, such as storms, flooding and fire. Succession also follows human interference with ecosystems, such as deforestation, slash-and-burn agriculture and ploughing.

Primary succession

A good example of a primary succession is an island that suddenly appears as a result of volcanic activity. In 1963, a volcanic eruption off the southern coast of Iceland created the island of Surtsey. Two years later it was declared a nature reserve for the study of colonisation and succession. Other examples of new land to be colonised are the moraine exposed by retreating glaciers (see Figure 13.18), the silting up of wetland areas and the sand that is continually deposited on beaches to form sand dunes. Over time the area of land that becomes colonised gradually changes and goes through several stages (see Figure 13.19). Each of these stages is known as a seral stage.

Figure 13.18 This glacier in the Glacier Bay National Park in Alaska is receding. The land that is exposed is colonised first by pioneer species and then by low-growing plants, which you can see in the foreground.

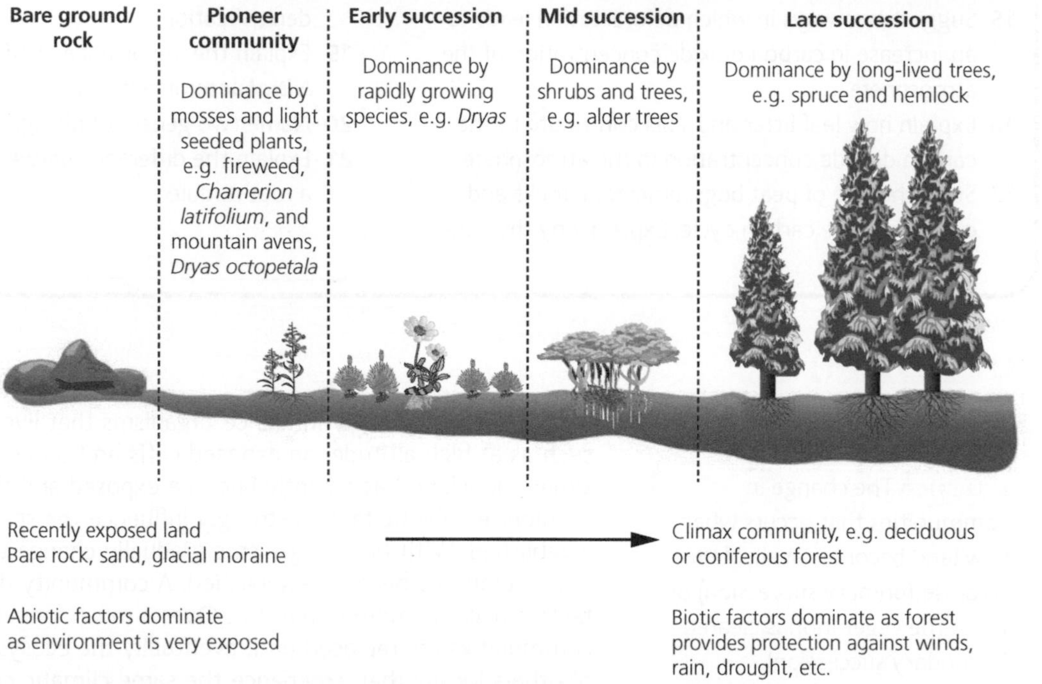

Bare ground/ rock	Pioneer community	Early succession	Mid succession	Late succession
	Dominance by mosses and light seeded plants, e.g. fireweed, *Chamerion latifolium*, and mountain avens, *Dryas octopetala*	Dominance by rapidly growing species, e.g. *Dryas*	Dominance by shrubs and trees, e.g. alder trees	Dominance by long-lived trees, e.g. spruce and hemlock

Recently exposed land
Bare rock, sand, glacial moraine

Abiotic factors dominate as environment is very exposed

Climax community, e.g. deciduous or coniferous forest

Biotic factors dominate as forest provides protection against winds, rain, drought, etc.

Community		Seral stage			Climax
		Pioneer	Early succession	Mid succession	
Biotic features	Dominant plants	Mosses, lichens, some herbs	Herb species, grasses	Shrubs, alder trees	Trees
	Number of plant species	11	17	18	31
	Niches	Few	More	More	Most
	Food chains	Very short	Short	Longer	Longest
	Food webs	Simple	Becoming more complex		Complex
	Biomass	Very low	Increasing	Increasing	Very high
	Primary productivity	Very low	Highest	High	High
Abiotic features	Soil	None (rock, sand)	Shallow soil	Deep with topsoil and subsoil	
	Soil pH	7.2	7.0	6.8	3.5
	Soil organic matter (humus)	None	Some	More	Large quantity
	Soil nitrogen / g m^{-2}	3.8	5.3	21.8	53.3
	Recycling	Rapid	Slower	Slow (as much dead material is woody)	

Figure 13.19 The features of the primary succession at Glacier Bay, Alaska.

The succession at Glacier Bay has been followed for the past 200 years since the position of the glacier was first recorded. If you stand at the edge of the glacier on recently uncovered moraine and walk away from the glacier you can walk through areas that were once covered by ice. It's a long walk: the glacier has retreated about 100 km over 200 years. The walk takes you first over rocks and boulders, but this inhospitable place is already colonised by microorganisms, including photosynthetic bacteria and algae: the first producers to become established. Other pioneer species that need no soil are lichens and mosses. Small flowering plants colonise this environment. These species are opportunists, they produce many small seeds that are easily carried in the wind and will germinate where there is little soil. They are also often self-pollinated as there may be few or no pollinators, such as insects, in this environment. With time, the presence of these plants changes the physical conditions so that other plant species can become established. Soil begins to form from wind-blown particles and decaying plant material; this makes it possible for deeper-rooted plants, such as grasses, to grow.

The moraine that is exposed by the retreating glacier has almost no fixed nitrogen in any form that plants can absorb. This lack of fixed nitrogen limits the growth of many plants so if seeds arrive here the plants do not grow. The exceptions are those with nitrogen-fixing bacteria in their roots. Examples are *Dryas octopetala*, often known as mountain avens, which grows into low-growing shrubs, and alder trees, *Alnus sinuata*. *Dryas* forms a carpet over the moraine and alder trees grow into thickets. These plants now change the environmental conditions even more significantly, especially by increasing the nitrogen content of the soil when leaves die and are decomposed. The roots begin to hold more soil and the features of the soil begin to change. The plants of the early seral stages are outcompeted and die. Now plant species with more demanding requirements can become established and these in turn outcompete mountain avens and alder. Seedlings of tree species can become established and these grow into the dominant species in the community. No further changes in the succession occur and the hemlock trees, *Tsuga heterophylla*, spruce trees, *Picea sitchensis*, and all the other species that coexist with them form the climax community.

Each successive plant community leading up to the climax community creates new conditions that lead to its replacement by plants more competitive under those new conditions. Conditions, such as light and moisture availability, the organic content of the soil (humus) and the mineral ions in the soil all change, making it possible for other plants to become established. As the succession proceeds the plants provide habitats for animals that invade the area. With time, a greater variety of microorganisms becomes established, particularly the decomposers and nitrifying bacteria that are important for the cycling of carbon and nitrogen.

During a succession, the production of biomass increases until the climax community is established; then biomass will be constant. Primary productivity increases to a peak in the early succession stages and then decreases to remain constant. Species diversity increases throughout a succession until it too remains constant when the climax community is reached.

The type of primary succession that occurs differs according to the nature of the land that is colonised. The succession at Glacier Bay is a lithosere as it begins with bare rock. There are differences in the types of plants that are in each seral stage. The names given to the types of succession are listed in Table 13.7.

Key term

Climax community The community that is the final stage of a natural succession.

Table 13.7 Types of primary succession.

Type of new land	Name of succession
Wetland that is silting up	Hydrosere
Sand dunes	Psammosere
Mud deposits in estuaries	Halosere
Bare rock	Lithosere

Deflected succession

Key terms

Deflected succession A succession that does not progress to the natural climax community; factors that cause this are grazing, mowing and fire.

Plagioclimax The community that becomes established when a deflected succession occurs. Chalk grassland is an example of a plagioclimax.

Around 10 000 years ago people began cutting down forests throughout Britain and began grazing livestock and planting crops. Loss of forest would naturally be followed by secondary succession, but farming maintains an early seral stage (see Figure 13.20).

When succession does not reach the climax it is a deflected succession. In the UK, moorland, heathland, chalk grassland, lawns and playing fields are examples. Grazers reduce the competition from fast-growing grasses so that many other species can survive. The grazing also prevents the growth of shrubs and trees so the later communities do not become established. Mowing is very like grazing although much less selective. Instead of progressing through the seral stages to a climax community, the succession remains at an early seral stage which often has higher productivity than the climax community. The seral stage that is maintained is a plagioclimax.

Figure 13.20 Chalk grassland, a deflected succession. This view of Uffington on the Berkshire Downs, is typical of much of southern England. Without grazing by sheep, this ecosystem would change to hawthorn scrub and then forest.

Activity

Primary succession on sand dunes

Only a few species are adapted to live in areas where sand is constantly deposited by the wind. Many sand dune ecosystems (Figure 13.21) are eroded by trampling visitors or have been destroyed by development.

Some students investigated the distribution of plants along a section of a sand dune at Winterton sand dunes on the north Norfolk coast. They took readings at 10 stations at intervals from the strandline. Table 13.8 shows the distribution and abundance of the plant species and Table 13.9 shows the data for the abiotic factors that they collected.

Figure 13.21 Sand dunes, showing progression from yellow dunes in the foreground to grey dunes in the background.

Table 13.8 The percentage cover of selected species at 10 sampling sites on a sand dune.

Distance/m	0	2	4	8	16	32	64	128	256	512
Marram grass	0	0	0	0	30	24	15	0	0	0
Sea couch grass	11	8	9	35	6	18	4	0	0	0
Lyme grass	0	0	0	0	0	8	13	0	0	0
Red fescue grass	0	0	0	0	0	0	0	1	26	0
Lesser hawksbit	0	0	0	0	0	10	15	2	1	1
Bell heather	0	0	0	0	0	0	0	0	40	0
Bramble	0	0	0	0	0	0	0	0	0	16

Table 13.9 Measurements of abiotic factors at 10 sampling sites on a sand dune.

Distance/m	0	2	4	8	16	32	64	128	256	512
Soil pH	8	8	8	7	8	8	8	8	8	8
Wind speed/m s^{-1}	3.9	4.0	3.4	1.7	3.5	3.1	4.9	5.8	6.3	2.6
Air temperature/°C	17.1	17.5	16.8	18.1	16.2	18.6	15.8	18.0	19.0	15.9
Vegetation height/m	0.23	0.14	0.12	0.91	1.10	0.40	0.76	0.12	0.09	0.25
Soil moisture reading (0–10)	1	2	2	2	1	1	1	1	1	1

1 Present the data as a series of bar charts aligned to show the distribution and abundance of the different species across the dunes.
2 Explain why it would be difficult to draw a kite diagram to show the results of this investigation.
3 Suggest an explanation for the distribution of sea couch grass and marram grass in the sand dunes.
4 Explain why this ecosystem is an example of a primary succession.

5 The area is a deflected succession as behind the dunes there is a council-owned grass car park and a field of cattle. State three factors that prevent the succession continuing to a climax.
6 State some of the limitations of the results shown in Tables 13.8 and 13.9.

Tip

If writing about a succession it is good to use the general principles about changes to biomass, productivity and biodiversity.

Test yourself

22 Use Figure 13.19 to make graphs or charts to show the changes in abiotic factors in the succession at Glacier Bay.
23 What are the abiotic factors that influence a sand dune community? List as many as you can.
24 Many molecules of biological significance contain nitrogen. List as many as you can.
25 Explain how sheep maintain deflected successions.
26 How do nitrogen-fixing bacteria help legumes to succeed early in primary successions?

Exam practice questions

1 Which is a biotic factor?

   A Disease

   B soil pH

   C Salinity

   D Temperature of a lake   *(1)*

2 Freshwater fish excrete ammonia. This can cause health problems for fish after they are put into a new tank of water. The graph shows the changes that occur in the concentrations of ammonium ions (NH_4^+), nitrite ions (NO_2^-) and nitrate ions (NO_3^-) in a newly-established fish tank for 50 days after adding fish.

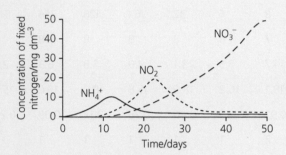

Which process is responsible for the changes in the concentrations shown in the graph?

   A Deamination

   B Denitrification

   C Nitrification

   D Nitrogen fixation   *(1)*

3 The table shows the energy flow through grassland. All the figures are kJ m^{-2} year^{-1}.

Photosynthetically active radiation (PAR) that strikes plants	1046000
Energy for transpiration	523000
Energy reflected	165000
Energy in new growth	21000
Energy loss from respiration	2000
Heat energy absorbed by the soil	335000

Which is the best estimate of the net primary productivity as a percentage of the PAR?

   A 1%

   B 2%

   C 8%

   D 10%   *(1)*

4 A field is fertilised by farmyard manure (FYM) before it is sown with wheat.

   a) Explain what happens to fixed nitrogen in the organic compounds in FYM so that it becomes available in chloroplasts to make proteins in the wheat seedlings.   *(8)*

FYM is a good source of biomass for detritus food chains. Ground beetles are carnivores in these food chains.

   b) Describe how you would estimate the population of ground beetles in a woodland ecosystem.   *(4)*

   c) Describe how you would investigate the activity of ground beetles in an area of grassland at different times over a 24 hour period.   *(4)*

5 As a glacier retreats it leaves behind a mass of rocks of different sizes known as moraine. The graph shows some of the changes that occur as the area becomes a woodland ecosystem over a period of 100 years.

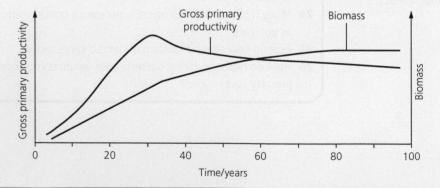

a) Suggest the units that should be used for gross productivity. *(2)*

b) Explain the changes to gross productivity during the period shown in the graph. *(5)*

c) Explain why not all of the gross productivity is available to primary consumers in the ecosystem. *(3)*

d) Suggest why the biomass does not change after 80 years. *(2)*

e) State and explain what happens to the biodiversity over 100 years. *(4)*

6 Badgers live in groups known as clans. Each clan feeds in a territory centred on a series of burrows or sett. An investigation into the diet of badgers in the UK examined faecal samples and found that 97% of badger faeces around each sett contained the remains of earthworms. The study also showed that 53% of the individual food items that the badgers consumed were earthworms.

a) Suggest how researchers could discover how much energy badgers derive from earthworms in different parts of the UK. *(6)*

b) The table shows the numbers of badgers in each clan and the relative abundance of earthworms in the territory occupied by each clan.

Sett number	Earthworm biomass per badger territory/kg	Number of badgers in the clan
1	1680	4
2	1240	3
3	42362	8
4	104470	11
5	37132	4
6	5812	2
7	31079	6

  i) Draw a graph to show the results. *(4)*

  ii) Analyse the data in the table to find out if there is a correlation between the earthworm biomass and the number of badgers in each clan. Show your working. *(5)*

  iii) Discuss the conclusions that can be made about the results. *(3)*

7 Chalk grassland in the UK has a high biodiversity with over 60 species of flowering plant found only in this ecosystem. Many of the plants have root nodules that contain nitrogen-fixing bacteria. The plant diversity supports many invertebrate species, some of which are specialist consumers.

a) Chalk grassland is an example of a deflected succession. Explain why. *(3)*

b) Suggest the meaning of the term *specialist consumer*. *(1)*

c) Simpson's index of diversity is used to assess the species diversity within ecosystems. Outline how you would collect the data necessary to calculate this index of diversity. *(6)*

d) How would you carry out an experiment to discover the importance of plants with root nodules to the productivity of chalk grassland? *(8)*

Stretch and challenge

8 Mayfly nymphs are a juvenile stage of an invertebrate animal that lives in freshwater. Some students investigated the distribution and abundance of three families of mayfly nymphs in riffles and pools along a stream in the Exmoor National Park. The tables show the data that they collected.

Sample site	Ecdyuronidae		Baetidae		Ephemeridae	
	Riffles	Pools	Riffles	Pools	Riffles	Pools
1	15	0	5	0	1	35
2	27	0	3	0	0	13
3	28	4	0	0	1	23
4	17	1	0	0	4	9
5	35	0	3	0	0	10
6	20	1	0	0	0	68
7	8	0	3	0	4	20
8	40	0	3	0	0	17
9	41	0	0	0	0	37
10	42	8	9	0	0	5

Sample site	Dissolved oxygen/% of maximum		Velocity of water/m s^{-1}	
	Riffles	Pools	Riffles	Pools
1	92	85	0.42	0
2	93	89	0.82	0
3	91	91	0.28	0
4	92	90	0.45	0
5	93	92	1.52	0
6	93	91	0.38	0
7	92	92	0.49	0
8	92	91	0.77	0
9	92	89	0.65	0
10	93	90	0.42	0

271

a) Suggest why the students counted nymphs at the family level rather than at the species level.

b) Suggest how the students collected the data. The students thought that there was a relationship between the abundance of the Ecdyuronidae found in the riffle habitat and the velocity of the water.

c) Carry out a statistical analysis to test this hypothesis.

d) Suggest an explanation for the results of the students' investigation.

9 An experiment was carried out to investigate the effect of limpet grazing on a rocky shore on the Isle of Wight by creating limpet exclusion areas using plastic-coated wire fencing. Each area was 0.25 m². The limpets (*Patella vulgata*) were carefully removed from each area and the total percentage cover of each species of alga and numbers of colonising limpets within the quadrats were counted each month. Control areas were also established by fencing the corners of other quadrats. This was done to determine the effects of the wire fences. Measurements and counts were also made within marked unfenced areas of the shore. Each treatment was replicated three times and was monitored for 2 years. The results are shown as three graphs, A–C.

a) Describe the changes that occur to the two species of alga and to the limpets in each of the three areas.

b) Explain the effect of excluding limpets from areas of rock and use the data in the graphs to support your statements.

c) Discuss the ethical issues in carrying out experiments of this type.

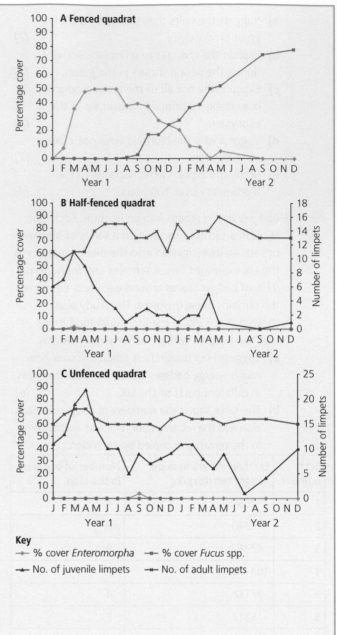

Key

→ % cover *Enteromorpha*   → % cover *Fucus* spp.

→ No. of juvenile limpets   → No. of adult limpets

Chapter 14

Populations and sustainability

Prior knowledge

- Microorganisms can be cultured in liquid nutrient media in either aerobic or anaerobic conditions.
- Yeast reproduces asexually by budding.
- The difference between abiotic and biotic factors.
- Examples of abiotic and biotic factors that influence ecosystems.
- The difference between interspecific competition and intraspecific competition.
- The reasons for conserving biodiversity.
- An ecosystem is a specific area where a community of interdependent organisms interacts with the environment.
- A population is all the individuals of one species that live in the same place at the same time.
- A niche is the role of a species in an ecosystem in terms of the habitat it lives in, its interactions with other species in the community and its effects on the environment.
- The flow of biomass and energy between trophic levels can be determined by taking measurements of the productivity of organisms.
- Natural selection acts on the phenotypes of individuals.

Test yourself on prior knowledge

1 Explain the difference between abiotic factors and biotic factors.
2 Give examples of abiotic factors and biotic factors.
3 Give an example of interspecific competition.
4 Explain why intraspecific competition can often be more intense than interspecific competition. Give examples of resources for animals and plants.
5 Suggest why intraspecific competition is a powerful agent of selection.
6 State four reasons for conserving biodiversity.

Factors that influence population size

When some organisms of the same species enter a new environment with plenty of resources it is highly likely that the population of that organism will increase. The population growth may well be exponential.

An example is the growth of a yeast population. A small sample of yeast was put into a fermenter with a nutrient solution and incubated at a constant temperature for 18 hours. Samples were removed from the culture at hourly intervals and the dry mass of the yeast in each sample was determined. The results are shown in Table 14.1.

Table 14.1 Growth of a yeast population.

Time/h	Dry mass of yeast population/mg cm^{-3}
0	9.6
1	18.3
2	29.0
3	47.2
4	71.1
5	119.1
6	174.6
7	257.3
8	350.7
9	441.0
10	513.3
11	559.7
12	594.8
13	629.4
14	640.8
15	651.1
16	655.9
17	659.6
18	661.8

Tip

This type of growth appears as an S-shaped curve when plotted on linear × linear graph paper. Look at Chapter 16 before answering Test yourself question 1.

Key terms

Logistic growth S-shaped population growth in which exponential increase is limited by resources, so environmental resistance sets in and carrying capacity is reached.

Limiting factor Any biotic or abiotic factor that restrains the growth of a population.

The results in Table 14.1 are plotted in Figure 14.1. This pattern of growth is known as logistic growth.

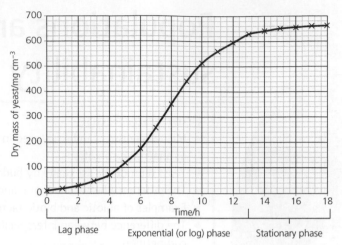

Figure 14.1 The population growth curve for a yeast culture (data from Table 14.1).

At the beginning, the environment in the container has plenty of resources: a respiratory substrate, such as glucose or sucrose, in excess, dissolved oxygen and mineral ions. Over time these resources become limiting factors as the substrate, the oxygen and the mineral ions are used up. Respiration may become anaerobic in which case ethanol is produced, which may inhibit enzyme action in the yeast cells. In addition, as the population increases in size the intensity of competition between the yeast cells increases. The concentrations of substrate, oxygen and ethanol are abiotic factors that limit the rate of growth of the yeast. Competition between individuals of the same species for resources is intraspecific competition, which is the only biotic factor that influences the growth of yeast in this simple, single-species ecosystem. Other biotic factors, such as interspecific competition, disease and predation, are absent.

Competition between individuals in the same species tends to be more intense than competition between different species. This is because they all have the same resource requirements (Figure 14.2). Competition is usually indirect in that individuals do not engage in combat. Those that lose out in indirect competition tend to grow slowly if they eat little food or die of starvation or migrate. Direct competition occurs in combat for territories and for mates. Often such combat is ritualised, consisting of displays without any physical contact so no damage is done to either competitor.

Abiotic and biotic factors, described in Chapter 13, determine the growth of a population and its final size. Because these factors can limit the growth of a population when it reaches a certain level, they are called limiting factors. The intensity of the biotic factors on populations increases as the density of a population increases. The intensity of abiotic factors, such as extreme cold weather, drought and flooding, remains constant whatever the density of the population.

The pattern shown in Figure 14.1 is repeated whenever a simple organism is placed in a confined space in a single-organism ecosystem. Is the same pattern repeated in the wild where more complex organisms enter an ecosystem occupied by a community? The best examples of this occur when an organism invades a new habitat, possibly as a result of migration. Sometimes the invader becomes established in the new area, and shows logistic growth, as in the case of kudzu in the USA (Figure 14.3).

Figure 14.2 Puffins nest in burrows on cliffs. There is competition between the birds for the best burrows. The less successful puffins get burrows at the edge of the breeding colony where they are at a greater risk of predation.

Figure 14.3 Kudzu, *Pueraria* spp., an invasive plant in the USA. Kudzu kills the trees by smothering them and blocking out the light. In 2012 and 2013 kudzu almost disappeared as it was eaten by a type of stink bug. But in 2014 the population of kudzu had grown back to cover the same areas as in previous years.

But in many cases there is no niche for the invader and it cannot compete successfully with species that are well adapted to the environment.

Logistic growth also occurs after species have experienced some catastrophe, such as an epidemic. Populations of wildebeest in East Africa increased like this following an epidemic of rinderpest in the early 1960s.

However, in a real environment there are many biotic and abiotic factors that limit population increase until it decelerates and reaches a maximum size. A population is unlikely to remain at exactly the same size for long; instead it will fluctuate about a mean population, as in the example in Figure 14.4. As population numbers increase the population encounters factors that prevent any further growth. The combined effect of these factors is known as environmental resistance. The maximum population density that the ecosystem can support over lengthy periods of time without harm to the environment is the carrying capacity.

Eventually a population reaches a maximum density. This is the stationary phase (the plateau of a growth curve; see Figure 14.1). The carrying capacity for a population of plants is determined by the factors that limit growth such as grazing, disease and competition for space and resources. The carrying capacity for populations of animals is determined by predation,

> ### Tip
>
> There are many examples of species that have been introduced to different parts of the world and have 'escaped' into the wild. Because of the disastrous effects they usually have on the indigenous wildlife they are unfortunate examples of logistic growth.

> ### Key terms
>
> **Environmental resistance** The combined effects of the factors that prevent the further increase of a population.
> **Carrying capacity** The maximum population density that an ecosystem can support.

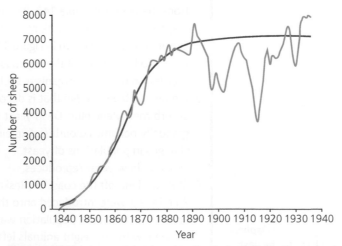

Figure 14.4 Sheep were introduced to South Australia in the early nineteenth century. Their numbers increased exponentially and then fluctuated around a mean value. The carrying capacity for sheep in South Australia is the maximum number that the ecosystem can maintain.

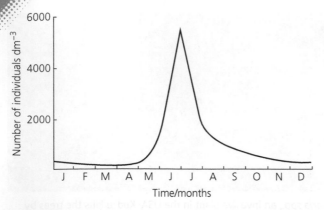

Figure 14.5 J-shaped population growth of a species of marine planktonic algae. This happens because nutrients become available for a short period of time in surface waters in the spring when the light intensity increases.

disease and competition for resources and space for territories, nesting sites, shelter, and so on.

Not all populations grow in this way. The populations of some protoctists, animals and plants increase very steeply when resources become available and often never reach the carrying capacity before numbers crash. Many planktonic algae and the small animals that feed on them show the type of population growth shown in Figure 14.5, which is known as a J-shaped curve rather than an S-shaped one.

Organisms that reproduce asexually are most likely to show this type of population growth as they exploit resources when they become available. Aphids that feed on crops, such as beans and wheat, can increase very rapidly when the plants reach a certain size and weather conditions are favourable.

These populations reach a peak and then crash rather than reaching a constant size. This is typical of species that are controlled more by abiotic factors than biotic factors and respond to the presence of light, mineral ions or food in their environment. They exploit resources that are only available for a short period of time.

Migration influences population size. When populations reach their carrying capacity, some individuals may emigrate to find new sources of food and other resources. The resulting immigration increases the size of the receiving population as happens to population of voles in the old field ecosystem described in Figure 13.11.

Tip

Aphids and *Daphnia* can both reproduce very quickly by parthenogenesis, a form of asexual reproduction. The advantage is that the population gains use of resources that may not be present in the ecosystem for very long. See Chapter 12 for more about parthenogenesis.

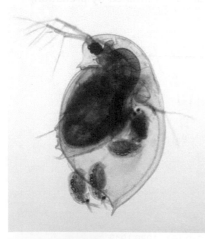

Figure 14.6 Population growth in *Daphnia* sp. This female is giving birth to the next generation by parthenogenesis – a form of asexual reproduction.

Test yourself

Questions 1–7 are about the data in Table 14.1 and Figure 14.1.

1 Determine the $\log_{10}$ of the biomass of yeast in Table 14.1. Plot a graph of $\log_{10}$ biomass against time. If you have log × linear graph paper then plot the biomass figures from Table 14.1 directly onto the graph paper. See Chapter 16 for help.

2 Explain what the $\log_{10}$ × linear plot shows about the growth of the yeast population.

3 Calculate the hourly increase in the dry mass of the population of yeast from the figures in Table 14.1. Record your answers in a table and plot a graph showing how the rate of increase changes over the 18 hours.

4 Explain the results shown in Figure 14.1 and the two graphs you have drawn.

5 Give two limitations of the dry mass method for determining growth in a population of a microorganism, such as yeast.

6 Methylene blue is a dye that is decolourised when reduced. Yeast cells absorb methylene blue. Dead yeast cells remain blue, while living cells gradually become colourless. Explain why this is useful when determining changes in populations of yeast.

7 Describe how yeast reproduces.

8 St Paul Island off the coast of Alaska is about 106 km² in area. In 1911, 25 reindeer were introduced onto the island. The population grew rapidly until 1938, when the population was just over 2000. The population then crashed, with only eight animals left in 1950. Sketch a graph to show the population change and explain these observations.

9 Explain how migration influences the size of populations.

Interspecific competition

The English bluebell, *Hyacinthoides non-scripta*, is thought to be under threat from the Spanish bluebell, *Hyacinthoides hispanica*, which was introduced into gardens in the UK in the seventeenth century (Figure 14.7). Since 1909, Spanish bluebells have been appearing in woodlands and they are now spread quite widely in the UK. There is believed to be direct competition between the two species and they also hybridise freely. Both the hybrid and the Spanish bluebell grow more vigorously than the English bluebell and both produce highly fertile seed that enables both to invade areas of bluebell woodlands. No one knows exactly how the three types of bluebell compete, but in 2013 it was reported that both invaders were flowering earlier than English bluebells, which might give them a competitive edge when colonising new areas.

Figure 14.7 (a) English bluebells and (b) Spanish bluebells. No one knows exactly how they compete with each other, but the more vigorous invasive Spanish bluebell is a threat to the native species.

Interspecific competition has been investigated in simple ecosystems in the laboratory and also in natural ecosystems. Figure 14.8 shows the results of field experiments on two species of barnacle.

Chthamalus stellatus and *Balanus balanoides* are two species of barnacle that live on rocky shores. The larvae of *C. stellatus* settle across the shore, but only grow into adult barnacles in certain areas as they seem to be in direct competition with *B. balanoides*. A researcher set up small wire cages to exclude predators on some rocks in areas where adult *C. stellatus* were never found. All individuals of *B. balanoides* were removed from some of the areas where the cages were fitted and were left in place in others. The numbers of adult *C. stellatus* under each cage were counted for the next two years. The results are shown in Figure 14.8.

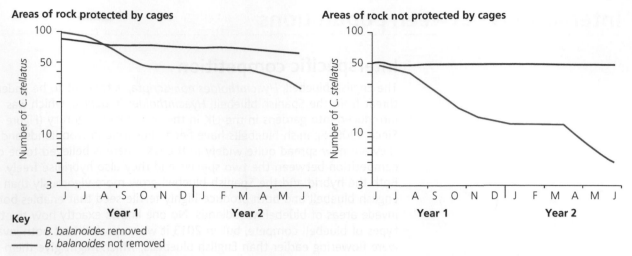

Figure 14.8 The results of field experiments with two species of barnacle: *Chthamalus stellatus* and *Balanus balanoides*. The red lines show the numbers where *B. balanoides* were removed. The blue lines show the numbers where *B. balanoides* were not removed. Note the log scale for the numbers of *C. stellatus*.

Key terms

Niche The role of a species in a habitat, including its trophic level, its interactions with the abiotic environment and interactions with other species, e.g. as a predator, a host or a symbiont, a parasite or pollinator.

Competitive exclusion When individuals of different species compete for the same resources, one species may succeed and the other be excluded from the niche and completely disappear.

Resource partitioning Different species competing for the same sort of resource or resources (e.g. food) occupy slightly different niches so that they avoid direct competition.

Character displacement Two species that compete with each other show differences in characters (features) where they exist in the same ecosystem, but do not show these differences where they do not coexist.

The numbers of *C. stellatus* in the areas with the cages decrease even though predators are not responsible. In the other areas the numbers decrease even more. The researcher noticed that there was direct competition between the two species with *Balanus* growing over *Chthamalus* and smothering it or growing under the barnacle and lifting it off the rock.

If two species have very similar niche requirements the competition between them will be intense, as with the two barnacle species. This does not mean that they will fight each other, as competition can be more subtle, but the less successful species will starve or not find anywhere to reproduce. This means that each niche in an ecosystem is occupied by one species. Other competing species find themselves excluded. This is the principle of competitive exclusion.

The results of competitive exclusion in the past can be seen in anolis lizards which occupy slightly different niches; for example, a study in a forest in the Dominican Republic found seven different species feeding on the same prey animals but avoiding direct competition with each other by having different places where they prefer to perch to catch insects and other small arthropods. This is known as resource partitioning.

When similar species exist in the same ecosystem they may avoid direct competition by character displacement. Two species of small marine snail, *Hydrobia ulvae* and *Hydrobia ventrosa*, live in mud flat habitats. Populations of these two species that do not occupy the same habitat (allopatry) have shells which are about the same size. However, *H. ulvae* always has the larger shells where the two species are sympatric, living in the same habitat.

Tip

For another example of coexistence of two similar species in the same habitat, see question 6 at the end of Chapter 15 of *OCR A level Biology 1 Student's Book*. This is about two species of cormorant in the waters around Nova Scotia in Canada and the way food resources are partitioned between them.

Grazers of different species could compete with each other for food. Many have evolved into specialist grazers. In tropical forests, there are grazers that feed on plants at different heights; there are some that specialise on eating fruits and others that eat leaves. In the African savanna, zebra and wildebeest coexist. Zebra graze the harder parts of the grasses whereas the wildebeest prefer softer, new growth. Zebra will move into areas of long grass before other herbivores and eat the grass down, thereby exposing new growth that is suitable for wildebeest.

Test yourself

10 Which biotic limiting factors do not affect a single-species ecosystem maintained in sterile conditions?
11 Distinguish between intraspecific and interspecific competition.
12 Calculate the percentage decrease in the four groups of *C. stellatus* over the two years of the study shown in Figure 14.8.
13 Explain how the results show that the decreases in numbers of *C. stellatus* are not simply the results of predation.
14 Define the terms *sympatry* and *allopatry*.
15 Give an example of resource partitioning.

Predator–prey relationships

Increases in populations of primary consumers, such as the aphids in Figure 14.9, give plenty of food for their predators, in this case ladybirds. The numbers of a predator usually increase following an increase in their food supply. As numbers of predators increase the intensity of predation increases, so then there is a decrease in the numbers of prey species. With fewer prey, the numbers of predators decrease as well.

Figure 14.10 shows the likely changes in populations of prey and predator in an ecosystem where there is one prey species for the predator and only one predator taking the prey species. This is an unusual situation, but the classic example of the predator–prey cycle was first described for lynx and snowshoe hares in the Arctic regions of Canada in the 1930s. Both species were trapped for the fur trade in the nineteenth and early twentieth century. Data collected from various sources revealed that numbers of the lynx (predator) and its prey, the snowshoe hare, fluctuated with peaks about every 10 years in populations across different regions of Canada.

Notice that in the lynx/snowshoe hare example (Figure 14.10) there is a predator–prey cycle in which population numbers of both species fluctuate at fairly regular intervals but that the peaks and troughs of the predator cycle lag slightly behind those of the prey. This can be interpreted as follows.

Figure 14.9 Ladybird beetles are a major predator of aphids.

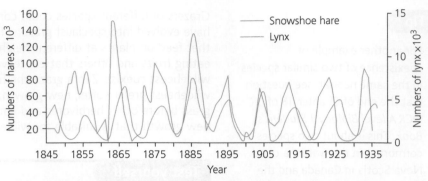

Figure 14.10 Fluctuations in numbers of lynx and snowshoe hares in Canada between 1845 and 1935. The data were taken from a variety of sources including the records of skins received from trappers by the Hudson's Bay Company.

- The prey population increases in number as there are few or no limiting factors.

- There is now more food available to the predator so its numbers increase after a lag period during which the predators reproduce.

- The increase in the predator population leads to more of the prey species being eaten.

- The increase in predation acts as a factor to limit the population of the prey and its numbers begin to fall as the numbers of predators continue to rise.

- As there are fewer prey animals, there is less food for the predators and their numbers fall as they have fewer offspring and some of them starve.

- The cycle repeats itself.

This explanation appears to work well for simple ecosystems where one prey animal has one predator and each predator has one main prey animal. However, it has been shown that the decrease in the numbers of the prey animal is more often due to increased intraspecific competition for food and starvation of the prey, rather than being due to simply to predation. Competition for food is often more important in limiting the growth of a prey species rather than the effects of predation.

The predator–prey example described here is in a simple ecosystem with one prey species and one predator. In most ecosystems, however, there are more complex feeding relationships where each predator preys on several species and each prey species has several different predators. When the numbers of any one prey species decrease the predators can feed on other species. Similarly, the predator that takes the larger proportion of any one prey species may change over time.

Tip

In many cases control of the prey species is bottom up, not top down, because the main determining limiting factor was shown to be food supply not predation.

Activity

Gulls eat crabs eat mussels

Part of a common food chain on rocky shores is

mussel → crab → gull

Crabs are predators of mussels, *Mytilus edulis*, a type of mollusc. In an investigation of predator behaviour, adult male crabs of a single species with carapace ('shell') widths between 160 and 180 mm were collected from the wild. They were kept individually in large tanks and maintained under constant conditions. Mussels were also collected and placed into five groups as shown in Table 14.2.

Table 14.2 Mussels divided by size.

Group	Diameter of mussel/mm
1	15–20
2	21–25
3	26–30
4	31–35
5	36–40

Twelve crabs, A–L, were fed *freely* with meat from mussels. Then they were starved for three days. Equal numbers of mussels chosen from two of the groups in Table 14.2 were spread randomly on the bottom of a seawater tank containing one crab. After 24 hours uneaten mussels were removed and measured. The crab was not fed for the next 24 hours and then the mussels were offered again. The procedure was repeated until 10 trials were completed for each crab. The whole procedure was repeated to give the results shown in Table 14.3.

Table 14.3 Results of the experiment.

Mussel groups offered as prey	Crabs	Mean number of mussels eaten		χ^2 value	Probability
		Smaller-size group	Larger-size group		
1 and 3	A, B, C	47	27	5.4	< 0.05
2 and 3	D, E, F	66	48	2.8	NS
3 and 4	G, H, I	35	29	0.5	NS
3 and 5	J, K, L	47	7	29.6	< 0.05

1 State the type of variation shown by the mussels.
2 State the independent variable in this investigation.
3 Suggest a reason for using crabs of the same species with specified carapace widths.
4 Explain why each crab was offered the same selection of mussels in all 10 feeding trials.
5 State the null hypothesis that would have been used in obtaining the value of χ^2 in Table 14.3.
6 What can you conclude from the results?

8 a) To what extent do these data in Table 14.4 support the hypothesis?
  b) What additional information do you need in order to carry out a statistical analysis of the data?
9 The researchers noticed that gulls often fly away from the place where they pick up crabs to eat them somewhere else. How might this observation explain the data in Table 14.4?

7 This was considered to be a very well-designed investigation. Explain why.

In a separate investigation the predatory behaviour of gulls on crabs was studied. The hypothesis was made that gulls do not discriminate by size when feeding on crabs. Measurements were made of the carapaces of four crab species in an ecosystem. The feeding behaviour of the gulls was watched carefully and the carapaces of dead crabs were measured. The results are shown in Table 14.4.

Table 14.4 Data on crab size selection by gulls.

Crab species	Mean carapace width of living crabs/mm (± SD)	Mean carapace width of crabs predated by gulls/mm (± SD)
A	30 (± 3.5)	33 (± 2.7)
B	65 (± 6.4)	48 (± 6.2)
C	40 (± 3.9)	42 (± 4.1)
D	27 (± 2.8)	25 (± 2.4)

10 Explain how you would investigate the flow of energy in the food chain: mussels → crabs → gulls.

A predator–prey relationship

Owls are predatory birds. Their most common prey animals are mice, shrews and voles. Owls regurgitate pellets containing the bones and fur of their prey. Owl pellets are collected and dissected to find out what they have been eating.

A long-term study was carried out to investigate the relationship between populations of voles and owls in a forest ecosystem. Researchers counted the number of owls in the forest and used Longworth traps to estimate the vole population in a small sample area within the forest. The results are shown in Figure 14.11.

1 Define the term *population*.
2 Describe the changes in the vole population over the 10 year period.
3 Suggest reasons, other than predation by owls, for the decreases in the population of voles.
4 State the evidence from the graphs that support the idea that the vole population is not the only factor controlling the population of owls.
5 Explain how evidence could be collected to investigate the idea that voles form the largest proportion of the diet of owls.

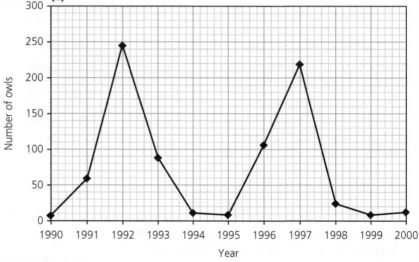

Figure 14.11 Changes in populations of (a) voles and (b) owls over a 10 year period.

Answers

1 A population is a group of organisms of the same species that live in the same area at the same time.
2 The vole population in the sample area fluctuated over the 10 year period reaching peaks of 1120 in 1993 and 620 in 1998. The population was between 180 and 300 between 1990 and 1992 and reached its lowest numbers in 1996 (<100) and 2000 (<50).
3 There could be a lack of food so the voles starve. This could be due to an increase in the population of competitors, such as mice, or an increase in the population of another predator, such as weasels. The lack of food could be due to a change in the weather so plants have not grown very well and have produced less food for voles; for example, fewer seeds. The voles might have been infected by a disease for which they had no resistance. A large proportion of the vole population might have migrated away from the study area.
4 The population of owls increases before the increase in the population of voles in 1992-3 and 1996-8.

If the population of voles controlled the owl population, then the increase in numbers of owls would happen later as in the typical predator-prey relationship. The decrease in the owl population occurs when vole numbers are increasing. This decrease may be due to smaller populations of other prey species or to other factors that cause owls to die or to migrate from the woodland ecosystem.

5 Owls regurgitate pellets containing fur and bones. These pellets could be collected and analysed to find the number and size of the animals that the owls have eaten. The energy content of the whole diet could then be estimated. The mass of the voles could be analysed and put into size classes. The energy content of the voles can be estimated and the proportion of energy consumed as voles calculated as follows.

$$\text{proportion} = \frac{\text{energy in voles}}{\text{total energy intake of owls}} \times 100$$

Conservation of biological resources

Tip

In Chapter 13 of *OCR A level Biology 1 Student's Book* we discussed ways to maintain biodiversity by *in situ* conservation, the establishment and management of protected areas of natural habitat, and by *ex situ* methods of conservation, such as transferring endangered organisms into protected environments away from their natural habitats and preserving gametes, embryos and seeds in gene banks.

Key terms

Conservation The maintenance of biodiversity and sustainability of biological resources using different methods of *in situ* and *ex situ* management.

Preservation The protection of ecosystems, habitats and species for the future without allowing any use by humans.

Conservation and preservation are subtly different. Preservation is maintaining species and habitats *as they are now* so that they will continue to exist in the future, and protecting them from any use by humans. Conservation is the management of habitats and species taking into account that both will change over time in response to environmental changes and allowing managed use of wildlife and habitats by humans.

In this section we are concerned with conserving the biological resources that we take from our environment, such as fish that we harvest from the sea and lakes, and the trees that we harvest for timber. These resources are *sustainable* if we can go on taking these resources in the future without depleting them.

The economic reasons for conserving biological resources are to provide

- sources of raw materials for industries; for example, timber products from natural and plantation forests for the construction and paper industries

- food security: a continuous supply of food, such as fish, that is sourced from the environment rather than reared in fish farms

- employment for people in industries involved in processing raw materials and manufacturing; for example, it has been estimated that the timber trade in the UK directly employs some 167 000 people

- employment for people in transport, marketing and retailing; ecotourism also provides employment for many people in places with high biodiversity

- a source of national income from exporting biological resources such as timber and paper products and fish; the UK is a net importer of timber and fish, but exports of fish, including shellfish, are worth over £1 billion

- environmental services, such as the regulation of climate, water supply, control of pests by their natural predators and disease control; the huge stores of carbon in peat bogs and forests are examples of such a service.

There are a variety of social reasons for conserving biological resources

- human communities in rural areas with little industry or commerce gain stability from the employment provided by fishing and forestry

- areas set aside for forestry provide spaces for recreation; the Forestry Commission and private forestry owners manage their forests as an amenity for people to enjoy the visual appeal of forested areas, the opportunities to take exercise and to observe wildlife.

The ethical reasons for conserving biological resources are

- the duty that we have to conserve resources for the livelihoods and well being of future generations

- the support of indigenous peoples across the world who maintain their traditional ways of life relying on certain biological resources; for example, the Inuit in the Arctic rely on seals and whales and the peoples of the forests of the Amazon and Central Africa depend on foods that they harvest from the forest.

Tip

You can read about the reasons for conservation of biodiversity in Chapter 13 in *OCR A level Biology 1 Student's Book*.

Management of resources in a sustainable way

A sustainable resource is a biological resource that is renewed by the activity of organisms. This means that there is always sufficient for us to harvest from the environment, and the resource does not run out. Our survival on Earth requires that the biological resources that we depend on for food, raw materials and the provision of ecosystem services are maintained in a sustainable fashion. Sustainability is also used to cover the idea that areas surrounding ecosystems should not be harmed and that biodiversity in areas that are harvested is not harmed. For example, coppiced woodland encourages biodiversity, particularly ground-level plants that prefer high light intensities. This encourages many insects, such as butterflies. The encouragement of biodiversity is another reason why some timber is classified as from a sustainable source. Sustainable management is the way in which these goals are achieved.

Fish stocks

Fish stocks around the world have been severely depleted. Some have collapsed entirely (see Chapter 13 in *OCR A level Biology 1 Student's Book*). Fish stocks can be conserved by using a variety of different methods that control where, when and how fishermen can catch fish. This involves restricting fishing in certain areas and at certain times and regulating the types of methods that fishermen use for catching fish. The organisation that regulates fishing needs to determine the maximum sustainable yield (MSY), which is the number or biomass of fish that can be caught without reducing the potential of the fish stock to regenerate itself each year. If a fish stock is labelled as sustainable then it may mean that reproduction of the fish will replace all the fish that are harvested and/or that it is claimed that fishing methods do not damage the environment.

Ways to regulate fishing include

- setting up exclusion zones where fishing is banned: these areas are often spawning grounds where fish reproduce or nursery grounds where young fish develop; fish migrate from these areas into areas where fishing is permitted

- limiting the number of boats that can fish a particular species or in a particular area

- banning fishing at certain times of the year, most notably during spawning

- having rules on the type of fishing gear that fishermen can use; for example, regulating the mesh size of nets and the hook size of line fishing so small fish are not taken and can survive long enough to breed

- issuing quotas so that fishermen cannot take more than a certain number or mass of fish each year

- inspecting the catches landed at port by fishing boats and using fisheries protection vessels to police regulations at sea

- restocking the sea with young fish.

Traditionally, methods of regulating fishing have depended on estimates of the size of fish stocks, concentrating on the success or failure of individual species to breed to produce enough fish to harvest without driving the species to extinction. A different approach, pioneered in the control of fishing in Antarctic waters, is the ecosystem approach (see below, in the section on Antarctica). The aim of this approach is to understand the structure and dynamics of the whole ecosystem including the position

Sustainable resource A biological resource required as a food or as a raw material that is renewed by the activity of organisms so that there are always sufficient stocks to take from the environment.

Sustainable management Aims to provide for the needs of an increasing human population without harming ecosystems, both natural and artificial, and their ability to provide materials and services for us.

Tip

You can read about the threats to fish stocks in Chapter 13 of *OCR A level Biology 1 Student's Book*.

of fish populations within the food web and their interactions with other populations that sustain them. This includes giving protection to seabed habitats within the ecosystem that are at risk of damage from trawling, which involves dragging a net along the sea bed. It has also been discovered that fat, older female fish are the best spawners and that taking steps to conserve them is one of the best ways to restock and then maintain a fish stock. Most fishing methods remove older fish (as they are easier to catch) and rely on younger fish to reproduce. One way in which this can be done is to create 'no catch zones' where fishing vessels are excluded. The fish reproduce in these areas and when the population increases young stages migrate to areas outside where fishing vessels are permitted to work.

There are many organisations across the world that regulate fishing activity and try to prevent illegal, unreported and unregulated fishing.

Some fish stocks have recovered significantly following the introduction of one or more of the methods outlined above. Other fish stocks have not recovered. For example, three areas totalling 17 000 km² on George's Bank in the Gulf of Maine in the North-West Atlantic were closed to fishing in the mid-1990s. Five years later all fish stocks had recovered except cod. Predation of young cod by other fish may be the reason for this, but there may well be a set of complex interactions that are responsible. In other areas of the Atlantic cod stocks are currently in good condition and well managed.

Timber

In the UK only about 13% of the land surface is wooded, but it is on the increase. The heavy requirement for timber during the two World Wars in the twentieth century resulted in less than 5% of the land area of the UK being covered in forest in 1945. It has taken a long time to restore the situation. Woodland can be managed sustainably to provide a source of timber for use in a variety of industries. Tree crops can be divided into two

- fast-growing coniferous trees, such as Scots pine, Sitka and Norway spruce, larch and Douglas fir: these are grown for many commercial processes such as paper making and for the construction industry

- slower-growing broad-leaved trees, such as oak, beech, hazel, alder and sweet chestnut.

Forests can be managed in different ways according to the types of tree and the timber products required. Four ways in which timber may be taken from forests are clear felling, selective felling, strip felling and coppicing.

Clear felling involves cutting trees over a certain area. This takes all the trees, usually of the same age, leaving the area clear of any tree canopy. This is usually seen as a very destructive practice as there is disruption to the biodiversity of the area clear felled and the soil is exposed to the weather, which may lead to soil erosion and loss of valuable nutrients, such as nitrate ions. Its advantage is that it is an economic method. In managed woodlands clear felling is followed by replanting (see Figure 14.12).

Selective felling is a way to minimise the damage to the forest ecosystem by cutting down some mature trees, diseased trees and unwanted species. Other trees are left in place. The removal of a few trees leaves space for natural regeneration from tree saplings or for the planting of young trees that have been raised in nurseries. The removal of single trees from forests inevitably involves some disruption to the ecosystem as the trees have to be removed and that requires access by machinery. This method also allows the best trees time to grow to maturity when they have higher economic value.

Figure 14.12 Sustainable forestry: the mature trees in the background are ready to be felled for the timber trade. The young trees were planted as replacements for a previous harvest and have another 20 years to grow before they are ready to be felled.

Figure 14.13 A coppiced woodland in Kent. This is coppiced sweet chestnut, *Castanea sativa*. Many broad-leaved species can be coppiced, but conifers cannot.

Figure 14.14 A pine forest. Some coniferous forests in the UK provide important habitats for endangered species, such as the red squirrel and the pine marten.

Strip felling is the clearance of small patches or strips of forest leaving adjacent areas of forest untouched. The cleared strips of land are replanted and adjacent areas are cut after the trees have grown to a harvestable size. Large areas are not felled all at the same time, so there is less disruption, less chance that biodiversity in the forest will be adversely affected and little or no soil erosion.

Coppicing is a method of managing broad-leaved woodlands. This involves cutting trees down to ground level and leaving the stumps to regrow. When this happens several stems grow from the buds on the stump. These grow more rapidly than newly planted saplings because they have well-developed root systems. After five years or more the shoots are cut down to the stumps. The wood is used for making a variety of products including hurdles, poles, posts, firewood and charcoal. Tree species that are suitable for coppicing are hazel, ash, sweet chestnut, oak, willow and alder (Figure 14.13).

Coppicing can be repeated indefinitely. Small strips or patches of woodland are cut in different years, providing a variety of habitats and giving high biodiversity. There is renewed interest in this method because willow can produce very large quantities of wood in a few years. The wood is unfit for construction work, but excellent for paper manufacture or to make into wood chips to burn in domestic heating systems, and in power stations to generate electricity. In many coppiced woodlands there is a mosaic of areas that are at different stages in the rotation. Oak is a slow-growing tree and so is cut on a 50 year rotation. Forestry is an industry which plans for the long term.

As trees grow, they take carbon from the atmosphere and store it as wood. The faster they grow, the faster they store carbon. Generally speaking, conifers are the fastest-growing trees in the UK (Figure 14.14). Furthermore, when that timber is used for durable goods, such as in construction, the carbon continues to be stored and not released back into the atmosphere. So, the more wood and wood products we make and use the more we are contributing to carbon sinks.

Effects of humans on the environment

This section covers two important areas

- the effects of human activities on plant and animal populations in ecosystems that are specially sensitive to environmental changes and the ways in which these activities are controlled

- managing the conflicts between the preservation of species and their future conservation with the demands of an increasing human population.

Table 14.5 lists some specific issues and the human activities that threaten ecosystems in seven regions of the world. In such a short space it is not possible to deal in any detail with the range of issues that face each area. To gain a fuller picture you should read more about each of these regions to gain an appreciation of the reasons why they are important to conservation. You should also study a suitable example somewhere near to where you live to learn more of the impact of human activities on sensitive ecosystems and the ways in which these are managed.

Tip

There is lots of information available about the wildlife of the Galápagos Islands and Charles Darwin's time there. You may have to search a bit harder to find information on threats to the local ecosystems and the conservation measures that now exist to reduce the impact of human activities. Try the websites of the Galápagos National Park, The Charles Darwin Research Station and The Galápagos Conservancy for further information.

Table 14.5 Some special regions of the world, why they are of specific interest to conservation and one or more of the threats to those interests.

Region	Specific interest	Threats
Galápagos Islands, Eastern Pacific Ocean	Many species endemic to the islands including those studied by Charles Darwin	Invasive species, overpopulation with humans, the pressures of tourism
Antarctica	Very productive marine environment supporting huge biodiversity	Climate change, overexploitation, pressures of development for tourism and scientific research
Snowdonia National Park, Wales	Maintaining the scenic beauty of a mountainous area	Invasive plant species
Lake District, North-West England	Maintaining access for visitors	Environmental damage by pressure of visitors
Flow Country in northern Scotland	Peat bogs as carbon sinks, rare ecosystems	Development and forestry
Masai Mara National Reserve in Kenya, East Africa	Populations of large mammals characteristic of savanna ecosystem	Increase in local human population and tourism
Terai region, Nepal	Several flagship species of conservation	Poaching, habitat fragmentation

Examples of environmentally sensitive areas

Antarctica

Antarctica is a continent of 14 million km² with many outlying islands, such as South Georgia, the South Orkneys and the South Shetland Islands. The whole region is protected by an international treaty: the Antarctic Treaty reserves the continent for peace and science. As most of the continent is a frozen wasteland it is the marine ecosystems that are of most concern.

There is a very low productivity on land. Plant life is restricted to a few areas which are ice free and consists of low-growing mosses, lichens and liverworts (see Figure 14.15). In the water, however, there is huge productivity by the phytoplankton which make use of the high concentration of minerals in Antarctic waters during the summer months. The phytoplankton are eaten by krill, which in turn are fed upon by many marine animals (see Figure 14.16).

There are three main impacts of human activity on Antarctic ecosystems

- the effects of global warming as a result of an increase in the concentrations of greenhouse gases in the atmosphere
- ozone thinning caused by the action of chlorofluorocarbons in the atmosphere
- direct exploitation of ecosystems through over-harvesting and introduction of alien species.

The Antarctic Peninsula has experienced the effects of global warming. The area has warmed by over 3 °C in the last 50 years. Certain types of lichens, mosses and liverworts have expanded their geographical range into parts of the Peninsula that were previously too cold or covered by ice. The

Figure 14.15 Antarctic vegetation is restricted to the Peninsula and to the surrounding islands. The grass in the foreground is Antarctic hair grass, *Deschampsia antarctica*, one of only two flowering plants endemic to some of the Antarctic islands and the Peninsula. Alien species, such as blue grass, *Poa annua*, from the USA have invaded the Peninsula.

Figure 14.16 Antarctic krill, *Euphausia superba*, is a species of crustacean that grows to a length of 60 mm and forms the main food source for many predators including penguins and marine mammals.

Figure 14.17 Tourists, researchers and film crews are all likely sources of alien species, including possible pathogens. How will they affect Antarctic ecosystems in the future?

primary productivity of the area has increased as a result of the warmer temperatures and an increase in areas not covered by snow.

Visitors to Antarctica have been found to be sources of alien plant and animal species, particularly those from the Arctic. These species are becoming established as the Antarctic Peninsula and the surrounding islands are becoming warmer.

The numbers of tourists visiting Antarctica are increasing. For example, there were approximately 7500 visitors in 1996–1997 and over 37 000 in 2013–2014 (see Figure 14.17). The overall trend towards continued increases in tourist numbers could have localised and/or wider regional impacts on the Antarctic environment in the future.

The seas around the Peninsula, particularly on the western side, have warmed significantly and this has caused a 40% decline in the area covered by sea ice over the past 30 years. The impact of this on the marine ecosystem has been significant. Krill is an important part of the marine ecosystem in Antarctic waters (see Figure 14.16). These crustaceans graze phytoplankton from the underside of winter sea ice when they are young. Krill form the main food source for many species of penguins, crabeater seals and baleen whales, such as humpback and blue whales, which strain krill from the water. The decreases in populations of Adélie penguins and chinstrap penguins are thought to be linked to reductions in the sea ice and the associated decrease in phytoplankton available for krill to feed on.

The loss of sea ice has also affected the breeding of Adélie penguins. They have lost their breeding grounds when the sea ice has melted too early in the summer or where their breeding grounds have become wet due to the increase in snow associated with the warmer, moister climate conditions. Emperor penguins, which breed on the sea ice surrounding Antarctica, have also experienced a decline in numbers during recent decades. Reductions are up to 50% in places and one of the most northerly colonies of emperor penguins on a small offshore island close to the Antarctic Peninsula is now thought to have disappeared completely.

Long-term monitoring of penguin populations on the Antarctic islands shows a much more complex picture, suggesting that not all change is for the worse. The numbers of gentoo penguins have increased. On South Georgia the population of macaroni penguins has declined from 2.5 million breeding pairs in the 1970s to just about 1 million today; however, king penguins, which were hunted heavily by humans in the early twentieth century, have increased from a few hundred in the 1920s to over 450 000 today. Changes to populations of secondary consumers may have an impact on tertiary consumers such as bird species and killer whales, which in the Antarctic prey largely on crabeater seals.

The Convention on the Conservation of Antarctic Marine Living Resources (CCAMLR) came into force in 1982 as part of the Antarctic Treaty System. It was established mainly in response to concerns that an increase in catches of krill by fishermen in Antarctic waters could have a serious effect on populations of krill and other marine life, particularly birds, seals and the fish, which also depend on krill. The aim of the convention is to conserve marine life, although this does not exclude harvesting so long as the catches are sustainable (in both meanings of the word). Under the convention CCAMLR was set up and has pioneered the development of the ecosystem approach to the regulation of fisheries. An ecosystem approach does not concentrate solely on the population dynamics of the species fished, but also seeks to reduce the chances of any harm to the whole ecosystem of which harvested species are a part.

Tip

Among the many websites devoted to Antarctica are: Discovering Antarctica (www.discoveringantarctica. org.uk), the Australian Antarctic Division (www.antarctica.gov.au) and the British Antarctic Survey (www.antarctica.ac.uk).

Further protection is afforded by the Protocol on Environmental Protection to the Antarctic Treaty that came into force in 1998. This protocol establishes clear measures and environmental regulations for all activities in the Antarctic including the designation of specially protected areas and exclusion zones where no scientific bases can be established or tourist activities permitted.

Snowdonia National Park

In Wales, Snowdonia National Park (Figure 14.18) covers just over 2000 km². One of the major problems in the park is the invasive *Rhododendron ponticum* that forms large bushes or shrubs (see Figure 14.19). Rhododendrons were introduced as a horticultural plant and to provide cover for pheasants for shooting about 100 years ago. Since escaping from cultivation rhododendrons have become established over 2000 hectares in Snowdonia. This has very negative consequences for wildlife. The bushes can grow to heights of 3 m or more and they have dense evergreen leaves which make the ground underneath so dark that few plants can survive beneath them. Rhododendrons have killed the local vegetation wherever it has colonised. The plants do not support many other species as their leaves are poisonous to most invertebrates and mammals.

The bushes are long-lived; when they do eventually die and collapse, the branches grow adventitious roots and establish new plants by vegetative reproduction. *R. ponticum* is an effective coloniser because it produces vast numbers of seeds: a single large bush can produce 1 million tiny seeds each year. The seeds need moist conditions to germinate successfully and although many fail to become established it only takes a single seed from each bush to start growing to give a rapid rate of colonisation. The rate of spread has been exponential and as it is ideally suited to the conditions in Snowdonia it has not reached its carrying capacity.

The Flow Country

The Lake District

Snowdonia

Figure 14.18 The 15 National Parks in Britain (shaded areas). The map also shows the location of the Flow Country in Scotland (in blue) which is not a national park.

Figure 14.19 *Rhododendron ponticum*, an introduced species that has caused havoc throughout Snowdonia National Park and in the west of Scotland as well as in other parts of the UK including nature reserves. Here it is being cleared by burning.

Rhododendron leaves decompose very slowly so they accumulate and ruin both soils and bodies of freshwater. *R. ponticum* hosts the plant pathogen *Phytophthora ramorum*, a threat to conifers, especially Japanese larch, *Larix kaempferi*, an important commercial tree. The disease appears as wilted withered shoot tips and the leaves turn black. There is no way to treat infections by *P. ramorum* so the only method of control is to fell infected trees before they become a source of spores that then spread to surrounding trees. This means that *R. ponticum* is a threat to the forests on which many people in the park rely for employment. Unfortunately, this pathogen does not seriously threaten rhododendron's survival, making control of this invasive plant even more urgent. Due to its highly invasive nature, *Rhododendron* infestation makes otherwise suitable habitats unsuitable for grazing animals, as it is unpalatable and toxic, thus reducing the amount of land available to farmers.

In the last 30 years, a considerable amount of effort has gone into trying to get rid of *Rhododendron* bushes by cutting them down and burning or chipping them (see Figure 14.19). It is essential that regrowth is then sprayed with herbicides approximately 18 months after cutting. Failure to do so would mean that the plants would grow back even more vigorously than before. The best way to control scattered bushes is by stem injection. This involves drilling holes into each stem of the bush and applying herbicides. All areas that are in management schemes start with three phases of work over five years to control the plants, with another ten years of management needed as a minimum.

The Snowdonia National Park Authority, the National Trust, Gwynedd Council and Natural Resources Wales formed the Snowdonia Rhododendron Partnership which works closely with private landowners concentrating on controlling *R. ponticum* in the Nant Gwynant/Beddgelert area. In several places the plant is being managed successfully; in others the 'war' against this invasive plant has yet to start.

Managing the conflict between human needs and conservation

The Flow Country

The Flow Country is an expanse of blanket bog that extends over an area of 4000 km² of Caithness and Sutherland in the north of Scotland (see the map in Figure 14.18). Blanket bog is an area of high rainfall and high humidity on poorly drained soil. The continuous flow of water leaches out any bases in the peat so it remains acidic. It is poor in nutrients and lacking in detritivores and decomposers. The climax community is not one of large bushes and trees, but of low-growing heather and, where it is very wet, *Sphagnum* moss. The region is important because of the considerable abundance of large continuous areas of *Sphagnum* carpets and hummocks, including the species *S. fuscum*, *S. imbricatum* and *S. pulchrum*, and for its numerous pools of water and small lochs (lakes) as you can see in Figure 14.21. The vegetation is mainly cross-leaved heath (*Erica tetralix*), ling (*Calluna vulgaris*) and *Sphagnum papillosum* as well as deergrass (*Trichophorum cespitosum*), bog myrtle (*Myrica gale*) and hare's-tail cottongrass (*Eriophorum vaginatum*).

The size and diversity of this ecosystem makes it unique in Europe. The Flow Country is the largest peat mass in the UK and is three times larger than any other similar area in either Britain or Ireland. Dead plant material does not decay, so the carbon compounds are not respired to carbon dioxide as they would be in aerated soils. The peat in the Flow Country stores 400 million tonnes of carbon and is a major carbon sink.

Figure 14.20 The light green shrubs are rhododendrons that have invaded this area of moorland in Ireland.

Figure 14.21 The Flow Country of north Scotland. Permanent pools of water and cotton grass, *Eriophorum vaginatum*, are visible in the foreground. Areas of *Sphagnum* bog and these pools extend for several square kilometres.

The major threats to this ecosystem are drainage to 'improve' the ground for cultivation and afforestation (that is, planting trees and changing the habitat). When the land is drained it dries out and the natural vegetation dies. Many trees were planted between 1979 and 1987. Tens of thousands of hectares of the blanket bog were planted with non-native conifer trees, driven by inappropriate forestry grants and tax breaks. Eventually, the tax breaks were removed and steps were taken, by the Royal Society for the Protection of Birds (RSPB) among others, to protect the area from future development. Some of the land has been restored to its original condition by removing trees and blocking drains. The RSPB set up a nature reserve at Forsinard Flows which is known for its red deer population and for bird species including dunlins, golden plovers, greenshanks and golden eagles. Other areas in the Flow Country are protected as they are designated as Sites of Special Scientific Interest (SSSIs) (see Chapter 13 in *OCR A level Biology 1 Student's Book*).

The Terai region of Nepal

The Terai region lies along the southern edge of Nepal below the foothills of the Himalayas and extends across the border into northern India. The natural vegetation of the area is savanna, marshy grasslands and forest (Figure 14.22). The Terai has seen large-scale deforestation over the past 100 years.

The map in Figure 14.23 shows the position of the two main national parks in Nepal that are situated in the Terai region. Together with the neighbouring Banke National Park, the coherent protected area of 1437 km² represents the Tiger Conservation Unit (TCU) Bardia-Banke that extends over 2231 km² of alluvial grasslands and subtropical moist deciduous forests.

Figure 14.22 The natural vegetation of the Terai region of Nepal.

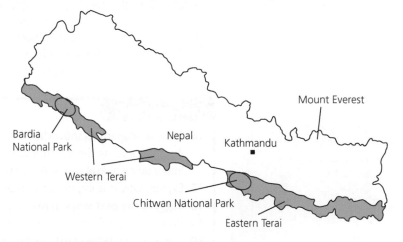

Figure 14.23 The Terai region of Nepal showing the location of the Bardia and Chitwan National Parks.

The Terai is known for three large mammals that live in the region. These three species, with their International Union for Conservation of Nature (IUCN) classifications, are

- the Bengal tiger, *Panthera tigris tigris* (endangered)

- the one-horned rhinoceros, *Rhinoceros unicornis* (vulnerable)

- the Indian elephant, *Elephas maximus* (endangered).

These three flagship species of conservation, and many more, are protected by 14 reserves and parks across India and Nepal to form the Terai Arc Landscape. This is one of only two places in the world where these three large mammal species coexist.

The countries of the region cooperate to conserve the Bengal tiger. In 2012 and 2013 intensive sampling involving transects and camera traps throughout the Terai in Nepal and India estimated the population to be 239 individual adult tigers of which 89 were adult males and 145 were adult

Figure 14.24 One-horned rhinoceros have been heavily poached in the Bardia National Park. In 1996, a study found only about 1500 of these animals left in the wild, but by 2007 the number had increased to about 2500. Poaching and habitat fragmentation remain threats to this species which nearly became extinct in the early 1900s.

Figure 14.25 The human pressures on the Terai: the region has a growing population which needs towns, cities, farmland and infrastructure, such as modern roads.

females (the sex of five tigers could not be determined). There are estimated to be 120 of these tigers in the Chitwan National Park and it also has the world's second largest concentration of one-horned rhinos (Figure 14.24). All three of the species listed above have been heavily poached. Body parts of rhinos and tigers are used in traditional medicine in Asia, particularly in China and Vietnam. Elephant tusks are used in the ivory trade, which flourishes in Thailand and elsewhere in South-East Asia. Thanks to the efforts of the Nepalese government and the army, poaching has decreased in Nepal with many soldiers patrolling the national parks to protect the animals.

Preservation might involve maintaining populations of these mammals. However, they cannot exist in isolation except in zoos. To be conserved they need to be in their natural habitat and they need all the ecosystem 'services' that we have discussed. Particularly difficult to conserve *in situ* are tigers, as they are top predators and require a large area to hunt to obtain sufficient energy. Not only are large predators fierce, but they also tend to be very rare anyway, even without human interference. To conserve them an ecosystem approach is needed to ensure that a fully functioning ecosystem exists with sufficient land area to provide enough prey for its top predators. There are advantages of this approach as studies have shown that removal of predators, especially large ones like tigers, decreases biodiversity at all trophic levels.

However, conserving large areas to provide enough space for one species inevitably involves conflicts with the human population (Figure 14.25), which uses forests as a supply of fuel and savanna for grazing livestock. Another problem is that top predators such as Bengal tigers need large habitats and few protected areas are large enough to support viable populations. In the Terai region wildlife corridors have been established to join up parks and reserves into a network of protected areas to allow tigers and other large mammals the space that they need. Wildlife corridors need to be broad enough for the animals to use them without threat. They also need to be maintained, and protected from human interference.

> ### Test yourself
>
> 16 Draw a diagram to show the flow of energy in the waters around Antarctica using the information in this chapter. Use it to explain the importance of krill to the Antarctic ecosystem.
> 17 Explain why it is important to maintain wildlife corridors.
> 18 Suggest the best ways to maintain these corridors.
> 19 Why are large, fierce animals, like the Bengal tiger, rare?
> 20 Suggest how large predators maintain biodiversity at all trophic levels.
> 21 Suggest ways in which the alien species *Rhododendron ponticum* can be controlled successfully in Snowdonia.
> 22 List the different threats from human activities to sensitive ecosystems across the world.

> ### Tip
>
> The seven places listed in Table 14.5 are given as examples of the effect of humans on the ecosystems and the way the conflicting demands of conservation and humanity can be resolved and managed. Some of these have been described. However, you may know other National Parks in the UK or perhaps have been on a scientific visit with an organisation like Operation Wallacea. You can use this knowledge when answering exam questions, such as Exam practice question 6, below.

Exam practice questions

1 A species is introduced into a new habitat and increases exponentially. Which statement explains why the population of this species increases?
   A There are many predators.
   B There are no limiting factors.
   C There is competition with other species.
   D There is environmental resistance. *(1)*

2 Which is not a good reason for using a logarithmic scale rather than a linear scale when plotting population changes?
   A A wider range of population size can be displayed on a single pair of axes.
   B Population fluctuations at low densities are more clearly displayed.
   C Population growth is always logarithmic.
   D The actual numbers of individuals can be read easily from the graph. *(1)*

3 There are many insect pests of stored food products. The red flour beetle, *Tribolium castaneum*, and the confused flour beetle, *T. confusum*, are two species that infest stores of flour. These two species gain all their energy, nutrients and water from flour. In an investigation of competition, the two species were kept together in containers of wheat flour under different environmental conditions.
   • Many containers were used for each set of environmental conditions.
   • The containers were all identical and contained the same mass and volume of flour.
   • The same numbers of male and female flour beetles of the two species were put into each container at the start of each experiment.

The numbers of beetles were counted regularly. The containers were left until only one species remained as the 'winner' in the competition. The table shows the results.

Environmental conditions	Percentage of containers in which T. castaneum was the 'winner'	Percentage of containers in which T. confusum was the 'winner'
A, hot and wet	100	0
B, hot and dry	10	90
C, warm and wet	86	14
D, warm and dry	13	87
E, cold and wet	29	71
F, cold and dry	0	100

a) i) Suggest suitable controls for this investigation. *(2)*
   ii) State the dependent variable and the derived variable for this investigation. *(2)*
b) Identify the environmental conditions most preferred by the two species of *Tribolium*. In each case give some evidence from the table to support your answer. *(6)*
c) Explain why there was a 'winner' in each container. *(3)*
d) Insect pests, such as flour beetles, not only eat the flour but also deposit nitrogenous waste in urine and faeces into the flour. This leads to the growth of bacteria and fungi in the flour. Describe how nitrogen in the nitrogenous wastes of insects is recycled for use by plants. *(7)*

4 Carabid beetles are one of the largest families of animals; many live in forest ecosystems. Researchers in Poland investigated the distribution and abundance of carabid beetles in forest plots of different ages. The percentage of beetles associated with the climax forest trapped in plots of the same age was calculated. The Spearman's rank correlation for this percentage and the age of the forest plots was $r_s = 0.696$; $p < 0.001$.
a) Explain what you can conclude from this result. *(3)*
b) Explain how you would compare the biodiversity in the youngest and oldest forest plots. *(5)*
Dung beetles belong to another family of the beetles, the Scarabaeinae. These beetles have three different lifestyles
• dwellers live inside dung where they feed and lay their eggs
• tunnellers dig burrows in the soil underneath piles of dung
• rollers make dung into balls and roll them away to bury in the soil.

Researchers investigated the numbers and biomass of each of the three types of dung beetle in two fields: one grazed by sheep and the other grazed by cows. The results are in the table.
c) The researchers did not kill all the beetles to obtain the biomass figures. Suggest how they derived the results in the table. *(3)*
d) Explain the role of dung beetles in grassland ecosystems grazed by large herbivores. *(4)*
e) Comment on the results. *(5)*

Type of dung beetle	Field A, grazed by sheep				Field B, grazed by cows			
	Numbers	%	Dry mass / mg	%	Numbers	%	Dry mass / mg	%
Dwellers	1486	40.6	7 206	14.2	8 968	82.7	4 4217	37.7
Rollers	201	5.5	10 437	20.5	105	1.0	11 903	10.2
Tunnellers	1976	53.9	33 185	65.3	1 764	16.3	61 017	52.1
Totals	3663	100.0	50 828	100.0	10 837	100.0	117 137	100.0

5 The collared dove, *Streptopelia decaocto*, spread from India across Europe to reach Britain in 1955 when four birds were reported. The graph shows the estimated numbers of *S. decaocto* between 1955 and 2009.

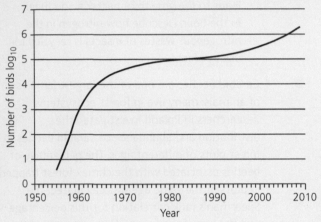

a) Describe the change in the population of the collared dove in the UK since 1955. *(4)*

b) Suggest reasons to account for the changes you have described in (a). *(4)*

In 1928, Oscar Richards investigated the growth of yeast under four different conditions (A–D). The medium in which the yeast was growing was changed in A every 3 hours, in B every 12 hours and in C every 24 hours. The medium was not changed at all in D.

c) Predict and explain what would happen to the growth of the yeast populations in each culture medium over 140 hours. *(5)*

6 Use examples in answering these two questions.

a) Explain how ecosystems can be managed to balance the conflict between conservation and human needs. *(6)*

b) Outline the effects of human activities on animal and plant populations in environmentally sensitive ecosystems and explain how they are controlled. *(6)*

7 The pipistrelle bat, *Pipistrellus pipistrellus*, is a night-flying, insectivorous bat that uses sonar to detect its prey. Researchers investigated the hypothesis that these bats select their prey rather

than catching insects at random. They used suction traps to collect samples of flying insects on successive nights. The insects were identified to family level and parts of the insects' exoskeletons were crushed to help identify the contents of the bats' faecal pellets. Bats were caught in nets and kept in bags until dawn when they were released. The contents of the faecal pellets collected from the bats were then compared with the exoskeleton fragments to identify the prey species that they had eaten. The researchers identified and counted 34 798 insects from the suction traps and identified 24 250 insects from faecal pellets. The table shows the relative abundance of the ten most common families of flying insects in the suction traps and in the faecal pellets over the 18 nights of trapping.

Insect family	Relative abundance of insect families	
	Percentage of total catch in the suction traps	Percentage in all the faeces collected over 18 nights
Chironomidae	53.8	56.4
Ceratopogonidae	7.8	4.1
Psychodidae	3.6	2.1
Cecidomyiidae	2.0	2.3
Glossosomatidae	23.5	23.4
Tipulidae	0.9	2.2
Hydroptilidae	0.9	0.9
Molannidae	0.9	0.7
Scarabaeidae	0.8	0.1
Mycetophilidae	0.7	0.7
Other families	5.1	7.1
Total	100.0	100.0

a) Use the Spearman's rank correlation test to analyse the data in the table. Show your working. *(5)*

b) State the conclusion that the researchers would make based on your answer to part (a). *(3)*

The researchers carried out the same analysis for the insects caught in the suction traps and present in the faeces for five of the 18 nights when trapping occurred. The results are given in the table below.

Night	Spearman's rank correlation coefficient	p
4th	0.848	*
5th	0.973	**
12th	0.884	**
15th	0.714	NS
18th	0.705	NS

$* = p < 0.05$; $** = p < 0.01$; NS $= p > 0.05$

The graphs below are scattergraphs to show the percentages of three orders of insects – Nematocera (many species of flies such as crane flies and mosquitoes), Neuroptera (lacewings) and Coleoptera (beetles) – in the suction traps and in the bats' faeces.

Nematocera

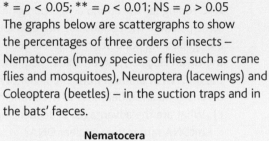

Neuroptera

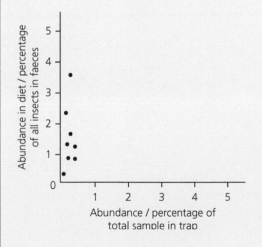

Coleoptera

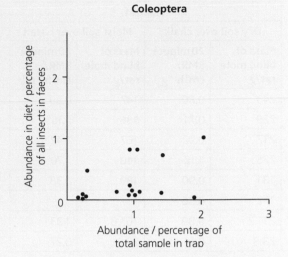

c) Explain whether the data in the table and the scattergraphs support the hypothesis that pipistrelle bats select their prey. *(4)*

Mexican free-tailed bats, *Tadarida brasiliensis*, emit sounds to jam each other's sonar that they use for detecting insects.

d) Discuss other ways in which intraspecific competition occurs. *(5)*

Some potential prey species of bats also jam the sonar of their predators.

e) Describe some other ways in which prey species avoid predation. *(4)*

f) Discuss how predation acts as an agent of natural selection. *(5)*

g) Suggest how insectivorous bats of different species could coexist in the same habitat. *(3)*

Stretch and challenge

8 Blind mole rats are found across the Middle East. The distribution of blind mole rats appears to be influenced by the type of rock underlying the soils in which they live. In one area in Galilee in northern Israel there is a sharp dividing line between two groups of mole rats in the species *Spalax galili*. One group lives in dry soils above chalk and the other in moister soils above basalt. The two areas have quite different vegetation. Researchers determined the basal metabolic rates (BMRs) of blind mole rats from the two groups.

Habitat			
Dry soil over chalk		Moist soil over basalt	
Mass of blind mole rat/g	20 minute BMR/ $cm^3 h^{-1} g^{-1}$	Mass of blind mole rat/g	20 minute BMR/ $cm^3 h^{-1} g^{-1}$
243	0.80	147	0.99
229	0.71	186	1.16
217	0.92	167	1.42
225	1.12	140	1.70
181	0.90	189	1.38
172	0.86		

Mean	211.2	0.89	165.8	1.33
SD	28.3	0.14	22.2	0.27

a) Suggest how the researchers determined the 20 minute BMR.

b) The researchers used a statistical test to find that the difference between the BMRs of the two groups is statistically significant. Comment on the results in the table in light of this conclusion.

The mitochondrial gene *ATP6* codes for part of the proton channel in ATP synthase. The researchers sequenced the gene in 28 mole rats from the basalt habitat and 14 from the chalk habitat. All the sequences from animals from the chalk had the codon AGC in position 185 coding for the amino acid serine; all the sequences from animals from the basalt habitat had the codon AAC coding for the amino acid asparagine in the same position. The codon AAC was also found in a neighbouring population of *Spalax golani* from the Golan Heights, also on basalt. Sequencing all the mtDNA of animals from each habitat revealed other significant differences between them.

c) What are the advantages of sequencing mtDNA rather than nuclear DNA?

d) Suggest a possible explanation for the results of the mtDNA sequencing.

e) Blind mole rats choose their mates based on the similarity of odour based on genotype. Suggest how this might affect the future of these two groups of *S. galili*.

Chapter 15

Maths skills

Most of the maths requirements are covered in Chapter 18 in *OCR A level Biology 1 Student's Book*. This chapter deals with two additional topics:

● using logarithms and log graphs

● using statistical methods to analyse data.

As everything in this chapter deals with data that are collected during different types of biological investigation, we will start by reminding you of the different types of variables involved in experimental work.

Variables

In all biological investigations variables are chosen, measured, controlled or derived. A variable is any factor that has an input to the investigation or is an output which is counted or measured. In this chapter, we will refer to four types of variable.

The **independent variable** is the variable that is the focus of the investigation. You choose its range and the intermediate values within that range which you will investigate. There are many graphs in this book and in *OCR A level Biology 1 Student's Book* that show the results of investigations with the independent variable always plotted on the *x*-axis. In some investigations there may be more than one independent variable; for example, if you choose to investigate the effect of light intensity and carbon dioxide concentration on the rate of photosynthesis (see Figure 15.1).

The **dependent variable** is the variable that is counted or measured in the investigation. As with the independent variable, there may be more than one dependent variable. Figure 15.2 shows measurements of the biomass of *Penicillium chrysogenum* and penicillin production in a fermenter like that

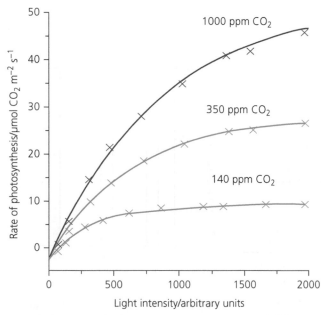

Figure 15.1 The graph shows the results of an experiment to investigate the effect of two variables: light intensity and carbon dioxide concentration. These are both independent variables.

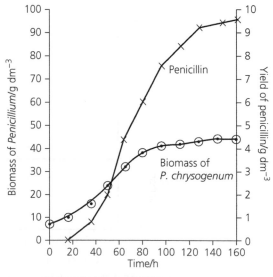

Figure 15.2 The biomass of the mycelium of *Penicillium chrysogenum* and the production of penicillin during a fermentation. Note the labels and units on the two *y*-axes.

in Figure 12.17. In these cases it is often necessary to have two *y*-axes: a primary axis on the left and a secondary axis on the right. When taking data from a graph like this make sure you look very carefully at the labelling and any key to the lines so that you take data from the correct axis.

A **derived variable** is calculated from the results of an investigation. Often results need to be processed before they can be analysed. For example, in investigations of respiration and photosynthesis final graphs often show how a factor influences the rate of the process. The rates often have to be calculated from the data collected, such as movement of a bubble or the volume of gas collected over a period of time. Rates of respiration and photosynthesis are therefore derived variables.

A **controlled variable** is any variable which is kept constant, or taken account of, during an investigation. For example, in the experiments that are plotted in Figure 15.1 the same temperature was used. Temperature is therefore a controlled variable.

> **Tip**
>
> All the mathematical skills described in this chapter are easily practised using a spreadsheet program.

Logarithms

Sometimes the independent variable or the dependent variable or both cover such a wide range that it is difficult to draw the results on a normal linear scale on graph paper. This is when a logarithmic scale may be used, where the intervals on the axis increase by an order of magnitude (×10). This would be appropriate when you have used a wide range of concentrations of plant hormones, say between 0.0001% and 1.0%. If the numbers are exact increases in order of magnitude, such as 0.0001, 0.001, 0.01, and so on. then these are plotted equidistant on the graph paper, as in Figure 15.3, which shows the effects of increasing concentrations of auxin on growth of shoots and roots. Figures for the dependent variable, or the derived variable, are rarely like this and intervening numbers need to be accommodated on the scale. Logarithmic scales are especially useful for plotting data on the growth of microorganisms, since under ideal conditions the numbers of individuals increase exponentially (see Figure 12.22 and Figure 15.6(b), below).

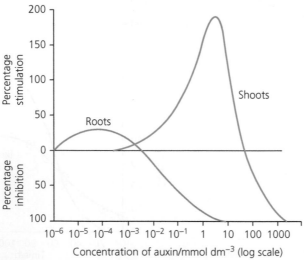

Figure 15.3 The stimulatory and inhibitory effects of auxin on shoots and roots. Notice that the *x*-axis is a logarithmic scale covering 11 orders of magnitude. (See Chapter 6 for information about the role of auxin in growth.)

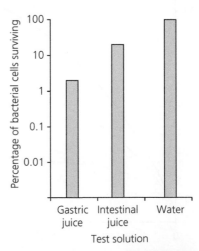

Figure 15.4 This graph shows the results of an investigation on the survival of cells of the probiotic bacterium *Bifidobacterium longum* in two different digestive juices.

If you are plotting numbers for the dependent variable on the y-axis then you can either plot directly onto log graph paper as in Figure 15.4 (though the grid is not shown) or calculate the logs of the numbers as shown in the Example on the next page, Straightening the curve, and then plot onto normal (linear × linear) graph paper.

Figure 15.5 shows some log graph paper. A cycle on the graph paper is one order of magnitude; for example, 0.1 to 1.0, 1.0 to 10.0, 10.0 to 100.0, and so on. Notice that the distance on the y-axis between 10 and 100 is the same as the distance between 1 and 10. You will see the advantage of this in Figure 15.6.

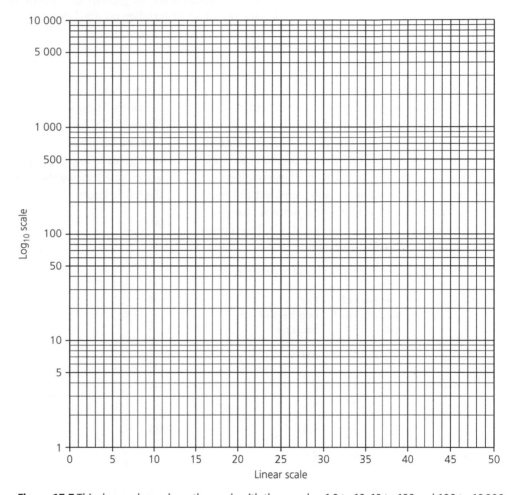

Figure 15.5 This shows a log scale on the y-axis with three cycles: 1.0 to 10, 10 to 100 and 100 to 10 000.

If you have intermediate numbers of the independent variable, then you have to use a log scale for the x-axis, as is the case when concentrations may be 0.0001 (1×10^{-4}), 0.0025 (2.5×10^{-3}), 0.0050 (5.0×10^{-3}), and so on.

If you leave plenty of space on the right of the graph paper it is possible to estimate the population by a certain time period in the future or predict by extrapolating the data. Taking logs of one or both variables usually straightens out a curve, as has happened in Figure 15.6(b). This makes extrapolating much easier.

If you have plotted onto linear graph paper, this means reading off the log and calculating from it the actual number. To do this with a calculator or computer, enter the log and then press the '10^x' button on your calculator or in Excel enter '=Power(10,A1)' where A1 is the cell where you have entered the log. In a very simple form this is what plant protection scientists do when predicting the likely number of pests during the summer growing season and advising farmers when it is appropriate to spray pesticides.

It is possible to have graphs with both the x-axis and the y-axis as log scales. This might happen if you investigate the effect of a serial dilution of nutrient medium on the growth of bacteria or investigate the effect of organic pollution or fertiliser on algal growth.

Example

Straightening the curve

Table 15.1 shows the numbers of bacteria kept in an aerated culture solution at 30 °C with plenty of nutrients for 10 hours. The numbers shown are those at the beginning and the end of the culture. You can work out the results from 3.0 to 8.5 hours. These numbers could be plotted on ordinary graph paper, but the wide range in numbers means that any significant changes when the numbers are small are not visible. The number of bacteria doubled every 30 minutes.

Table 15.1 Population growth of a culture of bacteria.

Time/h	Numbers of bacteria cm^{-3} ×10^3	Log$_{10}$ numbers of bacteria
0	1	3.0
0.5	2	3.3
1.0	4	3.6
1.5	8	3.9
2.0	16	4.2
2.5	32	4.5
3.0	...	...
...	...	...
8.5	...	...
9.0	...	...
9.5	524 288	8.7
10.0	1 048 576	9.0

The data in Table 15.1 are plotted on two graphs in Figure 15.6: first as a linear graph (a) and then as a semi-logarithmic plot on a log × linear graph (b).

The increase in numbers of bacteria is exponential for the whole 10 hours of the culture. It is not obvious in Figure 15.6(a) because the numbers for much of the time are too small to see this relationship. Converting the figures to logs (or plotting them directly onto log graph paper) shows that the growth is exponential for the full 10 hours.

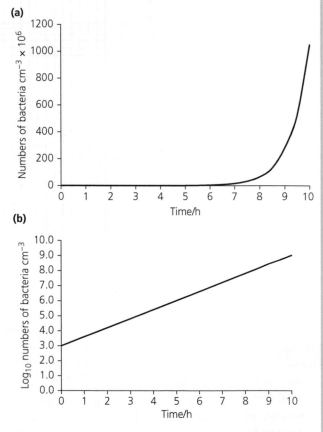

Figure 15.6 (a) A linear × linear plot of the data in the second column of Table 15.1. (b) A log × linear plot using the data in the third column of Table 15.1.

Test yourself

1 What is the difference between a controlled variable and a control experiment?
2 What does Figure 15.1 show about the limiting factors of photosynthesis?
3 What is the biomass of *Penicillium chrysogenum* at the end of the fermentation shown in Figure 15.2?
4 Explain why a log scale is used in Figure 15.3.

5 Why was water included in the investigation into survival of bacteria (see Figure 15.4)?
6 In an experiment the percentage survival of another species of probiotic bacterium in gastric juice was 0.013%. Make a sketch to show how that result would be plotted on the graph shown in Figure 15.4.

Using statistical methods to analyse data

Types of data

Table 15.2 shows the different types of variable and the data that you might collect in an investigation or be expected to process or display as a chart or graph or analyse in an examination paper.

Table 15.2 Types of data.

Type of variable		Type of data	Example	Common method to display data
Qualitative	Categoric	Nominal	Different phenotypes in a study of inheritance Presence or absence of roots on cuttings	Bar chart Pie chart
	Ordered	Ordinal (ranked)	Abundance scales in fieldwork studies of populations Arbitrary scales, e.g. 0–10	Bar charts
Quantitative	Continuous	Interval, having any value, e.g. 1.0, 2.5, etc.	Rates of respiration, photosynthesis and enzyme-catalysed reactions	Line graph or bar chart, depending on the independent variable
			Variation in length, height and mass	Histogram
	Discrete	Interval, integers only, e.g. 1, 2, 3, etc.	Number of heart beats per minute Number of plants in a quadrat	Bar chart, histogram or line graph, depending on the investigation

The type of investigation and the type of data collected determine the type of statistical method that you can use. The use of these methods in your analysis makes it possible to be much more precise with the conclusions that you are able to make.

Descriptive statistics

The first part of any analysis of data collected in an investigation is to simplify it, for example by calculating an average. You should recall from Chapter 18 in *OCR A level Biology 1 Student's Book* that there are three types of average: **mean**, **mode** and **median**. You should also recall that the dispersion of values around a mean is best shown by calculating the **standard deviation**.

In investigations, we often want to know whether the data that we have collected are 'good enough' to show a difference or to show a correlation. To do this we need to make use of descriptive statistics first.

Standard deviations should be calculated for replicates in such an investigation. If the SD is small, then you can have more confidence in the mean than if it is much larger. Remember that 64% of the replicates will be within ± 1SD of the mean and 95% will be within ±2SDs of the mean.

Calculating standard error and 95% confidence intervals

Standard error of the mean (SEM) can also be calculated. Standard error of the mean tells you more than the standard deviation. Standard error tells you about the mean of the population from which the sample has come. It is calculated by dividing the standard deviation by the square root of the number of readings.

$$SEM = \frac{SD}{\sqrt{n}}$$

The smaller the value for the standard error of the mean the more certain you can be about the relationship between the sample mean and the population mean.

Standard error of the mean is used to find the confidence limits within which the true value of the population mean is to be found with a probability of 95%. These are the **95% confidence limits** (95% CL). They are calculated by

$$95\% \ CL = \bar{x} \pm 2 \times SE$$

The difference between the confidence limits is the **confidence interval**. We can be 95% certain that the **population mean** is within ± 2 standard errors of the sample mean. The 95% CL are the best measure of the dispersal of the mean for error bars on graphs. They are also the best to use for comparing different sets of data.

In some investigations you may have two sets of data. For example, sprigs of Canadian pondweed, *Elodea canadensis*, may be exposed to blue light and to green light to find out how the wavelength of light influences the rate of photosynthesis. Although the independent variable in this investigation is qualitative – blue and green light – the results are quantitative as we can measure the volume of oxygen produced over a period of time and so calculate rates. This investigation was replicated 10 times and the results are shown in Table 15.3.

Tip

Standard deviation can be used to plot error bars on bar charts and line graphs. The error bar above the mean is equal to 1 SD; the error bar below the mean is also equal to 1 SD. Draw error bars as a ⊤ above and a ⊥ below the mean.

For details of calculating standard deviation see pages 323-324 in *OCR A level Biology 1 Student's Book*.

Tip

$\bar{x}$ (x bar) is the symbol for the mean.

Table 15.3 Rates of photosynthesis of *E. canadensis* in blue and green light.

Colour of light	Rate of photosynthesis/s⁻¹									
	1	2	3	4	5	6	7	8	9	10
Blue	31.3	20.8	27.8	25.0	23.8	27.0	21.7	30.3	24.4	22.2
Green	6.5	6.2	9.6	8.5	8.1	6.8	7.6	8.9	7.2	7.8

The results for the data in Table 15.3 are

Blue light: mean = 25.4; standard deviation = 3.4 (to 1 dp)

Green light: mean = 7.7; standard deviation = 1.1 (to 1 dp)

The standard error in each case is calculated by dividing the standard deviation by the square root of the total number of replicates, which is 10.

SEM for blue light = $3.4/\sqrt{10}$ = 3.4/3.16 = 1.08

SEM for green light = $1.1/\sqrt{10}$ = 1.1/3.16 = 0.35

The 95% confidence limits

Blue light = 2 × 1.08 = 2.16

Green light = 2 × 0.35 = 0.70

These are now added and subtracted from the mean to give the upper and lower limits. They can then be plotted as error bars on a bar chart of these results (see Figure 15.7).

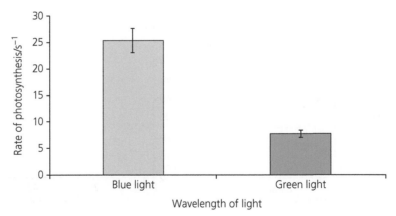

Figure 15.7 A bar chart showing the rate of photosynthesis of *E. canadensis* in two wavelengths of light. The error bars are 95% confidence limits.

You can see that the mean values for the two wavelengths of light are very different. The error bars do not overlap and this shows that the difference between the means is likely to be significant. We do not know that for certain, so have to use a statistical test. See the Student's *t*-test later in this chapter for an example of such a test. If the 95% CL error bars overlap then you can be certain that the difference between means is *not* statistically significant.

7 Students investigated a population of cuckoopint, *Arum maculatum*, in a wood. In summer, the bright orange fruits are on a long stalk. The students measured the lengths of the stalks of 20 plants in millimetres with the following results.

212, 231, 246, 254, 262, 268, 277, 281, 286, 289, 293, 297, 302, 305, 313, 318, 326, 333, 334, 372

Calculate the mean, standard deviation, standard error of the mean and 95% confidence intervals for the data.

8 What percentage of the sample is found between ± 1 SD of the mean?

9 What percentage of the sample is found between ± 2 SD of the mean?

10 State what you can conclude by calculating the 95% CL for the data on *A. maculatum*.

11 What can you state about a sample of 20 insects of the same species with a mean wing length of 38.3 mm and a standard deviation of 2.7 mm?

Types of statistical test

The data collected from A level practical investigations are usually one of four different types. When presented as graphs, they look like the four in Figure 15.8.

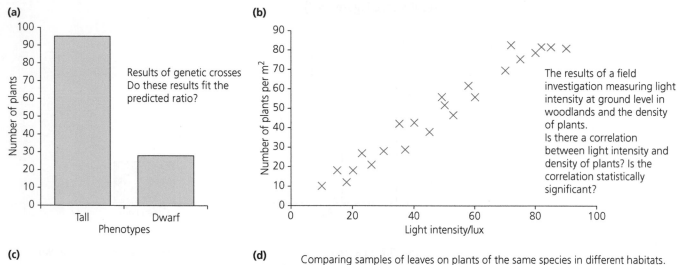

(a) Results of genetic crosses Do these results fit the predicted ratio?

(b) The results of a field investigation measuring light intensity at ground level in woodlands and the density of plants.
Is there a correlation between light intensity and density of plants? Is the correlation statistically significant?

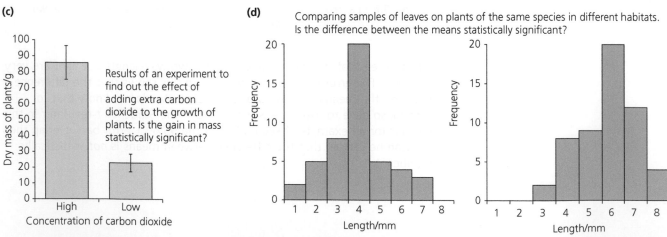

(c) Results of an experiment to find out the effect of adding extra carbon dioxide to the growth of plants. Is the gain in mass statistically significant?

(d) Comparing samples of leaves on plants of the same species in different habitats. Is the difference between the means statistically significant?

Figure 15.8 The data from most biological investigations at A level fall into one of the four categories shown here. Straight-line relationships, e.g. between enzyme concentration and rate of reaction, have been omitted.

Figure 15.9 This photograph shows leaves of common holly, *Ilex aquifolium*. Counting the numbers of variegated and non-variegated leaves gives categoric data; using a scale of 0–5 to assess the area of each variegated leaf that is yellow gives ordinal data; measuring the lengths of the leaves gives interval data and counting the prickles is data in the form of integers (whole numbers).

Tip

You should only use the chi-squared test when

- data are categoric
- the outcomes of an experiment can be predicted from theory or by assuming that they will be random events.

Tip

You should recognise at least five genetic ratios: $3:1$, $1:1$, $1:2:1$, $9:3:3:1$ and $1:1:1:1$. See Chapter 10 to find out how to interpret these ratios and what they tell you about the number of genes involved and the relationships between the alleles of those genes.

Sometimes you can see that there is a definite difference between results as is the case with graphs (a) and (c) on Figure 15.8. The results plotted for (b) show that the points look as if they fall on a line running from bottom left to top right. There is overlap between the histograms in (d), but is there enough difference between the means for the two populations to indicate something significant? Biological data are very variable and we often do not get such clear-cut answers to our questions.

As the data in (a) to (d) are in different forms, the statistical methods used for their analysis are different. For example, categorical (nominal) data – as in (a) – are analysed differently from interval data. Before you proceed take another look at Table 15.2.

There are three further statistical methods to describe

- the chi-squared test
- Spearman's rank correlation test
- Student's *t*-test for paired data and unpaired data.

The chi-squared (χ^2) test

When Gregor Mendel carried out his breeding experiments with pea plants he carefully counted large numbers of plants and/or seeds. He did not rely on very small samples. He also repeated his breeding experiments to check his results. If you repeat his experiments, for example with fast-cycling brassicas (otherwise known as fast plants) or with fruit flies or even model them using coloured beads or a computer simulation (as in the Hardy–Weinberg activity in Chapter 10), you are very unlikely to gain results that exactly fit any of the ratios described in Chapter 10. This does not matter: if the results exactly fitted the ratios we would be very suspicious. Biological data are very variable and in breeding experiments are at the mercy of chance effects such as the random fertilisation of gametes. In 1900, Karl Pearson (1857–1936) developed the chi-squared test as a 'goodness-of-fit' test to check the significance of differences between observed and expected results when using categoric data.

The formula for calculating the test statistic, χ^2, is

$$\chi^2 = \sum \frac{(O-E)^2}{E}$$

where Σ means sum of, O is the observed value and E is the expected value.

These are the steps to follow when doing a chi-squared test to analyse data from a genetics experiment.

1 Analyse the information to see what type of inheritance pattern there is: monohybrid, dihybrid, sex linked, and so on.

2 Determine the expected ratio in the results if that has not been done for you.

3 Write a null hypothesis. This should be something like 'there is no significant difference between the observed and predicted results'.

4 Draw a chi-squared table as shown in Table 15.5 in the Example on the next page. The number of columns is always six, but the number of rows depends on the number of classes of data. For a monohybrid cross with two phenotypes among the offspring you need three rows; for a dihybrid cross with four phenotypes among the offspring you need five rows in your table. (One row is needed for the headings of the columns.)

5 Complete the table by

   a) calculating the expected results using the predicted ratio

   b) calculating the difference between observed and expected results

   c) squaring the difference (to remove signs)

   d) dividing the square of the difference by the expected numbers for each class (this takes into consideration the size of the sample)

   e) adding up the results of the final column to give the test statistic, χ^2.

6 Decide on the degrees of freedom (df), which is calculated as one less than the number of classes (df = number of classes − 1).

7 Find the 0.05 probability (p = 0.05) in the table of probabilities for the appropriate degrees of freedom. It will be at the intersection of the row for df and the column for p = 0.05.

8 At the intersection, identify the critical value, χ^2_{crit}.

9 Use a table similar to Table 15.4 to decide whether the χ^2 value that you have calculated is equal to or greater than the critical value or less than the critical value. If the value calculated for χ^2 is equal to or greater than the critical value then the probability of getting the result is less than 0.05 (5%) and there is a statistically significant difference between the predicted and expected results. In this case, the null hypothesis is rejected. If the value calculated for χ^2 is less than the critical value then there is no significant difference between the predicted and expected results and the null hypothesis is not rejected.

10 Write a conclusion that states the value of χ^2, compares it with the critical value, χ^2_{crit} and states whether the null hypothesis can be rejected or accepted. Finish by saying whether or not the results contradict the Mendelian principle involved.

Always express your conclusions in terms of the probability of getting the result by chance, for example $p < 0.1$ but $p > 0.05$ and state that there is a statistically significant difference between predicted and actual results if that is the case.

Tip

The degrees of freedom when analysing data from genetics experiments is always one less than the number of classes. This is not always the case when using the chi-squared test to analyse data from other types of investigations.

Tip

In the exam you would only be given the top half of Table 15.4 as the bottom half tells you how to interpret the values for χ^2!

Table 15.4 Critical values for the chi-squared test with information about how to interpret the value of χ^2. Note that > means greater than, $\geq$ is greater than or equal to and < is less than.

Degrees of freedom	Distribution of χ^2							
	Probability, p							
	$\leftarrow$ Increasing values of p				Decreasing values of $p \rightarrow$			
	0.99	0.90	0.50	0.10	0.05	0.02	0.01	0.001
1	0.00016	0.016	0.46	2.71	3.84	5.41	6.64	10.83
2	0.02	0.10	1.39	4.61	5.99	7.82	9.21	13.82
3	0.12	0.58	2.37	6.25	7.82	9.84	11.35	16.27
4	0.30	1.06	3.36	7.78	9.49	11.67	13.28	18.47

$p > 0.90$		$p \geq 0.05$	$p < 0.05$	$p < 0.01$	$p < 0.001$
Result looks 'dodgy' = too good!		Result is not significantly different from expected outcome	Result is significantly different from expected outcome	Highly significant	Very highly significant

Example

Using the chi-squared test to analyse data from a monohybrid cross

Some pure-breeding fruit flies with long wings were crossed with pure-breeding fruit flies with vestigial (very small) wings. All of the F_1 generation of fruit flies had long wings. These F_1 flies were bred among themselves. Among the F_2 there were 390 long-winged flies and 114 vestigial-winged flies. The results should conform to the expected 3:1 ratio. Do the results differ significantly from the predicted results or not?

The null hypothesis is that there is no significant difference between the observed numbers of fruit flies of the two phenotypes and the expected numbers.

The chi-squared table is completed following the instructions above (see Table 15.5).

The value for χ^2 is 1.52 (2 dp). There is 1 degree of freedom (2 categories − 1 = 1). The critical value (χ^2_{crit}) from Table 15.4 = 3.84.

The calculated value is less than the critical value so the difference between the predicted and observed results is not significant. The probability of getting results like this is between 0.1 and 0.5, i.e. $p > 0.1$ but < 0.5, which means that the result could simply have occurred due to chance effects such as random fertilisation of gametes. The results of the F_2 are not significantly different from

the results predicted from the segregation of one pair of alleles in meiosis and their combination in diploid genotypes at fertilisation (see Chapter 10).

Figure 15.10 Breeding fruit flies with long and vestigial wings. This specimen tube contains larvae and pupae of *Drosophila melanogaster*, which live on the nutrient medium at the bottom. Once the adults emerge from the pupae they are anaesthetised, sorted and counted to give the numbers in the F_1 or F_2 generation.

Table 15.5 A chi-squared table for the data.

Category	O	E	O–E	(O–E)²	(O–E)²/E
Long wings	390	378	12	144	0.3809
Vestigial wings	114	126	−12	144	1.1428
Totals	504	504			$\chi^2 = 1.5237$

Test yourself

12 The results of certain types of experiment can be analysed with the chi-squared test. What do these experiments have in common?
13 How do you calculate degrees of freedom in the chi-squared test?
14 What does the value of χ^2 mean?
15 What is the importance of the 5% level (or 0.05)?
16 Tomato plants usually have indented leaves, known as 'cut'. Some tomato plants have leaves shaped like those of potatoes, known as 'potato'. Pure-bred tomato plants with cut leaves and purple stems were crossed with pure-bred plants with potato leaves and green stems. All the F_1 generation had cut leaves and purple stems. These F_1 plants were test crossed against tomato plants showing the recessive phenotype: potato leaves and green stems. The test cross offspring showed the following numbers of plants in each of four phenotypes.

Purple, cut	Purple, potato	Green, cut	Green, potato
70	91	86	77

Use the chi-squared test to analyse these data. What do you conclude from the results of your analysis?

Spearman's rank correlation test

In some investigations you may be looking to see if there is a correlation between two variables. Scattergraphs give a good visual indication of a relationship or association between variables. The data need to be quantitative or be ranked or ordered for you to plot them on graph paper. The relationships may be positive, negative or non-existent, as shown in the scattergraphs in Figure 15.11.

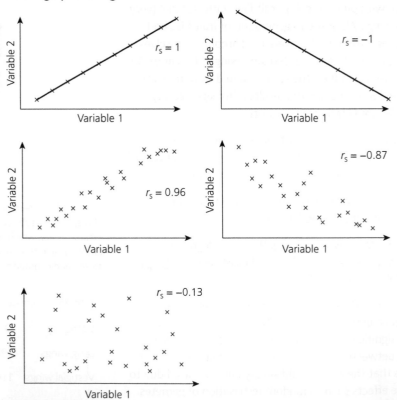

Figure 15.11 Scattergraphs. The strength of the correlation is the correlation coefficient, r_s. Correlation coefficients extend from + 1 to − 1.

Tip

You should only use the Spearman's rank correlation test when

- data points within samples are independent of each other
- the data are paired; for example, the soil moisture and the abundance of a certain plant at stations along a transect
- ordinal data or interval data have been collected
- a scattergraph indicates the possibility of an increasing or a decreasing relationship (this does not have to be a linear relationship, it could be one that increases and levels out).

In addition

- the number of paired observations should ideally be between 10 and 30: you can, however, use the test if there are more than five paired observations
- all individual results must be selected at random from a population; each individual must have an equal chance of being selected.

Tip

It does not matter whether you rank up or down; just make sure that you are consistent. If you give the lowest number in the first data set the rank of 1, make sure that you are consistent and use the same ranking for the second data set.

Tip

Remember that correlation does not mean that changes in one variable *cause* changes in the other variable.

The strength of the correlation is determined by finding the correlation coefficient. There are several ways to do this, but the only way you need to know is the Spearman's rank correlation coefficient.

The data do not have to be interval data. The data can be ordinal data. The abundance scales we met in Chapter 13 are examples of ordinal data: ACFOR, DAFOR and Braun-Blanquet. Data that have been ranked, for example by using an arbitrary scale such as 0–10, are also an example of ordinal data. An example is colour leaking from beetroot tissue at high temperatures. You can use an arbitrary scale if you do not have a colorimeter to take quantitative results. The test can be used if the relationship is non-linear, but it cannot be used if the relationship increases and then decreases or decreases and then increases.

The formula to calculate the Spearman's rank correlation coefficient is

$$r_s = 1 - \left(\frac{6 \times \sum D^2}{n^3 - n}\right)$$

where r_s is the Spearman's rank coefficient, $\sum D^2$ is the sum of the squares of the differences between ranks and n represents the number of pairs of items in the sample.

These are the steps to follow when calculating the Spearman's rank correlation coefficient.

1 Plot the results as a scattergraph to see if there may be a correlation. The easiest way to do this is to enter the figures in a spreadsheet and choose the scattergraph function. The relationship could possibly be a positive correlation or a negative correlation. It does not have to be linear; see Figure 15.12 for an example of a non-linear relationship.

2 Write a null hypothesis. It is usually going to say 'there is no significant correlation between …'.

3 Rank each set of data. The largest number in each set of data is given the rank of 1.

4 Calculate the differences in the ranks, D, by subtraction.

5 Square the differences in the ranks to give D^2. This removes the negative signs.

6 Add all the values for D^2 to calculate $\sum D^2$.

7 Calculate the Spearman's rank correlation coefficient by inserting the values for $\sum D^2$ and n into the formula above.

8 Interpret the value of r_s. The value of r_s may range from −1 to +1. A positive value indicates a positive correlation in which the values of variable y increase as the values of variable x increase. A negative value indicates a negative correlation in which the values of variable y decrease as the values of variable x increase. The closer the number is to 1, the stronger the correlation. A coefficient of 0 indicates that there is no correlation.

9 Interpret the significance of the value of r_s. The sign is ignored when comparing r_s to the critical value. Now look up the Spearman's rank coefficient in the table of critical values that correspond to the number of pairs of measurements in the results table (see Table 15.6).

10 Accept or reject the null hypothesis. If the value of r_s is equal to or greater than the critical value, then the null hypothesis is rejected. If the value of r_s is less than the critical value then the null hypothesis is accepted.

11 Write a conclusion that includes the type of correlation (positive or negative), the strength of the correlation (the value of r_s) and the degree of significance.

Table 15.6 The critical values of the Spearman's rank correlation coefficient (r_s) at $p = 0.05$ and $p = 0.01$.

Number of pairs of measurements	Critical values	
	$p = 0.05$ (5%)	$p = 0.01$ (1%)
5	1.000	–
6	0.886	1.000
7	0.786	0.929
8	0.738	0.881
9	0.700	0.833
10	0.648	0.794
11	0.618	0.755
12	0.587	0.727
13	0.560	0.703
14	0.538	0.679
15	0.521	0.654
16	0.503	0.635
17	0.488	0.618
18	0.472	0.600
19	0.460	0.584
20	0.407	0.570
30	0.362	0.467

Example

Finding a correlation

Researchers studied the River Wye to see if there was a correlation between the total dissolved solids and the diversity of macroinvertebrates by counting the number of different groups (taxa).

The researchers determined the total dissolved solids (TDS) at 13 sampling stations along the river and counted the number of taxa of macroinvertebrates. Table 15.7 shows their results.

The null hypothesis is: there is no significant correlation between the total number of macroinvertebrates and the TDS at the 13 sampling stations along the River Wye.

Figure 15.12 shows a scattergraph for the data in Table

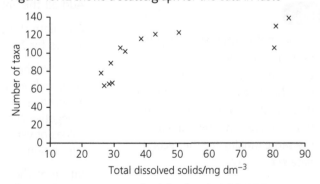

Figure 15.12 A scattergraph of the data in Table 15.7.

Table 15.7 Number of taxa of macroinvertebrates and total dissolved solids at 13 sampling stations along the River Wye.

Sample station	Total dissolved solids/mg dm⁻³	Total number of taxa of macroinvertebrates
1	27.0	64
2	28.5	66
3	29.5	67
4	26.0	78
5	29.0	89
6	32.0	106
7	33.5	102
8	38.5	116
9	43.0	121
10	50.5	123
11	80.5	106
12	81.0	130
13	85.0	139

The samples were then ranked as shown in Table 15.8.

Figure 15.12 shows a scattergraph for the data that shows there is a relationship. Table 15.8 shows how to calculate ΣD^2. Steps 3 to 6 are described on page 309. The value for the sum of D^2 and the total number of samples are inserted into the formula for calculating r_s

$$r_s = 1 - \left(\frac{6 \times 34.5}{13^3 - 13}\right)$$

$$r_s = 1 - 0.0948$$

$$r_s = 0.9052 = 0.91 \text{ (to 2 dp)}$$

The critical value (from Table 15.6) for 13 pairs of data is 0.560.

This value of r_s is greater than the critical value at $p = 0.05$ so there is a statistically significant correlation at $p = 0.05$. In fact it is greater than the critical value for 0.01, so we can say there is a significant correlation at $p = 0.01$.

As the value for r_s is greater than the critical value, we can reject the null hypothesis. The conclusion you can make is: it is 95% (or 99%) certain that there is a positive correlation with a strength of 0.91 between total dissolved solids and the diversity of macroinvertebrates. We cannot say that the concentration of total dissolved solids is responsible for the increase in diversity in the river; all we can say is that there is a statistically significant correlation between these two aspects of the river ecosystem.

Table 15.8 The ranked data.

> **Tip**
>
> * Step 3 Rank the data for the first column—rank the largest number as 1.
> ** Step 3 repeat for the data in the third column.
> *** Step 4 calculate the differences between the ranks.
> **** Step 5 square the difference.
> Step 6 sum all the numbers of D².

Total dissolved solids/mg dm⁻³	Rank*	Number of taxa	Rank**	Difference (D)***	D²****
85.0	1	139	1	0	0
81.0	2	130	2	0	0
80.5	3	106	6.5	−3.5	12.25
50.5	4	123	3	1	1
43.0	5	121	4	1	1
38.5	6	115	5	1	1
33.5	7	102	8	−1	1
32.0	8	106	6.5	1.5	2.25
29.5	9	67	11	−2	4
29.0	10	89	9	1	1
28.5	11	66	12	−1	1
27.0	12	64	13	−1	1
26.0	13	78	10	3	9
				ΣD^2	34.5

Tip

Remember that it is the Spearman's rank correlation test, not just 'Spearman's rank', which most people call it. It is a test of the strength of a correlation between two variables.

Tip

Student's *t*-test was devised in 1908 by William S. Gosset who worked at the Guinness Brewery in Dublin. He was interested in the performance of different varieties of barley used in brewing. The company would not allow him to publish with his own name so he gave himself the pseudonym of 'Student'. Just about everyone calls it the *t*-test and you can call it this in any answers that you write.

Tip

You should only use Student's *t*-test when

- interval data have been collected
- you want to see if there is a significant difference between the means of *two sets* of unpaired data, e.g. from two different populations of plants or animals
- the data in each set shows a normal distribution
- the sample is small (<30 measurements in each sample).

In addition

- the numbers in each sample may be different, e.g. 20 and 25.

Test yourself

17 Students investigated the abundance of blackfly larvae, *Simulium* spp., in a stream and also measured the velocity of water at each sampling point. The table shows their data.

Table 15.9

Sample stations	Number of blackfly larvae	Velocity of the stream/m s⁻¹
1	3	0.42
2	20	1.76
3	7	0.56
4	13	2.73
5	5	0.43
6	5	1.93
7	1	0.43
8	43	1.88
9	7	0.90
10	19	1.80

What conclusions can you make from the data in Table 15.9?

Student's *t*-test for unpaired data

The *t*-test is used to find out whether the means of *two sets* of data are significantly different. To use the *t*-test you have to be sure that the two sets of data both show a normal distribution.

The formula for the *t*-test for unpaired data is

$$t = \frac{|\bar{x} - \bar{y}|}{\sqrt{\frac{(s_X)^2}{n_X} + \frac{(s_Y)^2}{n_Y}}}$$

where $\bar{x}$ equals the mean of sample X, $\bar{y}$ equals the mean of sample Y, s_X is the standard deviation of sample X, s_Y is the standard deviation of sample Y, n_X is the number of individual measurements in sample X and n_Y is the number of individual measurements in sample Y. The vertical lines either side of the numerator indicate that the sign can be ignored; use the absolute difference between the two means.

The formula for calculating the degrees of freedom is

$$df = (n_X - 1) + (n_Y - 1)$$

Here are the steps to follow when using Student's *t*-test for unpaired data.

1 Write a null hypothesis. This is usually 'there is no significant difference between the mean of ... and the mean of ...'.

2 Calculate the means of the two sets of data.

3 Calculate the standard deviation for each set of data (see Chapter 18 in *OCR A level Biology 1 Student's Book*).

4 Square the standard deviations.

The hypothesis that you are
testing determines how you use a
t-test table like that in Table 15.10.
If the hypothesis states that there
is a significant difference between
the means of the two samples
(and the null hypothesis that there
is no significant difference), then
this is a two-tailed test and you
use the values of *p* given in the
table. If your hypothesis states
that the mean for sample A will be
greater or smaller than the mean
for sample B, then this is a one-
tailed test. In this case the critical
values for *p* = 0.05 are those given
in Table 15.10 for *p* = 0.1 and those
for *p* = 0.01 are those for *p* = 0.05.

5 Put the values you have calculated into the formula above.

  a) Find the difference between the means, ignoring the sign.

  b) Square the standard deviation for the first sample and divide
by the number of observations in that sample. Repeat for the
second sample.

  c) Add these two results from step 5(b) together and take the square root.

  d) Divide the difference in the two sample means from step 5(a) by the
answer to step 5(c). This is the value for *t*.

6 Calculate the total degrees of freedom (df) for all the data, using the
formula above.

7 Find the critical value at *p* = 0.05 for the degrees of freedom that you
have calculated.

Table 15.10 shows the critical values at different levels of significance and
degrees of freedom.

If the calculated value of *t* is less than the critical value then we can
conclude that the difference between the means could be due to chance
and there is no significant difference between the two samples. If the
calculated value of *t* is greater than the critical value we can conclude that
the difference between the means is statistically significant and reject the
null hypothesis.

Table 15.10 The critical values for the Student's *t*-test. Published tables give all the
intermediate degrees of freedom.

Degrees of freedom	Decreasing value of *p* →			
	0.10 (10%)	0.05 (5%)	0.01 (1%)	0.001 (0.1%)
1	6.31	12.71	63.66	636.60
2	2.92	4.30	9.92	31.60
4	2.13	2.78	4.60	8.61
6	1.94	2.45	3.71	5.96
8	1.86	2.31	3.36	5.04
10	1.81	2.23	3.17	4.59
16	1.75	2.12	2.92	4.02
18	1.73	2.10	2.88	3.92
20	1.72	2.09	2.85	3.85
25	1.71	2.06	2.80	3.73
30	1.70	2.04	2.75	3.65
40	1.64	1.96	2.58	3.29
	$p > 0.05$	$p < 0.05$	$p < 0.01$	$p < 0.001$
	Results are not significantly different	Results are significantly different	Results are highly significant	Results are very highly significant

The result of Student's *t*-test cannot prove that a particular theory is
correct. The result, however, can provide support for the theory.

Variation in limpets

The limpet, *Patella vulgata*, is a common mollusc on rocky shores. Some students investigated the variation in limpets on a shore that was exposed to wave action and on another shore that was very sheltered from the waves. The students sampled 10 limpets at random on each shore and calculated the length:height ratio of each limpet by dividing the length by the height. Table 15.11 shows their results.

The null hypothesis is: that there is no significant difference between the length:height ratios of the limpets on the sheltered shore and the exposed shore. Table 15.12 (copied from a spreadsheet) shows the calculations.

From the table, we can calculate

$$s^2/n + s^2/n = 0.014$$

$$\sqrt{s^2/n + s^2/n} = 0.119$$

$$\bar{x}_1 - \bar{x}_2 = 0.53$$

$$t = 4.47$$

$$df = (10-1) + (10-1) = 8$$

The value for $t = 4.47$. At 8 degrees of freedom the critical value at $p = 0.05$ is 2.31 (Table 15.10). The value for t is greater than this so there is a less than 5% chance of this result happening by chance. This means that we can reject the null hypothesis and accept the alternative hypothesis that there is a significant difference between the means of the length : height ratios for the limpets on the two shores. From this test we cannot explain the difference; all we can say is that we are more than 95% confident that the difference is statistically significant.

Figure 15.13 The common limpet, *Patella vulgata*.

Table 15.11 Length:height ratios of limpets on an exposed shore and a sheltered shore.

Length:height ratio of limpets	
Exposed shore	**Sheltered shore**
1.58	1.85
1.42	2.02
1.65	2.84
1.63	2.41
1.42	2.04
1.68	1.78
1.35	1.59
1.40	1.90
1.52	1.86
1.46	2.12

Table 15.12 Calculations for testing variation in limpets.

	Exposed shore			Sheltered shore		
	Length: height ratio	$x - \bar{x}$	$(x - \bar{x})^2$	Length: height ratio	$x - \bar{x}$	$(x - \bar{x})^2$
	1.58	0.07	0.0049	1.85	−0.19	0.0361
	1.42	−0.09	0.0081	2.02	−0.02	0.0004
	1.65	0.14	0.0196	2.84	0.80	0.6400
	1.63	0.12	0.0144	2.41	0.37	0.1369
	1.42	−0.09	0.0081	2.04	0.00	0.0000
	1.68	0.17	0.0289	1.78	−0.26	0.0676
	1.35	−0.16	0.0256	1.59	−0.45	0.2025
	1.40	−0.11	0.0121	1.90	−0.14	0.0196
	1.52	0.01	0.0001	1.86	−0.18	0.0324
	1.46	−0.05	0.0025	2.12	0.08	0.0064
$\bar{x}$	1.51	$\Sigma(x - \bar{x})^2 =$	0.1243	2.04	$\Sigma(x - \bar{x})^2 =$	1.1419
s	0.118			0.356		
s^2	0.014			0.127		
s^2/n	0.001			0.013		

Test yourself

18 Some students investigated the effect of adding fertiliser on the growth in height of seedlings of two cultivars of the same crop species. Table 15.13 shows their results.

Table 15.13

	Cultivar 1	Cultivar 2
Mean ($\bar{x}$)/mm	400	450
Number (n)	18	20
SD (s)/mm	40	45

Use Student's t-test to analyse the data. What conclusion can you make?

Student's t-test for paired data

In some investigations two (or more) sets of results are collected from the same individuals. For example, you may investigate the effect of exercise on heart rates or the effect of caffeine on the volume and/or concentration of urine. A different formula is used:

$$t = \frac{\bar{d}\sqrt{n}}{s_d}$$

Where $\bar{d}$ is the mean of the differences between the two data sets (e.g. before and after exercise), s_d is the standard deviation of these differences and n is the number of individuals in the investigation. The degrees of freedom is $n - 1$.

The steps to follow when using the Student's t-test for paired data:

1 Write a hypothesis. You might write one that states that the results will increase or decrease. Write an appropriate null hypothesis.

2 Calculate the difference between the two results (*before* and *after*) for each individual.

3 Calculate the mean difference ($\bar{d}$)

4 Calculate the standard deviation of the differences (s_d)

5 Put the values you have calculated into the formula above.

6 Use Table 15.10 in the same way as described on page 313, except that the degrees of freedom are calculated as $n - 1$.

The effect of light exercise on heart rate

A student investigated the effect of skipping on heart rate. Ten volunteers of the same age, sex, BMI and fitness level were fitted with a heart rate monitor. They all skipped for 30 seconds.

The null hypothesis is that there is no significant difference between heart rates before and after exercise.

The heart rates before and after exercise are shown in Table 15.14.

Table 15.14 Heart rates of ten volunteers before and after 30 seconds of skipping.

Volunteer	Heart rate / beats min⁻¹	
	Before skipping	**After skipping**
1	60	80
2	70	86
3	68	88
4	58	82
5	82	94
6	74	88
7	70	88
8	70	84
9	68	80
10	66	78

Table 15.15 shows the student's calculations.

Table 15.15 A student's calculations for the paired *t*-test.

Volunteer	Heart rate / beats min⁻¹				
	before skipping	**after skipping**	**differences ($\bar{d}$)**	**$d - \bar{d}$**	**$(d - \bar{d})^2$**
1	60	80	20	3.8	14.44
2	70	86	16	−0.2	0.04
3	68	88	20	3.8	14.44
4	58	82	24	7.8	60.84
5	82	94	12	−4.2	17.64
6	74	88	14	−2.2	4.84
7	70	88	18	1.8	3.24
8	70	84	14	−2.2	4.84
9	68	80	12	−4.2	17.64
10	66	78	12	−4.2	17.64
		mean ($\bar{d}$)	16.2		

Standard deviation $(s_d) = 4.16$ $\quad t = \dfrac{16.2 \times \sqrt{10}}{4.16} = \dfrac{16.2 \times 3.16}{4.16}$ $\quad t = 12.31$

At 9 degrees of freedom the critical value at $p = 0.05$ (for a two-tailed test) is 2.26.

The calculated value of *t* is greater than the critical value so the student can reject the null hypothesis and conclude that there is a significant difference between the two sets of results.

Choosing a statistical test

When answering a question or carrying out some practical work, you may have to choose the appropriate statistical method to analyse the data.

Use this checklist to help you to decide which test to use. Also see Table 15.2.

1 If you have categoric data then you can only use the chi-squared test. This is used to compare the actual data with predicted data.

2 If you want to compare the means of two samples then use the *t*-test if you are sure both sets of data are normally distributed (i.e. give a bell-shaped curve). If the data comes from two populations then use the t-test for unpaired data. If the two sets of data come from the same individuals then use the t-test for paired data.

3 If you want to see if two variables are correlated, then use Spearman's rank correlation test. The data can be ordinal data or interval data or both.

Index

Free online material

Answers for the following features found in this book are available online:

- Prior knowledge questions
- Test yourself questions
- Activities

You'll also find an Extended glossary to help you learn the key terms and formulae you'll need in your exam.

Scan the QR codes below for each chapter.

Alternatively, you can browse through all chapters at www.hoddereducation.co.uk/OCRABiology1

How to use the QR codes

To use the QR codes you will need a QR code reader for your smartphone/tablet. There are many free readers available, depending on the smartphone/tablet you are using. We have supplied some suggestions below, but this is not an exhaustive list and you should only download software compatible with your device and operating system. We do not endorse any of the third-party products listed below and downloading them is at your own risk.

- for iPhone/iPad, search the App store for Qrafter
- for Android, search the Play store for QR Droid
- for Blackberry, search Blackberry World for QR Scanner Pro
- for Windows/Symbian, search the store for Upcode

Once you have downloaded a QR code reader, simply open the reader app and use it to take a photo of the code. You will then see a menu of the free resources available for that topic.

1 Communication and homeostasis

3 Neuronal communication

2 Excretion as an example of homeostatic control

4 Hormonal communication

5 Animal responses

12 Cloning and biotechnology

6 Plant responses

13 Ecosystems

7 Photosynthesis

14 Populations and sustainability

8 Respiration

15 Maths skills

9 Cellular control

16 Preparing for practical assessment

10 Patterns of inheritance

17 Exam preparation

11 Manipulating genomes

Acknowledgements

The publisher would like to thank the following for permission to reproduce copyright material.

Photo credits:

p.1 © BSIP SA / Alamy; **p.2** © Stockbyte via Thinkstock/Getty Images; **p.6** t © EDWARD KINSMAN/SCIENCE PHOTO LIBRARY, b © Nature Photographers Ltd / Alamy; **p.7** © Andy_Astbury - iStock via Thinkstock/Getty Images; **p.8** © moodboard / Alamy; **p.9** © Adrian Schmit; **p.14** t © Richard Fosbery, b © MarieHolding - Thinkstock/Getty Images; **p.16** © Richard Fosbery; **p.17** all © Richard Fosbery; **p.21** tl © John Luttick and Laurence Wesson, James Allen's Girls' School, tr © John Luttick and Laurence Wesson, James Allen's Girls' School, b © Richard Fosbery; **p.22** © Richard Fosbery; **p.27** © DR. FRED HOSSLER, VISUALS UNLIMITED /SCIENCE PHOTO LIBRARY; **p.31** t © Richard Fosbery, b © Richard Fosbery; **p.33** © SCIENCE PHOTO LIBRARY; **p.37** t © Sebastian Kaulitzki / Alamy, b © CNRI/SCIENCE PHOTO LIBRARY; **p.41** © ANATOMICAL TRAVELOGUE/SCIENCE PHOTO LIBRARY; **p.54** t © DR MARK J. WINTER/SCIENCE PHOTO LIBRARY, b © BSIP SA / Alamy Stock Photo; **p.56** © Sebastian Kaulitzki - Hemera via Thinkstock/Getty Images; **p.57** © STEVE GSCHMEISSNER/SCIENCE PHOTO LIBRARY; **p.62** © May/BSIP/Corbis; **p.66** t © M.I. WALKER/SCIENCE PHOTO LIBRARY, b © Stephen Mcsweeny - Hemera via Thinkstock/Getty Images; **p.76** © M.I. WALKER/ SCIENCE PHOTO LIBRARY; **p.77** © Louise Howard, Dartmouth College, USA; **p.81** all © blueringmedia/Thinkstock; **p.82** © Richard Fosbery; **p.85** t © Craig Stephen / Alamy, b © mediagram – Fotolia; **p.86** all © SCIENTIFICA, VISUALS UNLIMITED /SCIENCE PHOTO LIBRARY; **p.87** © Margot Fosbery; **p.91** © Max Planck Institute for Plant Breeding Research; **p.98** t © JMP Stock / Alamy, b © Zoonar RF via Thinkstock/Getty Images; **p.99** t © Dr. Robert Calentine/Visuals Unlimited/Corbis, b © DR KARI LOUNATMAA/SCIENCE PHOTO LIBRARY; **p.100** © DR KENNETH R. MILLER/SCIENCE PHOTO LIBRARY; **p.118** © BILL LONGCORE/SCIENCE PHOTO LIBRARY; **p.119** Courtesy of Thermophile~commonswiki via Wikipedia Commons (http://creativecommons.org/licenses/by-sa/3.0/); **p.140** t © PHOTOTAKE Inc. / Alamy, b © paranoidplastic / Stockimo / Alamy; **p.151** t © Science VU/Dr. F. Rudolph Turner, VISUALS UNLIMITED /SCIENCE PHOTO LIBRARY, m © Dennis Kunkel Microscopy, Inc./Visuals Unlimited/Corbis, b © BSIP/UIG via Getty Images; **p.156** © David S. Goodsell and the RCSB PDB; **p.158** t © TIM VERNON / SCIENCE PHOTO LIBRARY, bl © KIbrahim1989 - iStock via Thinkstock/Getty Images, bm © Fuse - via Thinkstock/Getty Images, br © Anat0ly - iStock via Thinkstock/Getty Images; **p.159** from tl to br, © Getty Images/Hemera/Thinkstock, © GlobalP - iStock via Thinkstock/Getty Images, © Anette Linnea Rasmus – Fotolia, © Fuse - Thinkstock / Getty images, © nstanev - iStock via Thinkstock/Getty Images, © Fuse via Thinkstock/Getty Images; **p.185** t © 2015 Edvotek Inc. www.edvotek.com, b © Andrew Brookes/Corbis; **p.187** © WILDLIFE GmbH / Alamy; **p.189** © 2015 Edvotek Inc. www.edvotek.com; **p.192** © HANK MORGAN/SCIENCE PHOTO LIBRARY; **p.193** © Alcuin - iStock via Thinkstock/Getty Images; **p.195** © Genome Research Ltd. Francesca Gale; **p.196** © Genome Research Ltd. Francesca Gale; **p.218** © Bruce Watson; **p.219** © meryll - Fotolia.com; **p.222** © Mark Rieger, Dean and Professor, University of Delaware; **p.224** © khwanchai.s - Fotolia.com; **p.225** © Mark Reiger, Dean and Professor, University of Delaware; **p.226** © Andy Dean - Fotolia; **p.230** © arinahabich - iStock via Thinkstock / Getty Images; **p.231** © MONTY RAKUSEN/SCIENCE PHOTO LIBRARY; **p.233** © karichs - Fotolia.com; **p.246** © Dr Clare van der Willigen; **p.247** © Andrew Powell; **p.248** t © Andrew Powell, b © cheeseong - iStock via Thinkstock / Getty Images; **p.249** from top to bottom © Dr Clare van der Willigen, © Richard Fosbery, © Richard Fosbery, © Richard Fosbery; **p.250** © Richard Fosbery; **p.251** l © Richard Fosbery, r © AfriPics.com / Alamy; **p.252** © Richard Fosbery; **p.260** © Jochen Tack / Alamy Stock Photo; **p.261** © gabe9000c - Fotolia.com; **p.264** © DR JEREMY BURGESS/SCIENCE PHOTO LIBRARY; **p.266** © Elliot Hurwitt - iStock via Thinkstock/Getty Images; **p.268** t © Nigel Hicks / National Geographic Creative/Corbis, b © Gaynor Frost; **p.273** © Esme Mahoney-Phillips; **p.275** both © Dr Clare van der Willigen; **p.276** © micro_photo – Fotolia; **p.277** l © Sharpyshooter - iStock via Thinkstock/Getty Images, r © Richard Fosbery; **p.279** © Keith Naylor - Fotolia.com; **p.285** © Goodygreen - iStock via Thinkstock/Getty Images; **p.286** t © Richard Fosbery, b © Richard Fosbery; **p.287** t © Maria Jose Rosello, b © pilipenkoD - iStock via Thinkstock/Getty Images; **p.288** © Maria Jose Rosello; **p.289** © Cymdeithas Eryri Snowdonia Society; **p.290** t © Richard Fosbery, b © BOB GIBBONS/ SCIENCE PHOTO LIBRARY; **p.291** © Hemis / Alamy; **p292** t © Utopia_88 - iStock via Thinkstock/Getty Images; b © Lynn Winspear; **p.297** © Margot Fosbery; **p.305** © Dario Lo Presti - iStock via Thinkstock/Getty Images; **p.307** © GUSTOIMAGES/ SCIENCE PHOTO LIBRARY; **p.314** © Olivier Le Moal - Fotolia. com; Chapter 16 **p.1** © kasto80- iStock via Thinkstock/Getty Images; Chapter 17 **p.1** © zhudifeng - iStock via Thinkstock/ Getty Images.

t = top, b = bottom, l = left, r = right, m = middle

Text and figures:

p.12 Q6 figure, from Head, J.J., "Discovering Biology", fig.3.6, p.45 © Oxford University Press; **p.13** Q7 figure, Richard Fosbery, "Biology: Communication, Homeostasis and Energy", 2009; **p.19** figure 2.7, Human Health and Disease, University of Cambridge Local Examinations Syndicate, 21 Aug 1997. ©Cambridge University Press; **p.20** figure 2.9, Jean McLean, Richard Fosbery, Heinemann Coordinated Science: Higher, Heinemann;

p.21 figure 2.12, Richard Fosbery, Stuart LaPlace and Lorna McPerson, Unit 2 CAPE Biology Study Guide. 2012. Nelson Thornes; **p.32** figure 2.23, Richard Fosbery, Stuart LaPlace and Lorna McPerson, Unit 2 CAPE Biology Study Guide. 2012. Nelson Thornes; **p.33** Fig 2.24, Source: Retrieved from http://biology-forums.com/index.php?action=gallery;sa=view;id=9670; **p.38** figure 3.2, Lesson 2 - The Materialistic Mind - Your Brain's Ingredients, Neuroscience For Beginners. Retrieved from http://neuroscientist.weebly.com/blog/lesson-2-the-materialistic-mind-your-brains-ingredients; **p.41** figure 3.5, Source: http://commons.wikimedia.org/wiki/File:Leeftijdgehoordrempel.png; **p.44** figure 3.9, Freeman, Scott; Quillin, Kim; Allison, Lizabeth, Biological Science, 5th Ed., ©2014, p. 933. Reprinted by permission of Pearson Education, Inc., New York, New York; **pp.45-46** figures 3.10 and 3.11, A. L. Hodgkin; R. D. Keynes, "Active transport of cations in giant axons from *Sepia* and *Loligo*", The Journal of Physiology. © John Wiley and Sons Inc.; **p.53** Q7 figure, H. Ogawa, M. Sato, and S. Yamashita, "Multiple sensitivity of chorda tympani fibres of the rat and hamster to gustatory and thermal stimuli", The Journal of Physiology. © John Wiley and Sons Inc.; **p.57** figure 4.4, Diagram: © Copyright. 2011. University of Waikato. All Rights Reserved. Inset micrograph: Copyright Professor David Fankhauser; **p.67** figure 5.2, Source: http://www.paradoja7.com/diagram-of-the-nervous-system-for-kids/; **p.68** figure 5.4, Source: The Aneurysm and AVM Foundation (www.taafonline.org); **p.70** figure 5.7, Chapter 10, Central nervous system, spinal and cranial nerves. Retrieved from http://classroom.sdmesa.edu/eschmid/Chapter10-Zoo145.htm; **p.71** figure 5.8, Chapter 7: Ocular Motor System, Valentin Dragoi, Ph.D., Department of Neurobiology and Anatomy, The UT Medical School at Houston. Retrieved from http://neuroscience.uth.tmc.edu/s3/chapter07.html; **p.78** figure 5.17, Dr. Thomas Caceci, "Smooth and Skeletal Muscle", November 6, 2013; **p.84** Q7 figure, Ian Harvey (Editor), "Biological and Medical Illustrations: Set B", 15 Feb 2000. © Heinemann; **p.93** figure 6.11, Source: http://users.rcn.com/jkimball.ma.ultranet/BiologyPages/A/abscission_layer.jpg; **p.95** Q5, data © 2000. Botanical Society of America, quotation from "The House at Pooh Corner" © A. A. Milne, 2011; **p.116** Q5 figure, D. Voet, J. G. Voet and C. W. Pratt, Fundamentals of Biochemistry, John Wiley & Sons, Inc, New York, 2001. © Royal Society of America; **p.117** Q8 figure, Photosynthetic Carbon Metabolism, James A. Bassham, Proceedings of the National Academy of Sciences of the United States of America, Vol. 68, No. 11, pp. 2877-2882. © National Academy of Sciences; **p.135** Table 8.7, Freeland, P.W., "Problems in Practical Advanced Level Biology: With Specimen Results", Hodder & Stoughton; **p.138** Q7 figure, OCR Biology Practical Examination – Planning Exercise (2806/03) © Copyright Ocean Conservation Research; **p.138** Q8, Table 1 Supporting information for White and Seymour (2003) Proc. Natl. Acad. Sci. USA, 10.1073/pnas.0436428100, Retrieved from http://www.pnas.org/content/suppl/2003/03/04/0436428100.DC1/6428; Table1. html The paper is at http://www.pnas.org/content/100/7/4046. Full copyright (2003) National Academy of Sciences, U.S.A.; **p.167** Tables 10.2 and 10.3, Richard Fosbery, OCR A2 Biology Student Unit Guide: Unit F215 Control, Genomes and Environment. © Hodder Education; **p.168** Table 10.5, Richard Fosbery, OCR A2 Biology Student Unit Guide: Unit F215 Control,

Genomes and Environment. © Hodder Education; **p.170** Table 10.6, Richard Fosbery, OCR A2 Biology Student Unit Guide: Unit F215 Control, Genomes and Environment. © Hodder Education; **pp.170-172** Example, data © The Open University; **p.182** Q5 figure, R.J. Berry, M.E. Jakobson, Life and death in an island population of the house mouse. Copyright © 1971 Pergamon Press Ltd. Published by Elsevier Inc.; **p.192** figure 11.9, Mary Jones, Richard Fosbery, Jennifer Gregory, Dennis Taylor, *Cambridge International AS and A Level Biology Coursebook with CD-ROM*, Cambridge University Press; **p.194** figure 11.12, Muruganandhan J, Sivakumar G. Practical aspects of DNA-based forensic studies in dentistry. J Forensic Dent Sci 2011;3:38-45; **p.207** figure 11.22, Introduction to Chemistry: General, Organic, and Biological, Creative Commons by-nc-sa 3.0 license; **p.215** figure 11.25, Source http://beacon-center.org/wp-content/uploads/2013/07/Expanded_Genetic_Code.png; p.217 Q6 figure, Tamakoshi A, Hamajima N, Kawase H, Wakai K, Katsuda N, Saito T, Ito H, Hirose K, Takezaki T, Tajima K., Duplex polymerase chain reaction with confronting two-pair primers (PCR-CTPP) for genotyping alcohol dehydrogenase beta subunit (ADH2) and aldehyde dehydrogenase 2 (ALDH2)", Japanese Journal of CAncer Research. © John Wiley and Sons Inc.; **p.220** figure 12.3, Figure from Biology: A Modern Introduction by B S Beckett (2e, OUP, 1982), copyright (C) 1978, reproduced by permission of Oxford University Press; **p.222** figure 12.6, Illustration: Softwood cuttings - Prepare your cuttings © Te Aho o Te Kura Pounamu, Wellington, New Zealand, retrieved http://horticulture.tekura.school.nz/plant-propagation/plant-propagation-2/h1092-plant-propogation-2-study-plan/softwood-cuttings/. Used by permission; **p.227** figure 12.22, Source: http://www.genome.gov/pages/education/illustration_of_cloning.htm; p.230 Table 12.1, Effect of IBA on Rooting of Terminal Cuttings of Scented Geranium (Pelargonium graveolens (L.) Herit.) in Different Months, Ch. Pulla Reddy, R. Chandra Sekhar, Y.N. Reddy; **p.232** figure 12.16, Cambridge International Examinations Paper 0610/33 November 2011 © UCLES; **p.233** figure 12.17, Cambridge International Examinations Paper 0610/31 November 2011 © UCLES; **p.234** figure 12.19, P.W. Freeland, "Microorganisms in Action: Investigations", May 16, 1991; **p.235** figure 12.20, Jane Taylor, "Micro-organisms and Biotechnology", Nelson Thornes Ltd; P.W. Freeland, "Microorganisms in Action: Investigations", May 16, 1991.; **p.235** figure 12.21, Cambridge International Examinations Paper 0610/33 November 2014 © UCLES; **p.236** figure 12.22, Surinder Kumar, Vishwa Mohan Katoch, *Textbook of Microbiology*, Jaypee Brothers Medical Pub; **p.240** figure 12.25, Lowrie, Pauline, Wells, Susan, "Microbiology and Biotechnology", Cambridge University Press; **p.243** Q5 figure, Immobilization of penicillin acyclase from Escherichia coli on commercial sepabeds EC-EP carrier, p.38. © Faculty of Technology in Leskovac; **p.244** Q7 figure, Water stress and recovery in the performance of two Eucalyptus globulus clones: physiological and biochemical profiles. © Scandinavian Plant Physiology Society; **pp.253-4** Activity, Philip H. Smith & Patricia A. Lockwood, Ainsdale Hope School playing fields – further ecological studies, June 2014. © Ainsdale Community Wildlife Trust; **p.271** figure 13.4, David Slingsby, Ceridwen Cook, Practical Ecology, May 1986. © Macmillan Publishers Ltd; **p.255** figure 13.10, Richard Fosbery, Stuart

LaPlace and Lorna McPerson, Unit 2 CAPE Biology Study Guide. 2012. Nelson Thornes; **p.258** figure 13.11, Frank B. Golley, Energy Dynamics of a Food Chain of an Old-Field Community, Ecological Monographs, Vol. 30, No. 2 (Apr., 1960), pp. 187-206. © John Wiley and Sons Inc.; **p.261** figure 13.13, Copyright: University of Waikato. All rights reserved. http://sciencelearn. org.nz; **p.262** figure 13.15, Richard Fosbery, Stuart LaPlace and Lorna McPerson, Unit 2 CAPE Biology Study Guide. 2012. Nelson Thornes; **p.274** Table 14.1, ÜberGeschwindigkeit und Grösse der Hefevermehrung in Würze. Biochem. Z. 57, 313–334; **p.275** figure 14.4, Davidson J. 1938 A On the ecology of the growth of the sheep population in South Australia Trans. Roy. Soc. S. A., 62 (1): 11-148. © Cambridge University Press; p.276 figure 14.5, Sands, Margaret K., "Problems in Ecology", p.13. © Mills and Boons; p.291 figure 14.23, Utilizing a Multi-Source Forest Inventory Technique, MODIS Data and Landsat TM Images in the Production of Forest Cover and Volume Maps for the Terai Physiographic Zone in Nepal. *Remote Sens*. 2012, 4, 3920-3947; **p.293** Example, Sands, Margaret K., "Problems in Ecology", p.13. © Mills and Boons; Adapted from Cambridge International Examinations IGCSE Biology, 0610/33, November 2011; **p.294** Q5c, ©1928 Richards. Journal of General Physiology. 11:525-538. doi:10.1085/jgp.11.5.525; **p.294** Q7, Possible

incipient sympatric ecological speciation in blind mole rats (Spalax). PNAS, vol. 110, no. 7, 2587–2592, www.pnas.org/cgi/ doi/10.1073/pnas.1222588110; Table S4. Retrieved from www. pnas.org/lookup/suppl/doi:10.1073/pnas.1222588110/-/ DCSupplemental; **p.297** figure 15.1, Chapter 1 - Light use and leaf gas exchange from Plants in Action. © Australian Society of Plant Scientists; **p.297** figure 15.2, Swartz, R.W. (1979) The use of economic analysis of penicillin G manufacturing costs in establishing priorities for fermentation process improvement. Ann Rep. Ferm. Proc. 3 (75-110). © Elsevier; **p.298** figure 15.4, Paper 5 Cambridge International Examinations A level Biology Oct/Nov 2009 9700/52. © UCLES; **pp.310-311** Example, A survey of the Macroinvertebrates of the River Wye. Department of Applied Biology, UWIST, King Edward VII Avenue, Cardiff, p.26. © John Wiley and Sons Inc.; **p.313** Table 15.3, J. D. Boyd, J.W. Garvin, " Skills in Advanced Biology", 1990. © Nelson Thornes Ltd; **p.314** Example, OCR A level Biology - Environ-mental Biology 2805/03 June 2006, p.4. © Copyright Ocean Conservation Research.